Colloidal Dispersions and Micellar Behavior

Colloidal Dispersions and Micellar Behavior

K. L. Mittal, *Editor*

Papers from a symposium honoring
Robert D. Vold and Marjorie J. Vold
sponsored by the Division of
Colloid and Surface Chemistry
at the 167th Meeting of the
American Chemical Society
Los Angeles, Calif.,
April 2-5, 1974.

ACS SYMPOSIUM SERIES 9

AMERICAN CHEMICAL SOCIETY
WASHINGTON, D. C. 1975

Library of Congress CIP Data

Colloidal dispersions and micellar behavior.
  (ACS symposium series; no. 9)

  Includes bibliographical references and index.

  1. Colloids—Congresses. 2. Micelles—Congresses.
  I. Mittal, K. L., 1945- ed. II. American Chemical
Society. Division of Colloid and Surface Chemistry. III.
Series: American Chemical Society. ACS symposium
series; no. 9.

QD549.C62   541'.3451   74-34072
ISBN 0-8412-0250-8   ACSMC8 9 1-353

Copyright © 1975

American Chemical Society

All Rights Reserved

PRINTED IN THE UNITED STATES OF AMERICA

ACS Symposium Series

Robert F. Gould, *Series Editor*

FOREWORD

The ACS SYMPOSIUM SERIES was founded in 1974 to provide a medium for publishing symposia quickly in book form. The format of the SERIES parallels that of its predecessor, ADVANCES IN CHEMISTRY SERIES, except that in order to save time the papers are not typeset but are reproduced as they are submitted by the authors in camera-ready form. As a further means of saving time, the papers are not edited or reviewed except by the symposium chairman, who becomes editor of the book. Papers published in the ACS SYMPOSIUM SERIES are original contributions not published elsewhere in whole or major part and include reports of research as well as reviews since symposia may embrace both types of presentation.

CONTENTS

Preface ix

1. A Third of a Century of Colloid Chemistry 1
   Robert D. Vold and Marjorie J. Vold

2. Macrocluster Gas–Liquid and Biliquid Foams and Their
   Biological Significance 18
   Felix Sebba

3. Mechanical and Surface Coagulation 40
   W. Heller

4. Kinetics of Ultracentrifugal Demulsification 64
   Robert D. Vold and Alice Ulhee Hahn

5. Ultracentrifugal Technique in the Study of Emulsions 76
   K. L. Mittal

6. Electric Emulsification 97
   Akira Watanabe, Ken Higashitsuji, and Kazuo Nishizawa

7. Macromolecular Emulgents in Emulsion Stability 110
   S. N. Srivastava

8. Viscoelastic Properties of W/O Emulsions Containing
   Microcrystalline Wax 126
   H. Komatsu and P. Sherman

9. True Emulsion Polymerization of Acrylate Esters Using
   Mixed Emulsifiers 135
   A. R. M. Azad, R. M. Fitch, and J. Ugelstad

10. Kinetic Theory for the Slow Flocculation at the Secondary Minimum   145
    Pranab Bagchi

11. Stability of Sterically Stabilized Dispersions at High
    Polymer Concentrations 165
    F. K. R. Li-In-On, B. Vincent, and F. A. Waite

12. Stability of Aqueous Dispersions in Polymer–Surfactant Complexes.
    Ethirimol Dispersions in Mixtures of Poly(Vinyl Alcohol) with
    Cetyltrimethylammonium Bromide or Sodium Dodecyl Benzene
    Sulfonate 173
    Th. F. Tadros

13. Rheological Studies of Polymer Chain Interaction 193
    Robert J. Hunter, Paul C. Neville, and Bruce A. Firth

14. Nonaqueous Dispersion: Carbon Black in Heptane Solution of Manganese and Zinc Salts of 2-Ethylhexanoic Acid 211
    B. R. Vijayendran

15. Micelle Formation in Nonaqueous Media 225
    Ayao Kitahara and Kijiro Kon-no

16. Thermodynamics of Micelle Formation 233
    K. S. Birdi

17. Anomailes of Partially Fluorinated Surfactant Micelles 239
    Pasupati Mukerjee and Karol J. Mysels

18. Deuteron NMR Studies on Soap–Water Mesophases 253
    Håkan Wennerström, Nils-Ola Persson, and Björn Lindman

19. Influence of Electrolyte on Phase Equilibria and Phase Structure in the Binary System of Di-2-ethylhexyl Sulphosuccinate and Water .. 270
    Krister Fontell

20. Dissolution Mechanism of Water-Soluble Polymers (Partially Saponified Polyvinyl Acetates) 278
    Hironobu Kunieda and Kōzō Shinoda

21. Thermodynamic Aspects of the Mixing of Mono-Amides and Poly-Peptides with Water 288
    P. Assarsson, N. Y. Chen, and F. R. Eirich

22. Equilibrium Dialysis and Viscometric Studies on the Interaction of Surfactants with Proteins 299
    W. U. Malik and V. P. Saxena

23. Derivation of Equations to Account for the Variation of pH and Specific Conductivity of Gum Arabic Acid Sol with Dilution 316
    B. N. Ghosh

24. Electrokinetic Approach to the Selection of Blood Compatible Materials and Anticoagulant Drugs 322
    Supramaniam Srinivasan and Barry R. Weiss

Index 345

PREFACE

Professors R. D. Vold and M. J. Vold retired in June 1974 after more than 30 yrs at the University of Southern California. The annual meeting of the American Chemical Society in April in near-by Los Angeles provided an ideal opportunity to pay tribute to these two outstanding workers in colloid science by holding a symposium in their honor.

When the idea for this symposium was broached to me, I accepted it without hesitation as I knew of their global popularity and knew that such an event would elicit an enthusiastic response both from their former students and fellow colloid scientists. The response exceeded all expectations as 57 papers covering a wide variety of interfacial and colloidal phenomena by 98 authors from 14 countries were included in the program, and two papers were considered later on. Such overwhelming response certainly testifies to their popularity as well as versatility in this research field. In fact, this symposium turned out to be the second largest one at the meeting. In consideration of the contents of this symposium, it might be more appropriate to name it the International Symposium on Interfacial and Colloidal Phenomena Honoring Professors R. D. and M. J. Vold.

This volume of 24 papers chronicles part of the proceedings of this symposium. The other part is contained in a companion volume of 20 papers entitled "Adsorption at Interfaces." The papers in this volume deal with the formation, stability, and other aspects of a variety of colloidal dispersions and micelles in aqueous and nonaqueous media.

Colloid chemistry, more generally colloid science, has been aptly described by the Volds in their book, "Colloid Chemistry" (Reinhold, 1964) as "the science of large molecules, small particles, and surfaces." This description encompasses a variety of phenomena both in the animate and inanimate world.

Colloid science has experienced fluctuations of popularity among researchers. To some people, it is an academic stigma to study colloidal systems as they think that only researchers with cloudy thinking can engage in these cloudy systems. This reminds me of an anecdote which happened in 1966. After finishing my M.Sc. in India in 1966, I had decided to join the University of Southern California to carry out my graduate research in colloid chemistry. One day one of my professors asked me what kind of graduate work I would do at U.S.C. I replied, "Colloid science." He looked at me as though I had said something unpalatable and retorted, "Don't you have anything better to do?" Apparently, he never bothered to probe into the beauty of the ubiquitous colloidal systems.

Fortunately, the myths about colloid science are now being dispelled, and the field has gained respectability both in industrial as well as academic circles. In the last decade, the researchers have increased the interest in colloidal systems as is evidenced by the proliferation of journal articles, books, and collective volume series, and all available signals indicate that this trend will continue. With the advent of sophisticated instrumentation and the development of quantitative theories of colloids, much progress has been made in understanding colloidal systems. But it is still a long way before the World of Neglected Dimensions is fully understood.

The Volds have been active in a spectrum of research activities. He made his research debut in the study of the solubility of organic compounds in aqueous and nonaqueous media while she published her first paper on the mechanism of substitution reactions. After this, Robert worked in industry where he became acutely aware of the inadequacies of the theories of simple systems to practical industrial problems and decided to study the intricacies of colloidal systems. Subsequently, Marjorie became very interested in studying colloids. They actively pursued the phase rule studies of association colloids (soaps, greases, etc.) in aqueous and nonaqueous media employing surface chemical, electron microscopic, x-ray diffraction, and thermal analytical (DTA) techniques; adsorption at various interfaces; stability of colloidal dispersions; and calculation of van der Waals forces. More recently, their research interests have included understanding of the factors influencing emulsion stability using ultracentrifuge; use of computer in floc formation and calculation of the dimensions of coiling type polymers; dispersions of carbon black; phase behavior of lithium stearate greases; theories of colloidal stability in nonaqueous media; and the hydration of biopolymers like DNA.

Obviously, the Volds' research activities have run the gamut from less glamorous colloidal systems like greases to the more fashionable biopolymers. Their work on the phase behavior and properties of nonaqueous soap systems had a significant impact in the petroleum industry (cf. NLGI Spokesman **18,** 168 (1964)). Their research investigations have culminated in 136 scientific and technical publications. They have also written the book mentioned above. This small paperback is an extremely good exposition of the principles and the methods of study of colloidal systems. Owing to its popularity and utility it has been translated into Japanese.

Apart from their research contributions the Volds have rendered a great service to colloid science by popularizing it on a global basis, and they were very instrumental, along with Professors Adamson, Mysels, and Simha, in establishing an internationally acclaimed center for surface and colloid chemistry in the Chemistry Department at the University of Southern California. Robert Vold organized the Summer Conferences on Colloid Chemistry (1961–1964) at U.S.C. sponsored by the National Science Foundation.

The Volds are very meticulous researchers and highly stimulating teachers. I have found them very adept at inculcating good research habits in the minds of neophytes in research. No student can get away with sloppy record keeping. Furthermore, they know how to bring out the best in a student.

Robert D. Vold received his A.B. and M.S. degrees at the University of Nebraska in 1931 and 1932, respectively, and Ph.D. degree from the University of California at Berkeley in 1935. During 1935–1937 he was a research chemist with Proctor and Gamble in Cincinnati. He was a postdoctoral fellow with Professor J. W. McBain at Stanford University from 1937 to 1941. Since 1941 he has been engaged in teaching and research at the University of Southern California. In 1953–1954 he held a Fulbright Senior Research Fellowship with Professor Overbeek at the University of Utrecht, The Netherlands. In 1955–1957 he served as Visiting Professor of Physical Chemistry at the Indian Institute of Science, Bangalore, India, where he introduced new fields of research and helped to establish a Ph.D. program. In 1965 he served as a consultant for the Summer Institute at Jadavpur University, Calcutta, India, which is designed to improve the teaching of chemistry in Indian colleges. He was a member of the advisory board of the Journal of Colloid Science from its inception in 1946 until 1960.

Among the various offices he has held in National Associations include Chairman, Southern California Section, ACS, 1967; Committee on Colloid and Surface Chemistry of the National Academy of Sciences, 1964–1967; Chairman, California Association of Chemistry Teachers, 1961; National Colloid Symposium Committee, 1948–1953; and Chairman, Division of Colloid Chemistry, ACS, 1947–1948. He was awarded the Tolman medal of the Southern California Section meeting of the American Chemical Society for his research contributions and service to the profession in 1970.

Marjorie J. Vold received her B.S. in 1934 and Ph.D. in 1936 from the University of California at Berkeley at the unusually young age of 23. She was University Medallist (Class Valedictorian) at U.C. Berkeley.

After brief experience as a lecturer at the University of Cincinnati and the University of Southern California, she was a research chemist with the Union Oil Co. (1942–1946). Since then she has held various faculty appointments in the Department of Chemistry of the University of Southern California, with the title of Adjunct Professor for the last 14 yrs. She was awarded a Guggenheim Fellowship in 1953 which was taken at the University of Utrecht, The Netherlands. From 1967 to 1970 she served as a member of the Advisory Board of the Journal of Colloid and Interface Science.

She was awarded the Garvan Medal of the American Chemical Society in 1967. Among many other awards and honors include Los Angeles Times "Woman of the Year" 1966; National Lubricating Grease Institute Authors Award for the "Best Paper" presented at the Annual Meeting (in 1967), 1968; and she is listed among the 100 Outstanding Women of the U.S. by the Women's Home Companion 1969. Unfortunately, Mrs. Vold was struck with multiple sclerosis in the fall of 1958 and has been confined to a wheel chair most of the time since 1960. In spite of her poor health and this terrible handicap, she has shown admirable stamina to deliver advanced colloid chemistry lectures continuously for two hours.

On November 19, 1967 an article entitled "Science: A Tie That Binds" appeared in Los Angeles Herald Examiner which presented a few glimpses of the scientific and social lives of the Volds. A special love for science has been maintained in the Vold clan for almost a century. Individually speaking, the Volds are different in many respects. Mrs. Vold has put it succinctly, "Robert has a sound almost conservative judgment while I tend to go off half-cocked. We complement each other in our work and our lives."

Although the Volds are retiring from active duty, they have no intention of relinquishing their interest for colloid science—a discipline they have cherished for about 40 yrs—as they plan to write a text book on the subject.

May we join together on this occasion to wish them a very healthy and enjoyable retirement in San Diego.

Acknowledgments: First, I am grateful to the Division of Colloid and Surface Chemistry for sponsoring this event. I am greatly indebted to the management of the IBM Corp., both at San Jose and at Poughkeepsie, for permitting me to organize this symposium and edit the volumes. Special thanks are due to my manager, E. L. Joba, for his patience and understanding. The secretarial assistance of Carol Smith is gratefully acknowledged. Special thanks are also due to M. J. Dvorocsik and Elizabeth M. Ragnone for helping to prepare this volume for publication. The able guidance and ready and willing help of Paul Becher and K. J. Mysels is deeply appreciated. The reviewers should be thanked for their many valuable comments on the manuscripts. I am thankful to my wife Usha, for helping with the correspondence, proofreading, and above all for tolerating, without complaint, the frequent privations of an editor's wife. It would be remiss on my part if I failed to also acknowledge the enthusiasm and cooperation of all the participants, especially the delegates from overseas countries.

K. L. MITTAL

IBM Corp.
Poughkeepsie, N.Y.
November 19, 1974

A Third of a Century of Colloid Chemistry

ROBERT D. VOLD and MARJORIE J. VOLD
Department of Chemistry, University of Southern California, Los Angeles, Calif. 90007

Our work in colloid chemistry began in 1936 at the Procter and Gamble Company. The 38 years which have elapsed since have seen many changes, very considerable progress, and a great increase in our knowledge and understanding of colloidal systems. In this paper an attempt will be made to contrast the present state of the science with that existing in the middle thirties, and sketch some of the changes which have occurred. Although we arrived on the scene too late to meet the founders we were privileged to have a warm personal acquaintance with many of their immediate successors, since colloid and surface chemistry is still a relatively young science. Among these should be mentioned Holmes, Hauser, LaMer, Debye, Bartell, Jerome Alexander, Weiser, and most of all McBain, none of whom are any longer with us. These were great men as well as great scientists. We shall never forget Hauser's consideration at a Colloid Symposium in lunching with a couple of then unknown young sprouts, and introducing us to the leaders in the field, or McBain's talent for raising the self-esteem and bringing out the best in even the least gifted of his students.

It is difficult now to put one's self back into the environment of 1935. Then there was no BET equation, no general method for determining molecular weight by light scattering, no DLVO theory of stability, and considerable uncertainty as to whether colloidal behavior could be described completely by the usual relations of physical chemistry, or whether special new variables would have to be introduced. The ultracentrifuge and the electron microscope were in their infancy, and unavailable except in a very few specialized centers. In fact there was little commercially available instrumentation suitable for colloidal studies, as contrasted with the present time when the limitation is usually financial rather than lack of availability. In the old days one had to improvise and construct out of whatever components could be located. Now, provided one has funds, one can simply buy a differential scanning calorimeter or a zeta meter, or a BET apparatus, or a scanning electron microscope, etc.

Since it is impossible in the available space to cover the whole field, the present article will be primarily a personal odyssey describing research in fields in which we have been active at the Procter and Gamble Company to 1937, at Stanford to 1941, and subsequently at the University of Southern California except for Marjorie Vold's wartime employment at the Union Oil Company, a year in Professor Overbeek's laboratory in Utrecht, two years at the Indian Institute of Science in Bangalore, and nine months at the California Institute of Technology (RDV). At Procter and Gamble and at Stanford our concern was primarily with the phase behavior of soap-salt-water systems with peripheral involvement in the nature and properties of surfactant solutions. At U.S.C. we were first interested in the phase behavior of heavy metal soap-hydrocarbon systems, with a later expansion to study their colloidal structure and physical properties, involving gelation phenomena, and laying a foundation for understanding the behavior of lubricating greases. Simultaneously studies were commenced on adsorption of surfactants at the water-solid interface with particular reference to detergency, suspendability, and electrical double layer theory.

The year in The Netherlands focussed attention on the overriding importance of obtaining an understanding of the nature and magnitude of the attractive and repulsive forces operative between colloidal particles suspended in a medium, a problem which is still not completely solved but on which rapid progress is being made at the present time. Studies of adsorption of surfactants at the oil-water interface were then undertaken as a necessary prerequisite for investigating quantitatively the stability of emulsions. At the same time an active program on the stability of suspensions was undertaken, the zeta potential being the most important variable in the case of aqueous suspensions, with entropic and enthalpic effects the important factors with non-aqueous suspensions. The latter could be investigated most fruitfully through a study of flocculation kinetics, while a computer simulation of the sedimentation process further increased our understanding. With the availability of an ultracentrifuge for studying emulsion stability, it was also possible to undertake a study of the physical chemistry of DNA in magnesium salt solutions, with particular interest in the fate of the water of hydration of the magnesium ion on binding to the DNA, and to changes in conformation of the DNA molecule, resulting from the binding of magnesium ion or changes in the thermodynamic activity of the water.

Simultaneously we have been much concerned with the status of the teaching of colloid chemistry in American universities. Despite its importance in industry, biology, and fundamental theory, the field has been relatively neglected for the last several years because of the glamorous excitement attending recent developments in the unravelling of molecular structure, as has been eloquently pointed out by several authors (1,2,3). The mother is in danger of being devoured by her offspring. Material scientists, biophysicists, polymer chemists, theoreticians dealing with interparticle forces, catalyst chemists, and many others, are all dealing with colloidal phenomena but all too often have no concept of the central organization of the whole field which is provided by a series of coherent courses in colloid, surface and polymer chemistry. Many former centers, such as that at Columbia, have declined because of neglect due to running after the latest fashion. However, there is hope for the future as evidenced by the rapid recent growth of new centers such as those at Clarkson and Lehigh. But at most universities, unfortunately, colloid and surface chemistry are regarded as being too specialized and applied by the chemical purists, and too theoretical by the chemical engineers.

Returning to a chronological approach, the primitive state of the art in 1936 is strikingly illustrated by the fact that at that time space was still found in the Journal of the American Chemical Society for a paper of which the main point was simply that the ordinary laws of physical chemistry applied to soap systems. This was an off-shoot of my work at the Procter and Gamble Company attempting to apply the phase rule to systems of soaps and water. In certain composition regions liquid crystalline middle soap exists in equilibrium with isotropic solution, and hence the vapor pressure should be independent of composition. Although all now believe that the phase rule applies without exceptions the experimental demonstration left much to be desired, since the vapor pressure is near that of water, and variations due to changes in temperature tend to be larger than those due to variation in composition.

James W. McBain was one of the giants in the field of colloid chemistry at that time, so I jumped at the opportunity to spend four years in his laboratory as a postdoctoral fellow. It is of interest to recapitulate the problems under investigation in his laboratory in view of its reputation. These included development of an air-driven ultracentrifuge, determination of the surface area of soap curd by adsorption of methylene blue, attempts to measure the range of surface forces from the effect of different substrates on the strength of adhesive joints, direct experimental determination of the surface excess and comparison with the results predicted by the Gibbs equation, determination of phase diagrams of soap-salt-water systems, characterization of dodecylsulfonic acid as a colloidal electrolyte through measurement of properties such as diffusion constant, conductivity, osmotic coefficient, etc., and "solubilization" of dyes.

Tests of the Gibbs adsorption isotherm are perhaps particularly interesting. A "railroad" was built on which a spring-driven cart with a razor edge scoop collected a surface layer 0.05 mm deep from a trough filled with soap solution (1 ml from 1000 cm^2 of surface), with the concentration difference between bulk and surface solution determined by interferometry. This was an attempt to explain what was known as McBain's paradox. All curves of surface tension versus concentration of colloidal electrolyte known at the time went through a minimum, with the surface excess calculated from the Gibbs equation being necessarily zero at that concentration. But the surface tension of the water was lowered by 50 dynes/cm. or so despite the calculation that no soap was adsorbed at the surface. Various experimental difficulties, particularly the problem of preventing evaporation from the trough, made it impossible to obtain really definitive results. It was not until much later, and after several fanciful attempts by other authors to explain the minimum in the surface tension-concentration curve theoretically, that it was shown to be an artifact due to the presence of trace impurities, lauryl alcohol in the case of sodium lauryl sulfate solutions.

The electron microscope was a relatively exotic tool in 1937. It was with considerable excitement that Marjorie prepared a sample of the curd fiber phase of sodium laurate in water for McBain to send to a friend of his in the RCA research laboratories to be actually photographed with an electron microscope. The pictures were duly taken with a direct magnification of 8000 diameters and returned to us for interpretation. They showed a fibrous network with an angle between the branches invariably 62-65° which is just the angle to be expected if the fibers were crystalline in nature and grown together at junction points somewhat like crystal twins. Evidence was also found that their widths were simple multiples of the molecular length.

Our own work at Stanford was concerned primarily with the phase diagrams of the sodium soaps with water, and with salt and water, and with the polymorphism of the solvent-free soaps. On heating sodium palmitate in a sealed tube the first definite visible change is a sintering and increase in translucency at 256° C. This was often listed as the melting point in the early literature, although some of the more persistent investigators went right on heating to higher temperatures, and were rewarded by being able to observe true melting at 292° to 305° C., with a substantial increase in fluidity and transparency. So the question arose as to what was really going on. How many phases of sodium palmitate were there? Could they dissolve water? What is the binary phase diagram like?

We put samples of sodium palmitate and water in a dilatometer to observe changes in slope of the curve of volume versus temperature when phase boundaries were crossed, and found that the curve was full of wiggles. Even with anhydrous sodium palmitate flats were found at 117, 135, 208, 253 and 295° C., proving the existence of five polymorphic phases in addition to the isotropic liquid. Microscopic examination between crossed Nicols in sealed flat glass capillaries in a microfurnace confirmed the result. This was a rewarding kind of experiment in yet another way: the liquid crystalline structures that separate from the liquid are beautiful to behold.

The idea was born that the various polymorphic forms of the sodium soaps could be considered to result from the melting process taking place in steps instead of all at once at a single temperature. So we undertook to measure the heats of transition between the various phases. It turned out that some were independent of the length of the paraffin chain of the soap, indicating that the structural change involved in the transition was concerned with interactions of the polar head groups. Others were linear functions of the number of carbon atoms in the chain. The sum of the heats of all the transitions was very nearly the same as the heat of fusion of the corresponding fatty acid.

The method adopted for measuring the heats of transition was differential thermal analysis. Now one can purchase DTA apparatus but this was not the case in 1941. Sample and reference material were sealed in pear-shaped glass cells into which a thermometer well had been punched. The wells were filled with polyisobutylene to effect good contact between the glass cell and the thermometer bulbs. The two cells were placed side by side in a furnace made by wrapping resistance wire around a can insulated with asbestos paper. The rate of heating was controlled manually, readings of sample and reference thermometers were taken at frequent intervals on heating, and their difference plotted as a function of the temperature of the reference cell. Despite the crudity of the equipment, the thermograms obtained were not too bad compared to those obtained today with apparatus provided with an electronically controlled rate of heating and continuous automatic recording of the differential temperature.

In 1941 I joined the faculty of the University of Southern California where, among other things, I had the responsibility for organizing and teaching a course in colloid chemistry. Fortunately, while at Stanford we had attended McBain's lectures on the subject. McBain had used no text, drawing on his own wide knowledge and personal experience. He had a large number of excellent demonstrations but there was no laboratory associated with the course. I did not have sufficient self confidence to proceed in this fashion, and after much casting around finally settled on "Colloid Chemistry" by Harry B. Weiser for the lecture course, and Floyd Bartell's "Laboratory Manual of Colloid Chemistry" for the laboratory.

Looking at Weiser's book today is quite a shock. For example, of the first 113 pages devoted to adsorption, less than half is concerned with the adsorption of gases. Of these 43 pages only ten are concerned with adsorption isotherms. The Freundlich isotherm is given and the Langmuir isotherm, together with its derivation from elementary kinetic considerations. Polanyi's theory is mentioned but is essentially dismissed on the grounds that it requires van der Waal's forces to be effective over a considerable distance whereas "everybody knows" they can extend only over distances of the order of magnitude of molecular sizes, and usually the Freundlich or Langmuir isotherms generally fit the data better anyway. Heats of adsorption calculated from the Clausius-Clapeyron equation are presented but there is no treatment of their variation with surface coverage. Although the book was copyrighted in 1939, the BET equation, published in 1938, still had not found its way into the textbooks.

The Gibbs adsorption equation, relating adsorption and the concentration dependence of the surface tension, is given, accompanied by some results of McBain's microtome method purporting to show actual adsorption substantially higher than predicted by the equation. It is surmised that the Gibbs equation is to be viewed as a "limiting law".

One might well ask in view of the vast amount of work on adsorption since then what could be found to say about it thirty years ago. The answer is "plenty". There was the Lundelius rule about the least soluble materials being the most adsorbable; Traube's rule concerning the increase in sorbability as one ascends a homologous series; Paneth and Fajan's rules regarding the adsorbability of ions on ionic solids; and the Schulze-Hardy rule relating the coagulating power to the valence of the ion of opposite charge to that of the colloidal particle. There are a host of other intriguing relations, so that one is led to wonder a bit uneasily whether the newly acquired and highly touted rigor of modern colloid science has been attained at the cost of the loss of some of the beauty of the earlier qualitative observations. In many practical industrial situations a thorough understanding of the qualitative behavior may serve the perplexed chemist better than knowledge of the details of the mathematical derivation of an equation applicable only to simple or idealized systems. Actually, for most effective performance, familiarity with both approaches is necessary.

What is true about the treatment of adsorption is also true of other topics as well, such as stability of lyophobic sols, lyophilic systems including biocolloids, and association colloids. What has been discovered in the last thirty years or so makes the best of the treatments of 1941 seem very much out of date. There are not only sins of omission but also those of commission as well. For instance, Freundlich believed that the electric charge on the particles had to be neutralized by adsorption of ions of the opposite charge in order to bring about coagulation. Weiser surmized that charge neutralization did not have to be complete in view of the existence of a non-zero value of the zeta potential below which rapid coagulation occurred, and related the phenomenon to contraction of the double layer. But that was as far as he got; there wasn't any quantitative description.

An important feature of academic research in 1941 which should be borne in mind is that there was no National Science Foundation, no Atomic Energy Commission, no Environmental Protective Agency, etc. By and large most research in the universities was unsponsored. Some industrial firms, however, would give funds to the university for work under the direction of a given individual. There would be no strings on the nature of the research, but frequently there was an implied understanding that the professor concerned would be interested in the technical problems of the company. I was fortunate in having funds to support two such fellowships, one from the Lever Brothers Company, and one from the Colgate-Palmolive-Peet Company. So to a continuing interest in soap phases we began also to become interested in the characteristics of dilute soap solutions and the problems of detergency.

Understanding of the mechanism of detergent action involves many of the basic processes of colloid science, such as solubilization, adsorption, wetting, pore structure, the role of mechanical action, and the stability of suspensions. For solid soils the mechanical agitation has to be sufficient to separate the dirt from the substrate. The detergent acts by becoming adsorbed on the surface of both the dirt and the substrate, thus effectively preventing redeposition of the dirt when the agitation is stopped. In some of our simpler experiments we merely determined the effect of concentration of detergent on the amount of powdered manganese dioxide or carbon black remaining in suspension in the detergent solution after four hours settling time. The amount suspended passed through a maximum with increasing concentration of detergent but there was no correlation with the critical micelle concentration of the detergent solution. The results could, however, be rationalized in terms of attainment of a limiting zeta potential resulting from adsorption of the detergent. However, even without the added complexity of a fiber-liquid interface, there are still many unsolved problems.

For the work with solid sodium soaps we went to x-ray diffraction. Since single crystals of these substances recalcitranly refused to form, the x-ray was used essentially as a fingerprint device. Several polymorphs can be realized at room temperature, unimaginatively labelled as alpha, beta, gamma and delta. The alpha form, deposited from aqueous alcohol, turned out to be a hemihydrate. Martin Buerger of M.I.T. did succeed in growing it as a single crystal, and determined its space group.

About this time we became enamoured of the idea of studying the behavior of metal soaps (i.e., all the soaps except those of sodium and potassium) in oils by the techniques which had been so relatively successful in unravelling the mysteries of the soap-water systems. In the latter case it had been definitely established that such important use properties as formation of lather, tendency of a bar to crack on aging, et cetera, are determined by which of several polymorphs is actually present; moreover, the whole process of soap boiling could be described in terms of plotting a course through the phase map of the system. Lubricating greases are made by dispersing metal soaps in oil, or reacting fats or fatty acids dispersed in oil with alkali, diluting the resultant concentrate with more oil, and extensive milling. Just as the art of soap boiling could be reduced to a science by determination of soap-salt-water phase diagrams, it seemed possible that the art of grease making would be similarly clarified by determination of the phases present at each stage. By this time the Office of Naval Research had undertaken limited support of fundamental research - it was still prior to the date of the National Science Foundation - and underwrote our first ventures in this field.

Even prior to these studies it had become desirable to replace our makeshift differential calorimeter with something better. Differential thermal analysis had been used extensively in studying the dehydration of clays, with an apparatus consisting of a block of metal with holes bored in it to contain sample and reference materials with thermocouples in each well. But very rapid rates of heating were used - sometimes as high as 25° C. per minute - and we needed a much lower rate since phase transitions in the soaps can occur fairly slowly. And we needed vapor-tight containers both to prevent oxidation of the soap sample and to maintain constancy of composition in use with solvent-containing systems. We also wanted to automate the instrument.

The time-temperature controller we developed was very direct. The desired pattern for a given thermal program, say a steady rate of heating of 1 1/2° C./min., was plotted in polar coordinates and cut out on a thin aluminum cam. A lever arm followed the cam as it turned. The actual temperature of the furnace was measured by a thermocouple located in the reference sample, and the signal fed to both a circular chart recorder and to the controller. Here the difference between the actual furnace temperature and the programmed temperature activated an air valve and set a pneumatic motor to raise or lower an arm which was attached to a cord wound around the stem of a variable transformer supplying heating current to the furnace. The temperature difference between sample and reference cell was measured with a five junction thermopile, and recorded continuously on a strip chart recorder. For all its Rube Goldberg appearance, this apparatus, developed with the collaboration of Todd Doscher, worked very well.

With the use of this apparatus, and the polarizing microscope, systematic measurements were made of the thermal transitions of most of the heavy metal soaps, and the binary phase diagrams determined for lithium stearate in cetane and in decalin. The result of all this work was the definitive, although negative, demonstration that the use properties of greases could not after all be explained in terms of phase diagrams. The phase diagrams of soap in petroleum oil are extremely simple. Each of the successive phases of the solvent-free soap is somewhat soluble in oil, with the solubility varying from nearly zero at room temperature to nearly complete miscibility in some cases at higher temperatures, but the soap phases do not swell to an appreciable extent. The physical properties of a grease are determined primarily not by what phase or phases are present, but by the structure of the mix of colloidal soap fibers with the oil. Thus the problem of explaining such physical properties as resistance to shear breakdown, hardness as related to soap and additives present, resistance to syneresis, or consistency and other rheological properties, is as complicated as the phase diagrams are simple. It appeared that explanations would have to be sought in such properties as the type of interconnected gel network formed, the strength and nature of the "bonds" at the contact points, the anisometry of the fibers, the size and tortuosity of the channels, etc. These are difficult to observe and to quantify: an endeavor which is still far from complete although some progress has been made.

Study of such properties necessitated development and application of a variety of different techniques, including the penetrometer for measuring consistency, the electron microscope for observing fibers and "structures", the exudation of liquid from the gel under pressure, and the effect on properties and characteristics of differing thermal pretreatments. Although empirical correlations had been developed between the length-width ratio of the soap fibers recoverable from greases and measurable in the electron microscope, it seemed important to us to attempt to characterize the properties of the structural skeleton formed by these fibers in the soap-oil gel itself. At the time there was still considerable controversy even as to the existence of such a three-dimensional structure, since it was possible to "explain" most observations equally plausibly in terms of a solvate shell of variable thickness. The situation then is reminiscent of the present situation where phenomena otherwise difficult to explain are rationalized in terms of a change in the structure of water.

The reality of a three dimensional structure of soap fibers held together at contact points by van der Waal's forces was beautifully demonstrated by the observation by Browning that the oil in a grease could be replaced by other organic liquids with no change in the dimensions of the sample, and little change in its rheological characteristics. Peterson and Bondi confirmed and extended this conclusion by making an aerogel from the grease by a similar perfusion technique with liquid butylene and liquid ethylene successively, followed by heating under pressure above the critical temperature, and allowing the ethylene to escape. The resultant aerogel had the same volume and shape as the original blob of grease but was very hard and brittle, certainly a long way from a loose pile of separate soap crystallites. Electron microscopy showed the presence of an interconnected brushheap of soap fibers. Oil could be sucked back into the aerogel under vacuum, but the reconstituted grease was much harder than the original until it was subjected to mechanical working, when it regained its original consistency. The observations are consistent with the concept that oil or additive molecules separate the individual soap fibers at their junction points, and that van der Waal's forces act through the oil layer making flexible joints which deform when the grease is made to flow.

Some further progress was made by borrowing a technique from the biologists, i.e. study of thin sections in the electron microscope, working with Henry Coffer. By cooling a grease to dry ice temperatures it was possible to prepare sections of about 0.1 μm thickness by slicing with a microtome. Marked differences were found in the appearance of the fibers bound in the continuous network as contrasted with those seen in samples prepared by the more common solvent extraction and smear technique. The difference is a little like viewing an assembled skeleton rather than a pile of unsorted, partly broken bones, with the additional complication of inducement of right-handed or left-handed, twisted, rope-like aggreagates in the smears dependent on the direction of smearing. However, no one has yet been able to prepare sections of sufficient quality to permit any quantitative characterization of the colloidal structure.

A grease may be converted into a suspension of soap crystallites by diluting it with a 'solvent for the oil, such as benzene or heptane. The resulting suspension forms a very weak gel which slowly compacts to a final volume containing only a small amount of solids. This behavior suggested another approach to characterizing the gel structure through calculation of a theoretical value for the per cent solids in the settled gel. The method employed was a computer simulation of the settling process using the Monte Carlo technique. First the sediment volume was calculated for a sediment composed of spherical particles treated as cohering to each other on first contact. Then a parameter was introduced giving the particles a specified probability of cohesion on contact - analogous to the existence of an interaction potential between the particles -, and otherwise letting them roll over each other to achieve more efficient packing. For moist glass beads settling in a non-polar solvent the experimental value of the volume per cent solids in the sediment was 12.9% while the calculated value from the model chosen was 12.8%. The calculations became more difficult and less unambiguous as the model is altered to conform better to real systems by making the particles anisometric and introducing longer range attractive forces, although promising in the sense that these alterations reduce the solid content in the sediment to lower values coming closer to those found with real greases.

A book appeared in 1948 which changed the entire state of colloid chemistry with respect to understanding of colloidal stability: "Theory of the Stability of Lyophobic Colloids" by E.J.W. Verwey and J. Th. G. Overbeek. It set forth the mathematical foundation for showing that the tendency of fine particles to remain dispersed was due to the mutual repulsion of their electrical double layers being sufficient to overcome the van der Waal's attractive forces pulling them together. The addition of electrolyte reduces the repulsion, thus permitting the van der Waal's forces to dominate the interaction so coagulation occurs. It was just that simple. It turned out subsequently that the Russian investigators, Landau and Derjaguin, had had much the same idea, actually publishing the electrical double layer repulsion part in 1941, although it did not become known until after the war. So the total treatment is now known as the DLVO theory. Hauser had already understood this qualitatively in 1939 (4), but it had seemed very contrived to most chemists of that period since any desired prediction could be obtained by shifting around the position of attraction and repulsion curves on a potential energy diagram, and neither he nor we realized that they were quantitatively calculable quantities. In fact, so great was the scepticism that at one meeting of the Colloid Symposium Hauser laid a revolver on the table at the beginning of his talk with the remark that people were going to have to listen to him whether they wanted to or not.

At that time it was still difficult to accept the idea that van der Waal's forces could be significant at distances of the order of colloidal dimensions. It was in fact not new, having been suggested by Willstatter in the early thirties and worked out in detail for spherical particles by Hamaker in 1936-37. In actual fact it now turns out that the inverse sixth power attraction of atoms is not additive, so the Hamaker approach is now being replaced by the more complex but more nearly accurate Lifshitz treatment.

The proximate cause of our enlightenment was the Ph.D. thesis of Sparnaay working with Overbeek, which purported to give a direct experimental measurement of the magnitude of long range van der Waal's forces between macroscopic objects; two glass plates in a vacuum. The value of the Hamaker constant obtained, while a little too large, was of the right order of magnitude to fit with the Verwey-Overbeek theory. We found the work of Sparnaay and Overbeek exciting - so much so that we wanted to make a pilgrimage to Overbeek's laboratory to see with our own eyes, and perhaps to repeat and extend the measurements. It became feasible for us to go to Utrecht in 1953-54, thanks to Guggenheim (MJV) and Fulbright (RDV) Fellowships.

In Sparnaay's apparatus one glass plate was held on a three point support, movable for levelling purposes by air pressure, with a very fine micrometer screw to raise it toward a second plate suspended by a spring. The top plate was attached to a silvered disc which formed a part of a condenser, the capacity of which depended on the distance through which the spring was pulled down by the lower plate, thus permitting calculation of the effective force. The distance between the plates was determined from the interference colors.

The three fold enemies of the measurement were vibrations, dust, and stray electrification. Vibrations were minimized by mounting the apparatus on a tombstone on springs in the basement of the building, since the laboratory was only three blocks from the central railroad station. Dust was minimized by meticulous cleanliness and careful polishing. There was always some vibration. If it ceased abruptly one could be sure the plates had come to rest with a dust particle between them, necessitating that the set-up be dismantled, the plates cleaned once more, and the experiment started over again. An attempt - not always successful - to avoid stray electrification was made by including a radioactive mineral in the vacuum chamber to ionize the residual air so as to make it conducting, thus permitting dissipation of static charges.

Ten months proved to be too little time to reassemble Sparnaay's apparatus and obtain more accurate results worth publishing. The most that we accomplished experimentally was to convince ourselves that the attraction between the plates was real, and really was due to van der Waal's forces. Probably the greatest effect the year in Utrecht had on us, through the close collaboration and friendship of Professor Overbeek, was to redirect our thinking towards the importance in colloidal systems of the forces operative between colloidal particles on close approach or contact, and the effect of the size and shape of the particles on the properties of the system.

So far, this backward look over the growth of colloid chemistry has been rather narrowly confined to our own interest. What of the rest of the science? In 1954 alone there were some 715 papers dealing with colloid and surface chemistry, even excluding such prolific branches as adsorption of gases, high polymers, proteins, chromatography, ion exchange, and membrane phenomena. These 715 purported to be fundamental work and did not include a very extensive trade literature. So how does one pick out from these the more significant contributions? One way is to let the contemporary investigators do it for you - by examining to see to whom they gave various awards and for what kind of work.

Colloid chemists who have been awarded the Nobel Prize include Zsigmondy, Svedberg for the ultracentrifuge, Tiselius for his beautiful work on separation and characterization of proteins by electrophoresis, Langmuir for his work on films giving the first definitive measurement of molecular dimensions, and Debye. But Debye was recognized primarily for his work in a different field. Further insight into what was regarded as being important can be gained by looking to see to whom we gave our own Kendall Award, a task facilitated by the initiative of Karol J. Mysels, who recently arranged for the publication of the Award addresses (5) of the last twenty years. A paragraph quoted from its Introduction says it all.

"The variety of topics covered and techniques discussed is

enormous. Even when nominally the subject is the same, such as the contact angle, the earlier approach of Bartell has little in common with that of Zisman and, still later, Zettlemoyer touches upon it again from a different point of view. The solid surface has also been treated by Emmett, Brunauer, Halsey and Zettlemoyer, but with emphasis on various aspects of its interaction with gases, its function in catalysis, the importance of proper area determinations, the mutual interference of adsorbed molecules, and the characteristics of hydrophobic behavior. The liquid-gas surfaces are the subjects of Mysels and of Hansen, but one is concerned with thin films and the other with ripples on bulk liquid. Aqueous dispersions are studied by Matijevic and serve him as a tool to solve problems of inorganic chemistry. The first monodisperse aerosols were discussed by LaMer, and Kerker reported recently on advances made by light scattering and other modern techniques in their study. The theory of radiation scattering in general, and especially of x-rays, was the subject of Debye. Williams discussed the role of the ultracentrifuge and, later, Vinograd its improvement and uses in studying DNA. Muscular contraction was the concern of Hill, and biological problems also were very much on Scatchard's mind. Holmes ranges, in his review, from vitamin A to the contraction of the silica gels which others use as model surfaces. Ferry and Mason both discuss rheology, one from the macroscopic point of view. ---- the variety of personalities, areas of interests, approaches and points of view, as well as the gradual evolution of the whole field, are manifest from the printed page."

As has been well said in the preface to the book, during the last twenty years "we see the last vestiges of art leave the study of colloid and surface chemistry, the technological maturity is seen; mathematics, physics and the biological sciences become home ground for the colloid and surface chemists and their territories now extend to the whole world of science". It is a far cry from the opinion in the last century, quoted in J. Chem. Educ. $\underline{46}$, 690 (1969) from A. Conte., "Philosophie Positive" (1830): "Every attempt to employ mathematical methods in the study of chemical questions must be considered profoundly irrational and contrary to the spirit of chemistry. If mathematical analysis should ever hold a prominent place in chemistry - an aberration which is happily almost impossible - it would occasion a rapid and widespread degeneration of that science".

During the last ten or fifteen years my own work (RDV) has involved three major interests; stability of oil-in-water emulsions, stability of non-aqueous dispersions, and the interaction between DNA and magnesium ions. Fundamental work on emulsions became possible as a fringe benefit resulting from pressure on the Northern Regional Research Laboratory of the Department of Agriculture to do something to help the midwestern growers of flax when their market for linseed oil was threatened by the successful utilization of synthetic resin emulsions in paints for exterior use. Limited funds were made available for university research in addition to a large coordinated industrial program to solve the practical problems.

The first difficulty to be overcome was the lack of a reliable, rapid method for measuring stability quantitatively. Groot in our laboratory, Rehfeld at the Shell Development Company, and Garrett at the Upjohn Company, independently hit on use of the rate of separation and amount of oil separated from an emulsion in an ultracentrifuge for this purpose. This method, however, while useful for studying the factors affecting coalescence, is limited in its application to real emulsions, in which both flocculation and coalescence may be important, by the fact that in the ultracentrifuge the emulsion is already flocculated when measurements begin, and is in a physical state resembling a foam rather than an emulsion.

The basic question, still not unambiguously answered, is whether the rate-determining step in demulsification leading to coalescence and separation of the oil phase, involves the rheological properties of the adsorbed film of emulsifier, or the rate of drainage of the aqueous phase from the lamellae separating the deformed oil "drops". Experiments too extensive to summarize in a brief report were carried out with the assistance of an able group of collaborators (Robert Groot, Kashmiri Mittal, Ulhee Hahn, Milagros Acevedo, Miodrag Maletic) determining the dependence of the rate of separation of oil on the nature of the oil and the emulsifier, the concentration of the emulsifier, the temperature, the magnitude of the centrifugal field, and the concentration of added electrolyte present, augmented by determination of adsorption isotherms and surface area by adsorption and by drop size distribution. It was soon shown that the stabilizing action of the emulsifier reached its maximum value when the equilibrium concentration of surfactant in the aqueous phase reached the critical micelle concentration, and a good correlation was found between stability and fraction of the total oil-water interface covered by adsorbed emulsifier. Some of our most recent findings in this study are being reported in other papers in this symposium.

Another group (Paul Hiemenz, Pranab Bagchi, Victor Dunn) worked in turn on the problem of attempting to develop an experimental method capable of testing the various theories which had been proposed to account for the stability of non-aqueous colloidal suspensions. The unhindered action of van der Waal's forces, as in aqueous systems, would soon coagulate them, and in these cases the stabilizing effect of interpenetrating electrical double layers is absent. Possible mechanisms suggested initially to account for the stability of such systems involved, besides the possibility of small residual charges, the decrease in entropy which would result from decrease in the number of configurations available on overlap of the adsorbed layer of stabilizing molecules surrounding each particle. Later it was recognized that there might also be an enthalpic contribution to the stability because of the increased concentration of stabilizer molecules in the overlap region. Both theory and experiment are difficult in this problem, since it is necessary to make many assumptions in order to make theoretical calculations, and measurements attempting to characterize aggregates as contrasted with ultimate particles tend to give results varying with the experimental method. Moreover, in the case of "particles" in the micron size range mechanical action rather than Brownian motion is likely to be the dominant force causing collisions.

The method we employed essentially is that of determining the value of the stability ratio, defined as the ratio of the rate of flocculation of the unprotected sol to that of the stabilized system. Rates of flocculation were obtained by an optical method, and the results compared with those calculated from the Smoluchowski equation modified by an interaction term involving a theoretical calculation of the balance between a van der Waal's attraction and a repulsion arising from the decrease in entropy occurring on overlap of the adsorbed layers of stabilizer molecules. Obviously there are difficult problems as to the orientation of the adsorbed molecules, the thickness of the adsorbed layer, etc., in addition to the fact that the energy barrier to flocculation is a small term determined as the difference between the very large and somewhat poorly known attraction and repulsion energies. Our conclusion was that actual flocculation behavior could not be predicted by the entropic (or steric hindrance) theory alone, although kinetics of flocculation data could be used to calculate backwards to give a reasonable value for the van der Waal's constant. Use of polymeric stabilizers in these experiments greatly extend their versatility, since it is then possible to utilize the information available from the extensive work on polymer statistics and change of configuration with change of solvent. As is treated in one of the other papers in this symposium, however, there are here too difficult experimental problems of attainment of equilibrium and of relating the conformation of the adsorbed polymer to the thickness of the adsorbed layer.

Meanwhile Bheema Vijayendran, Basappa Chincholi, and Anton Havlik, successively, were studying the effect on DNA molecules of replacement of sodium ions by magnesium ions. Our earlier work on this problem involved use of the ultracentrifuge by the density gradient method of Meselson, Stahl and Vinograd to determine the buoyant density of magnesium DNA, and by the sedimentation velocity method to determine its sedimentation coefficient, related to its molecular weight. The results obtained indicated a qualitative change in behavior commencing at a water activity of about 0.90, which was tentatively ascribed to a change in the conformation of the molecule resulting from partial dehydration. Independent support for this hypothesis is being sought from light scattering measurements on the same DNA in magnesium salt solutions, since such a change in shape might be expected to give rise to a measurable change in the radius of gyration obtainable from a Zimm plot of the scattering data.

But what of the future? What are now regarded as being the most important topics for investigation in colloid and surface chemistry? Any even remotely complete listing would be impossibly long for this review. It seems best merely to summarize the conclusions of three authoritative statements which have appeared recently.

Professor Overbeek, reporting for a Science Research Council panel of The Chemical Society, recommended the following areas of study (6).

The effects of wetting and adsorption on the energy of surface layers on solids. Study of adsorbed layers by spectroscopic and optical methods.

Thermodynamics of colloids: further development of double layer theory and theory of interfacial layers, particularly of steric stabilizers.

Direct measurement of forces between surfaces and the effect of adsorbed layers on such forces.

Effects of hydrodynamic factors on stability, stability of emulsions and foams, and the effect of polymeric stabilizers.

Further development of optical and spectroscopic techniques for studying colloidal properties; determination of rheological properties of model systems.

Catalysis of chemical or biochemical reactions by colloids.
Further studies of liquid crystals and association colloids.
Chemical formation and stabilization of aerosols.
Study of membranes, and cell and particle adhesion.

There has just appeared (7) Volume 1 of Colloid Science in the new series of Specialist Periodical Reports of the Chemical Society. Chapters by specialists reviewing recent important literature on the following topics are included: Adsorption at the gas-solid interface; Adsorption at the solid-liquid interface, non-aqueous systems; Polymer adsorption at the solid-liquid interface; Capillarity and porous materials, equilibrium properties; Particulate dispersions; Emulsions; Non-aqueous systems. Additional volumes are contemplated to cover most of the important areas of colloid science.

Even in this country colloid science is once again becoming a more fashionable field of research, partly because of its involvement in numerous biological processes, and partly because of the energy crisis. Thus it has recently been spotlighted (8) by the National Science Foundation as one of the basic research areas that will receive special attention in the energy research and development program. Here attention has been directed primarily to research on both homogeneous and heterogeneous nucleation, evaporation and growth rates of aerosols, coagulation kinetics, chemical reaction in aerosols, the preparation and stability of colloidal dispersions, emulsions, and surfactants.

With all this on the immediate agenda it is surely an exciting time to be involved in colloid research. For ourselves, we would like to discover the Fountain of Youth, and start all over again.

Literature Cited

1. Ross, S., "Chemistry and Physics of Interfaces", pp. vii-viii, American Chemical Society, Washington, D. C., 1965.
2. Hansen, R. S., and Smolders, C. A., J. Chem. Educ. (1962) $\underline{39}$, 167.
3. Mysels, K. J., J. Chem. Educ. (1960), $\underline{37}$, 355.
4. Hauser, E. A., "Colloidal Phenomena", pp. 217-226, McGraw-Hill Book Company, New York, 1939.
5. Mysels, K. J., Ed., "Twenty Years of Colloid and Surface Chemistry - The Kendall Award Addresses", American Chemical Society, Washington, D. C., 1973.
6. Overbeek, J. Th. G., Chemistry in Britain (1972), $\underline{8}$, 370.
7. Everett, D. H., Ash, S. G., Haynes, J. M., Ottewill, R. H., Sing, K. S. W., and Vincent, B., "Specialist Periodical Reports. Colloid Science, Volume 1", The Chemical Society, London, 1973.
8. Anon., Chem. and Eng. News (1974), $\underline{52}$, (30), 14.

2

Macrocluster Gas–Liquid and Biliquid Foams and Their Biological Significance

FELIX SEBBA

Department of Chemistry, University of the Witwatersrand, Johannesburg, South Africa

Introduction

In a consideration of states of matter such as crystals or liquids, it is usual to assume that they are composed of units such as ions, atoms or molecules and to interpret the bulk property as a co-operative summation of the properties of the individual units. However, there are a number of physical systems which could be better understood by considering them as consisting of units which are themselves stable or metastable aggregates of smaller units, the molecules, and it is suggested that such systems should be named "macrocluster systems". (Unfortunately, the term "cluster" which would be suitable has already been appropriated by statistical mechanics with a very different meaning.) Examples of macrocluster systems are gas-foams, biliquid foams, scums, flocs, concentrated emulsions, living tissue, cell cytoplasm and probably biomembranes.

A macrocluster system is defined as a heterogeneous association composed of units, each capable of independent existencse but organised in such a way that they constitute a more advanced system which has physical properties additional to the properties of the individual units. Foams provide a useful example of a macrocluster system. It will be shown that gas foams consist of units which are bubbles of gas encapsulated in a thin film of surfactant solution. However, individual gas bubbles are not foams, and do not exhibit the semi-solid behaviour of some foams nor their flow properties. It is because these bubbles are held together by thin aqueous films, the adhesion being a consequence of such films, that the advanced system, the foam is produced which has properties which are more than those of the individual units.

As defined in this way, many composite materials would qualify as macrocluster systems, as would those alloys whose properties depend upon interphase deposits even if these be metastable states. On the other hand, emulsions, with the exception, perhaps, of very concentrated oil-in-water emulsions, are not macrocluster systems. This is because the physical properties of a dilute emulsion are essentially those of the continuous aqueous phase.

Gas Foams

Many authorities compare gas foams to emulsions claiming that the only difference is that the disperse liquid phase in an emulsion is replaced by a gas. This is incorrect and it will be shown that the structure of a foam must be very different from that of an emulsion. However, it is possible to make emulsions of a gas in water, but falling into the same trap, this author mistakingly called them "microfoams" (1). But, apart from the common factor that gas is the dispersed phase, this system has no foam properties and to avoid confusion, it is here proposed that, in future, this system be referred to as a "microgas emulsion". Another misconception is that a foam consists of gas separated only by a continuous aqueous skeleton, the lamella, (Figure 1). This naive model cannot explain some simple facts of foam behaviour as will be discussed.

It must be recognised that no matter what process is used to generate a gas foam, essential requirements must be fulfilled. Firstly, there must be a surfactant dissolved in the liquid, and secondly the gas has to generate a surface within the liquid and then break through a second surface. Consider the case of a foam generated by passing air through a narrow orifice into a dilute aqueous solution of the foaming agent. The first step is the formation of a gas-filled hole in the liquid. In the absence of a foaming agent, any such holes which collided, while still under the surface of the liquid, would immediately coalesce. This is because the larger hole thus formed would expose a smaller interfacial area than the sum of the two interfacial areas of the two holes from which it was formed. However, in the presence of a surfactant, coalescence is less probable because of the energy barrier produced by the orientated molecules at the interface and consequent double layer. The repulsion of the two double layers is the principal component of the so-called "disjoining pressure" which resists thinning of the aqueous film jammed between two colliding holes.

Under the influence of gravity, the gas-filled hole rises to the surface of the solution. Here it meets an obstacle to its progress, because it has to break through the surface. If the hole is very small so that it does not have enough kinetic energy to penetrate the surface, it may be reflected back into the interior of the liquid. Usually, however, it is large enough to have sufficient energy to penetrate the surface. This is not a simple break-through because the surface is already covered with a liquid expanded monolayer of the foaming agent accompanied by its own double layer. Therefore, repulsive forces will operate preventing the gas from disrupting the surface completely. If it did so, the gas would simply disperse into the outer atmosphere and a foam would not be formed. However, the buoyancy of the gas-filled hole is enough to form a "blister" on the surface, covered by a thin aqueous film which owes its stability to the disjoining forces induced by the proximity of the charged heads of the surfactant molecules, (Figure 2). The net result is that the gas encapsulated by the thin aqueous shell, floats on the surface very slightly immersed and it is proper, at this stage, to refer to it as a bubble which is a volume of gas encapsulated in a thin film sandwiched between two expanded monolayer surfaces, (Figure 3). Should the bubble be large enough, and there be enough momentum, it could leave the surface as a free floating bubble of the kind painted by John Millais, i.e., a sphere of gas encapsulated in a shell of surfactant solution with a monolayer of surfactant absorbed at both inner and outer surfaces of the shell. As the bubble floats on the surface, the question arises as to whether the bubble is complete. It would appear to be so, because if a free bubble from air is introduced to the surface of a surfactant solution, it adjusts itself to float on the surface seemingly in precisely the same way as a bubble generated directly in the bulk of the liquid. Thus the bubble sits in a hollow in the substrate liquid as shown in Figure 3.

When two such floating bubbles get close enough, they move towards each other with increasing velocity until they almost touch, being separated still by a very thin aqueous film. The existence of this film, which is critical to the behaviour of foams of all types, is dependent upon the disjoining forces which must be overcome before the film can be disrupted. The pressure driving the bubbles together is often referred to as the Laplace capillary pressure which exists because of the tendency for any liquid surface to expose the minimum surface area. This is the same pressure as that which causes two partially immersed plates to adhere, whereas two completely immersed plates show no such tendency. In the previous case, the pressure $P = \frac{2\gamma}{d}$ where γ is the surface tension and d is the distance between the plates. In the case of the bubbles which are not spheres, P has not been derived, but presumably will still vary inversely with distance apart. The final position of the bubbles, therefore, is determined by the balance of three forces, the Laplace forces which increase as the bubbles approach, the long range Van der Waals forces which tend to thin the intervening film and the double layer repulsive forces which resist these. The resultant of the last two produces the disjoining force which prevents the film thinning to disappearance.

The fact that Laplace pressure exist between bubbles on water leads to an important but unexpected conclusion. Because they owe their origin to the tendency of liquids to minimise their surface area, there must be an interface exposed, such as a meniscus. On the other hand, the liquid must adhere to another phase, otherwise it would simply roll up into a sphere. When

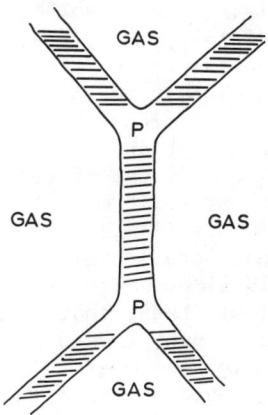

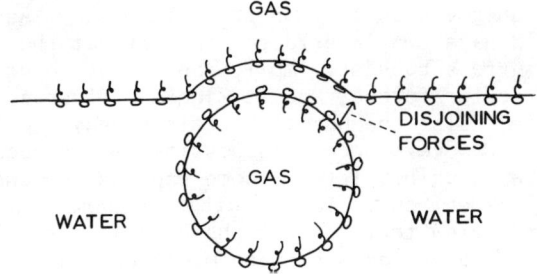

Figure 1. Conventional representation of foams. P represents plateau borders. Hatched areas represent thin aqueous film of surfactant with monolayer at gas interface.

Figure 2. Bubble approaching surface. ᵹ represents surfactant molecules.

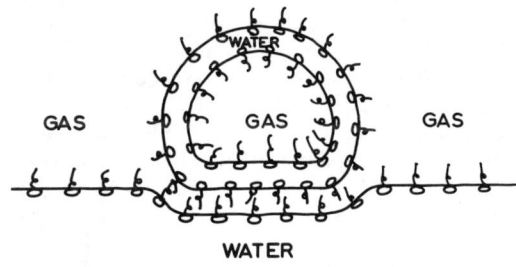

Figure 3. Bubble floating on surface. ᵹ represents surfactant molecules.

two plates are completely immersed in a liquid, there is no meniscus and thus no Laplace pressure. Immediately, the plates become partly immersed, a liquid-air interface exists and the tendency to minimise this interface pulls the plates together. However, the problem of attraction between two floating bubbles is not as simple. The fact that they do attract one another is obvious to anyone who has watched the bubbles on his cup of tea. The water meniscus cannot just adhere to the gas, so it must adhere to the film encapsulating the bubble. This film, clearly, must have different intermolecular attractions operating within it than does the bulk water otherwise it would simply merge with the bulk and the bubble would burst. It follows, then, that effectively the encapsulating film behaves as though it were a separate phase from bulk water. Though this might seem at variance with the usual definition of a phase, evidence already exists that the molecules of water close to an interface have different properties from bulk water. This has been explained as being due to a more ice-like structure because of the hydrogen bonding induced by the interface, and further evidence is provided by the fact that the surface of a surfactant solution has a very much higher viscosity than the bulk phase. As water in a thin film would have two interfaces, this effect should be enhanced. A surface boundary implies an abrupt change of intermolecular attractive forces, and it might at first seem strange that this could resist the normal kinetic energy tending to mix the molecules. However, once this possibility is accepted many phenomena to be described later become explicable, and there are two additional experimental facts, hitherto unreported which substantiate the view that surface phases have a reality. These are the existence of bridges between microgas emulsion bubbles and the existence of monolayer membranes. To avoid discontinuity these will be discussed as two subsections later.

For two equal solid spheres of radius r, and, with a drop of water between, the adhesive force is $2\pi r\gamma$ where γ is the surface tension of the intervening liquid, but this formula does not satisfy the conditions of the floating bubbles as they are flattened underneath and the undersurface is completely surrounded by a liquid. Also the spheres are not rigid, but deformable gas. However, as an approximation, it can be assumed that when they get close enough together, the approaching surfaces will be deformed enough to become parallel sides i.e. similar to two wet plates, partly immersed, so the adhesional pressure will be γ/d where d is the distance apart. This will be resisted by the disjoining forces. If the bubbles are very small, the internal pressure in the bubbles which is inversely proportional to their radii, will become greater, so distortion will be less than in the case of the larger bubbles.

The behaviour of monolayers of such bubbles is easily seen on a blackened Langmuir trough, by blowing air through a small jet under the surface. The sheet of bubbles is almost monodisperse. Though they adhere loosely, such floating bubbles do not yet constitute a foam. In order to convert them into a typical polyederschaum, it is necessary to crowd the bubbles, and this can readily be achieved by compressing them between barriers, when the bubbles will climb over one another forming a multi-layer of bubbles which soon distorts to a foam. More usually, a foam is formed because the vessel in which the bubbles are being formed has a limited area. A foam will not start to form until the surface is completely covered by a layer of bubbles. As the bubbles are unstable, the average life-time depending inter alia on the nature and concentration of surfactants and other solutes in the water, one of the prerequisites for the production of a foam is that the rate of generation of bubbles must be greater than the rate of destruction of bubbles moving from the point of generation to the walls of the vessel. The geometry resulting from the packing of a number of deformable spheres into a limited space causes the spheres to distort to polyhedra and the areas of closest approach of two polyhedra are planar. The regions of liquid confined by these planar interfaces constitute the lamellae. It should be noted that this model of independent bubbles forming a macrocluster system foam still retains the Plateau borders as regions of low pressure because of the curvature.

If the barriers which created a foam on the trough by compression are then moved apart, the foam breaks down, the bubbles slip back onto the water surface and the monolayer of bubbles is re-established. In other words, there is complete reversibility between spherical bubbles and the cells in a foam. It is difficult to visualise this happening if the lamella were a single uniform film. While it is conceivable that the outer surface of the two approaching bubbles could be destroyed under adhesional pressure and thus produce the homogeneous lamella depicted in textbooks, it is not easy to see that the reverse process could occur spontaneously. However, there is no such difficulty if the encapsulating film round each bubble remains unbroken, but this does mean that the lamella is complex and consists of two bubble films separated by a third thin aqueous film, (Figure 4).

Further evidence for this model of a foam is provided by the following experiment. If a viscosity increasing substance such as POLYOX is added to the surfactant solution, the foam stability is sufficiently increased for it to become possible to remove individual bubbles from the foam with a spatula, float such bubbles on a clean surface of the solution, and then replace them on the foam. A bubble re-inserted into its original site resumes its original shape. If a small amount of microgas emulsion is inserted beneath a foam, the small gas bubbles will rise through the foam bed, and if observed it will be seen that they never penetrate a cell surface directly, but always rise through the lamellae in spite of the fact that this often means a longer path or along paths that are only inclined slightly from the horizontal. The implication is that less work is needed to separate the bubbles than break cell walls.

When a foam whose foaming agent is an anionic surfactant is contacted with one produced using a cationic surfactant, the foams mutually destroy each other (2). Though this is difficult to explain on the conventional model of the single lamella, it is easy to see how two bubbles, because of the oppositely charged shells, will attract one another, i.e., will reinforce the Laplace pressures instead of resisting them as in a normal foam.

Finally, there is a well established technique for concentrating ions known as foam fractionation. This is inexplicable on the classical model of a foam but can be easily understood on the macrocluster model. The technique is simply a partition chromatography, the stationary phase being the shells round the bubbles, and the mobile phase the thin film of liquid between the bubbles.

Bridges Between Bubbles. Figure 5 is a microphotograph of a microgas emulsions (1), which is not a macrocluster systems. However, the photograph is of an emulsion which is sufficiently concentrated in gas bubbles that they actually touch and it will be observed that whenever two bubbles meet there is seemingly a bridge between them. They are always there no matter from what point they are illuminated and are, therefore, not optical illusions. The structure has the same appearance as the meniscus that would be seen if two wettable glass spheres had a drop of water between them. The microgas emulsion consists of gas-filled holes, with the surfactant monolayer at the interface between the hole and the aqueous phase. However, if this were so, there would not be the conditions for forming the meniscus. In some cases, the bubbles are seen to be distorted which means that there must be an adhesive force strong enough to achieve this. This adhesive force can only be the result of Laplace pressures, and such pressures can only exist if there is an interface between the surface phase and the bulk water phase, and the meniscus is formed by the surface phase. By analogy with two equal spheres of radius, γ, separated by a drop of liquid, where the adhesive force is $2\pi r \gamma$, the adhesive force will be $2\pi r \gamma_i$ where r is radius of bubble and γ_i the interfacial tension between surface water and bulk water. Since it is the surface water which is producing the meniscus, the attraction between its molecules must be greater than that between bulk water molecules. This is consistent with the suggestion that the surface phase is more hydrogen-bonded than bulk water. The interfacial tension between ice and water is of the order of 25 dynes cm^{-1} (3). The interfacial tension between surface phase and bulk phase must, therefore, lie somewhere between 0 and 25 dynes cm^{-1}. As the adhesive force is proportional to the radius of the bubble distortion of the bubbles is only observed in the larger bubbles.

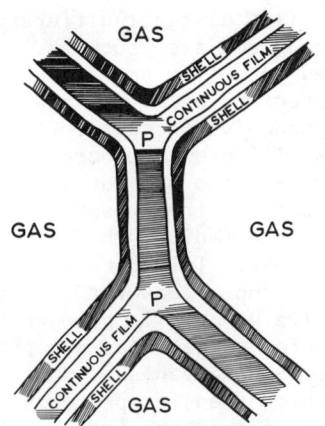

The interface between shell film and continuous film is effectivily a bimolecular film layer e.g.

Figure 4. Foam lamella

Figure 5. Microgas emulsion

It should be noted that there is a distinct difference between floating bubbles and microgas emulsion bubbles. This is because the former are encapsulated in a film bounded on both faces by monolayer and which is tenacious and elastic whereas the latter is not. It is, therefore, conceivable that in the case of the microgas emulsion the meniscus will draw the bubble surface with it thus forming a tube between the two gas bubbles. When two bubbles of a different size touch, as the pressure in the bubble is inversely proportional to the radius, the larger bubble grows at the expense of the smaller one. This growth can actually be observed under the microscope. However, the growth is not by a coalescence mechanism i.e. the walls between the bubbles do not collapse. The transfer of gas does not proceed as might be expected until the smaller bubble vanishes, but it seems to stop at a minimum size, after which transfer proceeds very slowly, if at all. The consequence is that the larger bubbles are observed to become coated by a large number of tiny bubbles, which because of their size, must be under higher pressure and, therefore, relatively rigid. This phenomenon might be caused by the connecting tube being pinched off below a certain size of bubble. This is an unusual example of a two unit macrocluster system.

Monolayer Membranes. The surface of a surfactant solution has sufficient lateral cohesion to form the basis of membranes which can be several centimetres across, and which are remarkably persistent. To make such a membrane a surfactant solution has to be placed upon an identical solution without disturbing the surface. Microgas emulsions provide one way of achieving this, because they are much less dense than water. If a microgas emulsion generated from a solution such as, for example, $5 \times 10^{-4}M$ sodium dodecyl benzene sulphonate containing $1 \times 10^{-3}M$ sodium sulphate is gently placed upon an identical solution, containing a small quantity of indicator dye, such as bromocresol blue, in a glass cylinder, it will float upon it. The gas slowly rises and escapes, until the upper solution becomes free of gas. However, it does not spontaneously mix with the lower solution, as can be seen by the existence of a boundary between the dyed lower solution and colourless upper solution. What is remarkable is the resistance of the boundary to disturbance such as tilting. One such membrane lasted three weeks and resisted transportation from one room to another. The membranes are slightly more easily made if the lower solution is made fractionally more dense than the upper, by incorporating slightly more salt. The salt appears to stabilise the membrane. The membrane is elastic and if a glass rod is made to disrupt the membrane to the depth of about 1 mm., on removal of the rod the membrane returns to its original position. As indicator dyes are themselves surface active, membranes have been made in which one of the solutions was coloured with iodine and still the iodine did not diffuse through the membrane. Though these membranes are impermeable to dyes, they appear to be slowly permeable to hydroxyl ions. This can be demonstrated by adding some alkali to the lower solution and a little phenolphtalein to the upper one, and observing the pink colour developing in the upper layer. However, the possibility of creep along the walls of the vessel past the membrane cannot be discounted. With care, sandwich layers can be made such as blue and red separated by a colourless solution. The nature of the membranes has not yet been determined and it is impossible to say whether they are monomolecular or bimolecular, nor is there any assurance that the air may not contribute to the stability. What is clear is that they are very impermeable so cannot serve as a model for biomembranes. What they do confirm is that membranes such as these resist mechanical disturbance and kinetic bombardment, reinforcing the suggestion that structures such as those postulated for foams can exist for long periods of time.

<u>Biliquid Foams</u>. In the following discussions, it becomes necessary to make a clear distinction between two very different types of surface active agent. On the one hand, there is the water soluble surfactant, usually ionised and typified by the alkali soaps and water soluble detergents. For brevity, this type will be referred to as WATSSA. On the other hand, there are the oil soluble surface active molecules, sparingly or insoluble in water, such as fatty acids, amines, long chain alcohols and cholesterol. These will be referred to as OILSSA. Biliquid foams, which differ from gas foams, only in that the holes are now filled with a liquid, owe their stability to similar forces to those operative in gas foams. The technique for their production has been described by Sebba (<u>4</u>). There are two distinct types depending on whether the hole contains an aqueous solution or an oil. In order to make a biliquid foam where the disperse phase is water encapsulated in a thin film of oil, advantage is taken of the fact that a microgas emulsion will float on an oil, so a microgas emulsion is placed on top of an oil such as Nujol or kerosine. As the gas leaves the microgas emulsion, its density slowly rises and when it is just above that of the oil it sinks through it. In doing so, it penetrates the two interfaces required for forming a foam, firstly the upper oil-water interface and then on leaving the oil, it gets its encapsulating oil film. This produces a biliquid bubble, and the macrocluster system, held together by the Laplace pressures in a thin water film between the bubbles, creates the biliquid foam.

It is possible with a hypodermic syringe to inject coloured solutions and various solutes into the water core of the bubbles, but there is no diffusion outwards, showing that the oil shell constitutes an impermeable barrier to diffusion. Thus, though it was at first thought that these foams might serve as a useful model for biological tissue, clearly the encapsulating oil film does not have the permeability properties of a bio-membrane.

The general behaviour of such water encapsulated biliquid foams resembles that of gas filled foams so closely that is reasonable to treat them as similar macrocluster systems. There is as yet no evidence to decide whether the oil film is coated with an aqueous surface layer in contact with a thin continuous water film or not. The aqueous phase contains WATSSA because its presence is essential for the production of the microgas emulsion. If a small amount of OILSSA is added to the oil phase, the character of the biliquid foam changes completely. The bubbles are very much smaller and more delicate and the adhesional strength is reduced so that the bubbles easily float away as oil encapsulated biliquid bubbles. It was this observation which led to the suggestion that something similar, with cholesterol as the OILSSA, was a factor in cancer, Sebba (5). It is not clear how the OILSSA reduces the adhesion, but as the Laplace pressure is proportional to the interfacial tension, and the OILSSA must reduce this, the explanation may be the simple one of reduction of the Laplace pressures and hence the adhesion.

The second type of biliquid foam is one in which the discontinuous phase is oil and the encapsulating phase is aqueous. To produce such a foam, a microgas emulsion is allowed to float on the oil, just as in the preparation of the first type. If, however, some more of the oil is poured onto the microgas emulsion, because the oil is heavier it will fall through as drops which get coated with the aqueous solution. At the bottom of the microgas emulsion, it passes through the second interface, and completes the requirements for a foam. This now consists of oil cells in a thin aqueous shell, the units, which have a formal resemblance to the gas filled soap bubble, being separated by a thin film of aqueous solution which resembles the continuous aqueous film in a gas filled foam. Thus, it is seen that these foams are more akin to gas-filled foams than is the first type of biliquid foam.

It is necessary to emphasize the distinction between oil-in-water emulsions and oil cell biliquid foams. In the former, there is a single interface between the oil drop and the continuous aqueous medium, the emulsifying agent at this interface providing an energy barrier which prevents coalescence. There are none of the conditions for a macrocluster system except in cases of very concentrated emulsions. In the biliquid foam, the oil cell is encapsulated by a thin aqueous film producing the unit for the macrocluster system, the adhesion now being provided by the Laplace pressures across the thin aqueous film which constitutes the continuous phase.

Biliquid foams of this sort are often met with in chemical and hydrometallurgical processes when oil as well as aqueous phases are involved. They have been looked upon as emulsions, but often prove resistant to destruction using techniques which should destroy emulsions. It would appear that the correct procedure would be to use techniques which are known to destroy foams, although this is admittedly an art and not always predictable.

It will be shown presently that oil encapsulated biliquid foams may provide a model for biomembranes and intracellular structures may consist of non-miscible liquids operating according to the rules for macrocluster biliquid foams.

Macrocluster Systems and Life. The behaviour of lenses of oil spreading on water surfaces has been fully dealt with by Langmuir (6) and Harkins (7). However, their treatment concerns only pure water and the behaviour of lenses when spread on dilute WATSSA solutions is very different. The interaction of two such lenses is an example of a two-unit macrocluster system, and an understanding of such behaviour may lead to biological insight.

A highly viscous oil like Nujol, if free of surfactant, does not spread on water or dilute surfactant solution. If, however, some OILSSA is incorporated into the Nujol, it will spread in a strange way. The nature of the spreading depends upon the concentration of WATSSA in the substrate and as all the conditions have not been fully investigated, a description of behaviour under only one set of conditions is given.

The substrate is water containing 0.2 g/l of sodium benzene sulphonate plus 0.1 g/l of sodium sulphate decahydrate. The spreading lens is Nujol containing 10% by volume of the TERGITOL 15-S-3. This is a non-ionic surfactant, an alkylpoly (ethyleneoxy) ethanol of HLB number 9 supplied by Union Carbide Corporation. The behaviour is more dramatic if the nujol is coloured by an oil soluble dye such as SUDAN 111 or WAXOLINE green, but the same behaviour can be observed in the absence of dye. The spreading behaviour is readily observed if the substrate is placed in a glass dish, such as a pyrex baking dish, which is illuminated from below by a powerful lamp. If a drop of the Nujol solution is placed on a clean surface, it spreads rapidly, but not as a circular lens but much more irregularly. If now, a second drop is placed on the surface, which presumably now has a film of OILSSA on its surface, it will spread with low contact angle more slowly to a fairly regular disc, a couple of centimetres in diameter. Spreading will continue for a few moments and then suddenly the lens will retract, reducing a few millimeters in diameter and with an obvious increase in contact angle. From this moment, spreading is much slower and very interesting. Sometimes, a rhythmic expansion and contraction will be observed, again within a few millimetres range of diameter. This may continue for a while, but is often so slight as to be unobserved. This is not a new phenomenon. It has been observed with cetyl alcohol on water (8), and was explained as due to a changing external surface pressure due to solution of the cetyl alcohol in the substrate. As the surface pressure dropped, the contact angle became lower, and spreading of the lens could proceed raising the surface pressure to a point at which the contact angle was raised sufficiently for spreading to stop. When that occurred the surface pressure dropped once again because of solution of the cetyl alcohol sufficiently for spreading to be recommenced. A similar mechanism will explain the rhythmic pulsing of the Nujol lens.

After a short time, the lens containing OILSSA begins to spread again, but this time it does not do so uniformly in all directions. Sometimes, many fuzzylike processes will appear giving the lens the appearance of a multiciliated organism. These "hairs" will gradually grow until after some minutes they will have extended from the central lens to a distance of many centimetres as filament-like processes often with branching, (Figure 6). The appearance very much resembles root-hairs. After a time, the filament begins to disintegrate into tiny globules of oil with high contact angle. There has thus been an enormous increase in interfacial perimeter contact. As it is almost impossible to avoid convection currents or similar disturbance, the pattern soon becomes very irregular and an infinite variety of patterns is obtained, the central lens departing markedly from its original circular pattern. Nevertheless, growth continues and often the pulsation is noticeable, so the odd "creature" floating on the water appears to be alive. It is possible to increase the rate at which the outgrowths extend by sucking off from the surface a considerable distance away, a trick often employed for cleaning contamination from surfaces. The effect of this is to reduce the surface pressure of the monolayer, thus increasing the rate of spreading of the lens. Sometimes, the lens does not grow "whiskers", but develops a puckered perimeter. As this occurs when the surface is older and contains a very large number of tiny globules, presumably the diminution of surface pressure is slower, being replaced from all the small drops, so spreading from the lens is reduced, and there is less tendency for outgrowths to appear. However, a complete description and explanation of these structures must await a separate communication as these are not themselves macrocluster structures, although they can be the units which build up to macrocluster structures.

The resemblance of some of the structures found in spreading lenses to cytological patterns may be more than a coincidence and of these the most striking are resemblances to the stellate cells in nerve tissue, the chromosomes and the endoplasmic reticulum systems. Each pattern is determined by a balance of a number of factors which have not yet been quantitatively determined, but the following are some of them:-
1) Nature of WATSSA
2) Concentration of WATSSA in substrate
3) Nature of OILSSA
4) Concentration of OILSSA in oil, which determines
5) Spreading pressure of lens
6) Equilibrium surface pressure of OILSSA on substrate which

is determined by
7) Rate of solubility or solubilisation of OILSSA in WATSSA solution
8) Nature and concentration of dissolved salts in substrate
9) Temperature
10) Viscosity of lens material
11) Convection currents in substrate or air currents above substrate
12) Free area between lens and neighbours or other obstacles
13) Size and shape of lens
14) Reaction to spreading forces
15) Age of WATSSA solution.

With respect to (14), the momentum of these spreading lenses is remarkable, and can themselves stir up currents of considerable magnitude which in turn can cause spiraling amongst the hair-like lenses. When these get close enough together, they form macrocluster systems which in appearance strongly resemble endoplasmic reticulum structures.

Perhaps, the most striking macrocluster system is that produced by two adjacent lenses, because if the conditions are correct, they can simulate the biological phenomenon of endocytosis to a remarkable degree. Endocytosis is the process by which a cell engulfs foreign matter and embraces pinocytosis and phagocytosis. In the following discussion the term endocytosis will be used to describe the phenomenon, although it does not here refer to living cells. The significant property which determines this behaviour is a difference of concentration of OILSSA in the two lenses. It is best observed by using different colours for the two lenses, one of which is called the gobbler and the other is the prey. For example, a good gobbler is a 10% by volume solution of TERGITOL 15-3-3 in Nujol, coloured red with a WAXOLINE dye. A good prey is a solution of Tergitol in Nujol less concentrated than the gobbler, coloured green with WAXOLINE green. A 2% solution works well, but Nujol without any OILSSA will also work. A lens of gobbler about 2 cm. diameter is laid on the WATSSA solution in water, and when it has almost reached a metastable equilibrium with the surface, i.e. spreads slowly, a small drop, about 2 mm. diameter of prey is placed a few millimetres away from the gobbler. They begin to attract one another, and each moves towards the other, but because of its smallness, the prey moves faster. As they get closer, the gobbler may extrude some tentacles, but these are not essential. The two lenses are being drawn towards one another by Laplace pressures modified by the fact that the prey, having a lower concentration of OILSSA, will absorb OILSSA from the intervening surface on the substrate, inducing a lower surface pressure i.e. higher surface tension, thus increasing the Laplace pressure. When the two lenses get very close, the Laplace pressure is high enough to deform the gobbler, so the gobbler envelops the prey which very soon is completely engulfed by the gobbler, and moves inwards towards the centre of the gobbler, but does not merge with it. This shows that there is an envelope surrounding the prey which protects it. In fact, it is often possible to see a thin residual strand of prey connecting it with the outside of the gobbler, (Figure 7). If the surface pressure outside the gobbler is carefully reduced by suction, it is possible to open up the cavity and the prey will re-emerge unscathed. Philosophers may muse on the fact that when the amoeba eats its victim, it is the victim which forces its way into the amoeba, thus happily committing suicide. Figure 8 shows schematically the mechanism for endocytosis.

If the concentration of OILSSA is the same in the two lenses, there is no attraction, and if they are placed close together, they may adhere but do not engulf. The contact angle seems to play a part as droplets which do not have a spreading contact angle i.e. is not less than 90° do not show the phenomenon. If there is no salt in the solution, it seems not to occur, but addition of $3 \times 10^{-4}M$ sodium sulphate enables it to proceed. This is understandable on the hypothesis that in order for the prey to distort the gobbler the Laplace pressure must be high i.e. the two lenses must approach within a minimum distance before it can overcome the repulsive forces caused by the double layer. Salts will reduce the thickness of the double layer and, therefore, enable the lenses to get closer.

The prey does not have to be a lens. If a platinum wire or thin glass rod is inserted into the substrate close to the gobbler, provided the contact angle is correct, the gobbler will start moving towards the solid and engulf it. Presumably even the small increase of surface area provided by the platinum or glass is enough to lower the surface pressure sufficiently for attraction to occur. Differences of spreading pressures may also offer a mechanism for the biological phenomenon of cytoplasmic movement. Lenses after spreading end up as innumerable small globules immersed in a monolayer of OILSSA, there clearly being equilibrium between the spreading pressure of the globules and the surface pressure of the monolayer. If to such a system, a drop of oil, Nujol or kerosine, low in OILSSA is added, extra-ordinary activity commences. It starts with the nearest small globules moving towards the low pressure oil, but this movement soon spreads and even the most distant globules begin to participate. As movement is rarely uniform in all direction, soon a streaming movement is observed and this may develop into a spiral movement. If there are still strand-like lenses on the surface, these follow the spiral movement, and wind-up like a coiled spring so that concentric layers of lens with the low pressure oil at the centre result. These strongly resemble the endoplasmic reticulum observed in cells. The motive energy is obviously the movement of OILSSA from the higher pressure on the substrate surface into the low pressure oil. What is interesting is that the small globules when they reach the centre oil lens, often rebound from

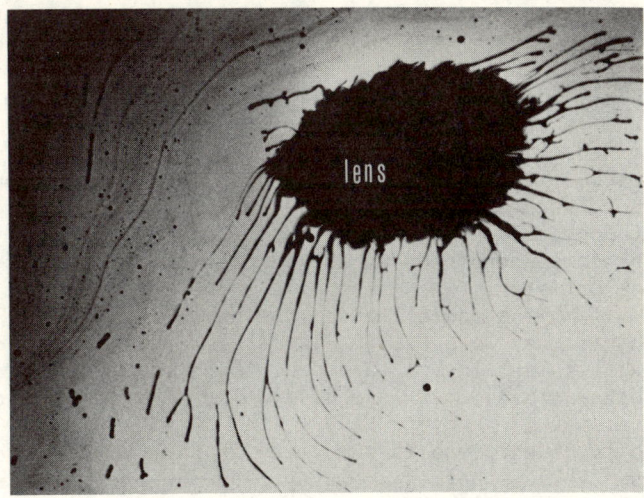

Figure 6. Lens spreading on water

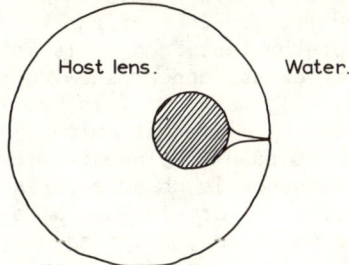

Figure 7. Model for endocytosis. Hatched circle represents victim lens. Note neck to periphery.

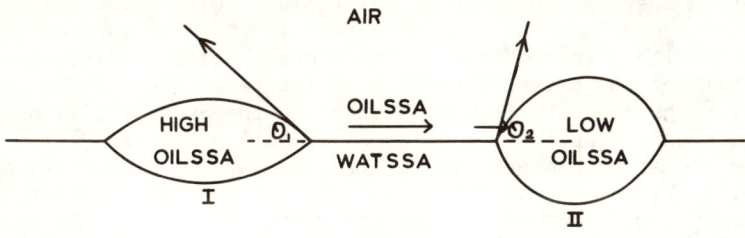

Figure 8. Mechanism of endocytosis. OILSSA moves from lens I to lens II thus tending to increase γ. This increases Laplace pressure which is $\propto \gamma$ so lens I and lens II are attracted. Spreading pressure of I > spreading pressure of II so when they touch I spreads around II. The higher the concentration of WATSSA, the less marked the effect of OILSSA on γ so attraction is reduced.

it, or encircle it but rarely merge into it. In other words, although a macrocluster system is formed, it is not a strong one. On the contrary, there is often a sudden spurt of activity at one point, and all the encircling small globules get swept to a distal point on the lens, from which point they again slowly form a ring around the lens. The reason for this explosive polarization is not obvious but it almost looks at though an enveloping bubble had burst. The general movement is remarkably rapid, and is an example of conversion of surface energy into movement. It is, of course, the same phenomenon as the camphor crystal darting about the surface of water. When there are thousands of little globules, all moving and all also showing reaction, i.e. jet propulsion, as they lose OILSSA, it is clear that the movement is complex, but it must in its turn contribute to the pattern and shape of other lenses which are developing or packing into macro- cluster systems.

Not only is endocytosis simulated in this way, but so too is exocytosis. This is better shown by oils not as viscous as Nujol, although the Nujol will show it more slowly. If some dyed kerosine, containing say 2% of TERGITOL OILSSA is spread on the aqueous substrate, it forms a typical lens with a fairly even circumference. This slowly deforms, islands of monolayer appear and dark zones on the circumference of the kerosine appear. This indicates thickening of the lens, and it is interesting that these regions of different thickness appear showing that pressure is not uniformly distributed, (Figure 9). If in the interior of such a lens, a drop of kerosine of different colour, but containing a higher concentration of OILSSA is placed, after a very short period the following sequence is observed. The new lens rapidly expands and then retracts to form islands of oil in a monolayer of OILSSA. These islands may or may not merge. They move towards the thinnest part of the lens where the monolayer, being at higher surface pressure than outside, causes a bulge and eventually breaks through carrying the added kerosine as a lens or lenses out with it. The lens then reforms to an approximately circular shape while the added lenses remain outside just touching the reformed lens which they then because of their higher OILSSA content attempt to engulf. What is significant is that the surface generated within the interior of the lens, in the readjustment now forms part of the external surface of the lens. This is precisely the opposite to what happens in the endocytosis, and simulates what happens in many types of cell, (Figure 10).

To summarise, interaction of two lenses, and presumably of two cells, is in essence the formation of a two unit macro- cluster system, with endocytosis occurring when the OILSSA concentration in gobbler is greater than that in the prey, and rejection of the cell (exocytosis) when the OILSSA concentration is greater in rejected cell than in the host cell. Fertilisation of an ovum by a spermatozoon is presumably a special case of endocytosis. It is easy to envisage a situation in which the

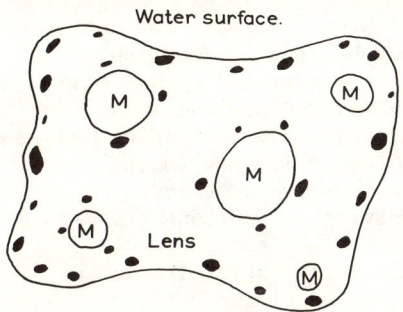

Figure 9. *Spreading lens. Lens breaking up. M represents monolayer. Rest of lens is thicker. Note sites of thickening at perimeter as lens retracts.*

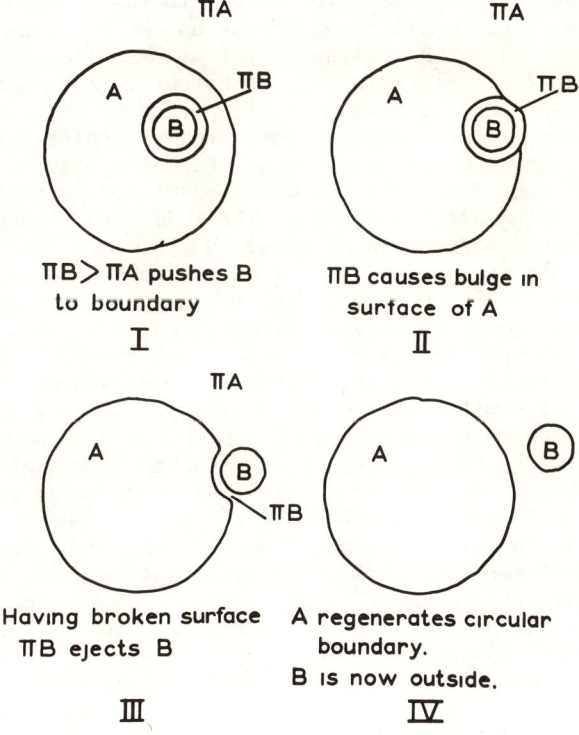

Figure 10. *Mechanism of exocytosis. Note that internal surface of A generated near B becomes part of the external surface of A after B has been ejected.*

spermatozoon introduces into the ovum some chemical which inactivates the OILSSA in the ovum. The effect of that would be to reduce the spreading pressure, and hence increase the contact angle, with the consequence that a second spermatozoon would not be attracted to enter the ovum. In a previous communication (5) it was suggested that cancer occurred because the concentration of cholesterol, a typical OILSSA, was above that in normal cells due to an induced genetic defect. It is easy to see now why such cells are invasive. It is because they behave as typical gobbler cells, normal cells having less OILSSA becoming the prey. If, however, these are not mobile, but the cancer cells are, the consequence would be infiltration by the cancer cell along the aqueous lamellae between the tissue cells.

<u>Two-Dimensional Biliquid Foams</u>. If beneath the surface of a dilute WATSSA solution, a stream of high viscosity oil such as Nujol containing some OILSSA is released, it does not have time to break into droplets but usually reaches the surface as a thread of oil. On penetrating the surface, the thread spreads to flat lenses, and these adhere to form two-dimensional biliquid foams, the conditions of breaking two surfaces applying in this case as well. A similar effect with smaller units is obtained by beating a lens into the substrate. The threads produced by an expanding lens will often adhere in the same way to form similar structures. These foams display some of the characteristics of single lenses. For example, the endocytosis process can occur. If the outside surface pressure is reduced by sucking off the surface, the individual lenses will move apart, and reform again on standing.

It is suggested that biomembranes are special examples of two dimensional biliquid foams. It is now believed that the Davson-Danielli model for a biomembrane is not satisfactory. Electron microscopy reveals the existence of pores. A two-dimensional biliquid foam, in which the lenses are very thin, and may even be bimolecular, seems to satisfy some of the conditions. One of the problems of biomembranes is how penetration is regulated. It is suggested that regulation could be achieved by metabolic control of OILSSA concentration, as illustrated in Figure 11 and 12. If the OILSSA concentration increases, the lenses will tend to expand and the OILSSA will spread to the inter-lens aqueous surface, until the surface pressure equals the horizontal component of the spreading pressure. If the OILSSA goes into the aqueous phase faster than it moves out of the spreading lens, the lenses will expand and the pores will close. If it is produced in the lens faster than it dissolves in the aqueous phase, the surface pressure will build up, and the lens will retract, opening the pores. In fact, there is similarity between this mechanism and the opening and closing of stomata except that in the latter case it is three-dimensional pressures in the guard-cells and not two-dimensional

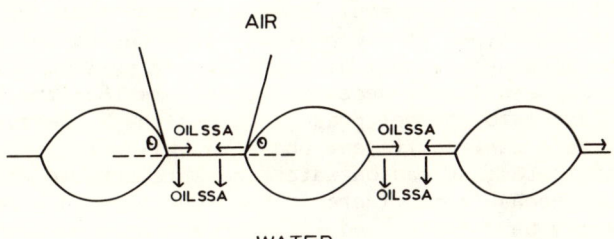

Figure 11. Opening of pores. OILSSA moves more slowly into aqueous phase so concentration on surface builds up. Effect is to decrease γ and Θ rises so lenses contract and pores open up.

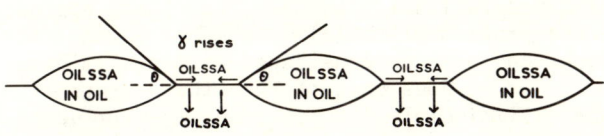

Figure 12. Closing of pores. OILSSA moves via surface into aqueous phase faster than it is spread onto surface. Effect is to increase γ and to compensate Θ drops, lenses expand and pores close.

pressures that are involved. In another publication, suggestions will be made as to the nature of the OILSSA in living systems and how its concentration is controlled.

Conclusions

In this paper, the attempt is made to show that the behaviour of many types of two phase systems can be better explained by considering them as macrocluster systems. In order to understand these, it is necessary to assume that the thin film of water adjacent to another phase, gas or liquid, behaves as though it itself was a different phase to bulk water. The behaviour of lenses, spread on water shows remarkable resemblance to cytological behaviour. There is no proof that this behaviour can necessarily be extrapolated into three dimensions, but it might be useful to assume it could. The problem is that when spread on water, the third phase is air, whereas presumably there is not gas in cytoplasm, though the question could be asked whether the absence of gas has been unequivocably proven. In cytoplasm, the three phases would be an oil-phase, an aqueous phase and the surface phase, but one of the fundamentals of any phase study is that thermodynamics is not concerned with the nature of the phases but only with the number of phases. However, there is one qualitative fact that would justify the extrapolation from two-dimensional to three-dimensional systems and that is that when experimenting with biliquid foams whenever three different fluid phases met, there was extraordinary motion between them resembling the simulated protoplasmic motion described in the two-dimensional model.

(No attempt has yet been made to explain the problems met with in industry in flotation, flocculation, and filtration in terms of the concepts introduced here, but these approaches might very well prove profitable).

This paper can only be considered as a preliminary investigation, but it does suggest that the secrets of life lie in the behaviour of immiscible liquids under the influence of their interfacial forces.

Acknowledgements

Thanks are expressed to Mrs. F. Horwitz who helped with the biliquid experiments and to Mr. T. Ambrose who helped with the photographs.

Literature Cited

1. Sebba, F., J. Colloid Interface Sci., (1971), _35_, 643
2. Sebba, F., Nature (London), (1963), _197_, 1195
3. Adamson, A.W., "Physical Chemistry of Surface" 2nd ed., pp. 348, 383, Interscience, New York, N.Y. 1967

4. Sebba, F., J. Colloid Interface Sci., (1972), 40, 468
5. Sebba, F., J. Colloid Interface Sci., (1972), 40, 479
6. Langmuir, I., J. Chem. Phys., (1933), 1, 756
7. Harkins, W.D., "Colloid Symposium Monograph", 6, p.24, Chemical Catalog Co., New York, N.Y., 1928
8. Sebba, F. and Briscoe, H.V.A., J. Chem.Soc., (1940), 114.

3

Mechanical and Surface Coagulation

W. HELLER

Department of Chemistry, Wayne State University, Detroit, Mich. 48202

Introduction

It is well known that colloidal suspensions may coagulate partially or even completely under the influence of agitation, such as stirring or shaking. In concentrated systems, particularly in those of originally uniform more or less pastelike or creamy consistency, such as concentrated latices or paints this may manifest itself primarily by the formation of microscopic or macroscopic lumps or crums and an increase in fluidity of the systems. In a more dilute system, separation into a more or less water clear upper layer and a concentrated lower layer will occur, the latter consisting primarily of coagulated colloid. This phenomenon of coagulation of colloidal systems by agitation, is generally referred to as "mechanical" coagulation. The present article is concerned with this type of coagulation.

Survey of the Literature

The first to point out that agitation should affect the rate of coagulation in a system which is already coagulating (as a result of the addition of coagulating electrolyte) was Smoluchowski ([1]). According to his theory, the collision number in a coagulating system is increased under the influence of flow and, therefore, the number of successful collisions also is increased. The actual effect of agitation upon a coagulating system was investigated systematically first by Freundlich and Basu who used CuO-sols of relatively low stability towards electrolyte ([2]). Figure 1 shows a sample of the quantitative data obtained by these authors on the coagulation of a CuO-sol after addition of a sodium sulfate solution so that the mixture contained 0.125 mmoles per liter sol. Curve 1 shows the progress of coagulation in absence of stirring and Curve 2 that observed on stirring the coagulating system. (The ordinate gives the actual sol concentration, in grams of Cu/liter sol, the abscissa gives the time in minutes elapsed since the electrolyte had been added.) It is apparent that stirring accelerates the coagulation considerably except in its final stage. Although no reference was made to this fact, it is clear from this figure that the effect of agitation was far too large to be accounted for on the basis of the Smoluchowski theory. The same authors found subsequently that addition of electrolyte made these sols susceptible to coagulation by agitation although the amount added was too small to produce coagulation in the system at rest.

Freundlich and Kroch (3) investigated for the first time this "pure" mechanical coagulation of CuO-sols. Typical results are reproduced in Figure 2. Two facts clearly emerge: (1) the coagulation by stirring is faster the higher the rate of stirring; (2) the residual colloid concentration declines linearly with the time of agitation suggesting a zero order reaction. The same linear relationship was found in a subsequent paper by Freundlich and Loebman (4) on the mechanical coagulation of Goethite (α-FeOOH)-sols. They were the second type of system found to be susceptible to mechanical coagulation. The ordinate in Figure 3, illustrating their results, gives the residual concentration in mg Fe/liter and the abscissa the time of agitation in hours. The curve refers to an α-FeOOH sol which had been stirred at 1000 rpm. An intriguing and as yet unexplained new fact, indicated by this figure, is that the coagulation by stirring alone may not go to completion but may stop after coagulation has progressed sufficiently far.

The original assumption that one was dealing with a purely mechanical coagulation did not readily agree with the finding already made by Kroch and extensively verified by Freundlich and Loebmann that the rate of coagulation increased roughly with the square of the velocity of stirring. The latter relation, they reasoned, would follow naturally, in the case of a surface coagulation if the water-air surface, created in the middle of the sample by the revolving stirrer, were cylindrical and if the height of this cylinder increased in direct proportion to the speed of the stirrer. (The actual contours of the liquid-air interface are more complex). This idea of a surface coagulation seemed to be reasonable also in view of the interesting effect found by Zsigmondy many years earlier (5) that shaking certain aqueous sols with inert organic liquids produced coagulation. Further support for this explanation of the nature of mechanical coagula-

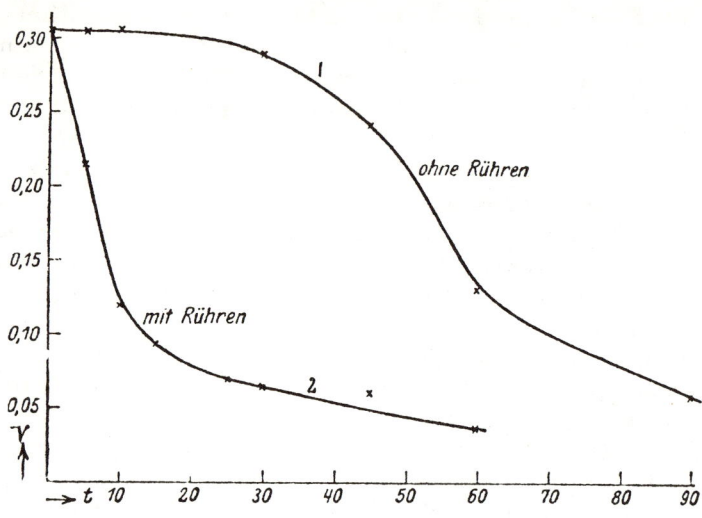

Figure 1. Acceleration by stirring of coagulation of CuO-sols by electrolyte (2). Ordinate: residual sol concentration (g Cu/l.). Abscissa: time, t, in minutes, elapsed since addition of Na_2SO_4 solution without stirring (curve 1) and with stirring during entire time, t (curve 2). Na_2SO_4 concentration: 0.125 mmol/l. sol.

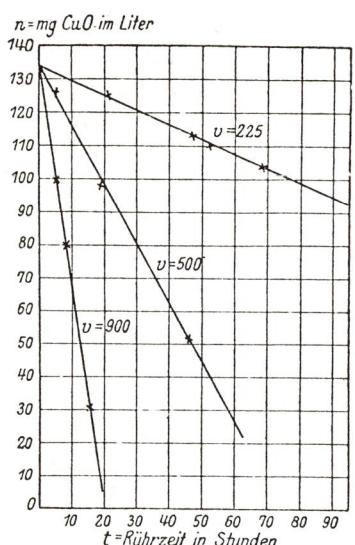

Figure 2. Coagulation of CuO-sols by stirring alone (3). Ordinate: residual sol concentration (mg CuO/l.). Abscissa: time of stirring in hrs. v: number of rotations of stirrer/min.

tion as a surface coagulation was provided by subsequent investigations by Freundlich and Recklinghausen (6). They compared the effect of stirring in presence and absence, respectively, of a liquid-air interface at appreciable rates of rotation (700-1000 rpm). The straight line A in Figure 4 shows that the CuO-sol, used again in these experiments did not coagulate at all if the stirring was done in absence of a water-air interface. Curve B on the other hand, obtained on admitting one single air bubble into the system, shows that coagulation occurred and progressed until the air bubble was removed (approximately 10 minutes after the start of the experiment) whereupon coagulation stopped completely.

Freundlich and Recklinghausen furthermore found in verification of Zsigmondy's qualitative finding (5) that coagulation occurs also, in absence of a liquid-air surface, on stirring a CuO-sol in presence of mechanically emulsified bromnaphthalene, i.e. in presence of a liquid-liquid interface only. Representative results are shown in Figure 5. Here Curve A pertains to 600 rpm and Curve B to 1000 rpm. The faster rate of coagulation in the latter case was explained by assuming the formation of smaller emulsion droplets (of bromnaphthalene).

Further support for the explanation of mechanical coagulation as a surface coagulation may be derived from independent work carried out about the same time by Stark. This author (7) investigated the coagulation produced by bubbling gas through especially prepared sols of ferric oxide. He explained the appreciable coagulation found not as a surface coagulation, but by assuming that the colloid was merely adsorbed at the surface of the rising gas bubbles and that coagulation took place after the gas bubbles burst, due to a critical increase, in the upper region of the sol, of the bulk colloid concentration. Figure 6 gives some representative results obtained by Stark with gas bubbles whose diameter was somewhere between 1-17 mm. The time indicated on the abscissa is the time elapsed after bubbling of gas (probably air) had started and the ordinate indicates the residual concentration of uncoagulated colloid in grams of iron per liter. Curve 1 represents the coagulation curve in absence of added electrolyte. The other curves show the effect of various concentrations of potassium sulfate on the rate of coagulation.

In the quantitative investigations of Freundlich and collaborators only two types of sols were coagulated "mechanically", those of α-FeOOH and CuO. Both of these, and also the iron oxide sols used by Stark, while perfectly stable at rest, required only small amounts of electrolyte to cause coagulation even in

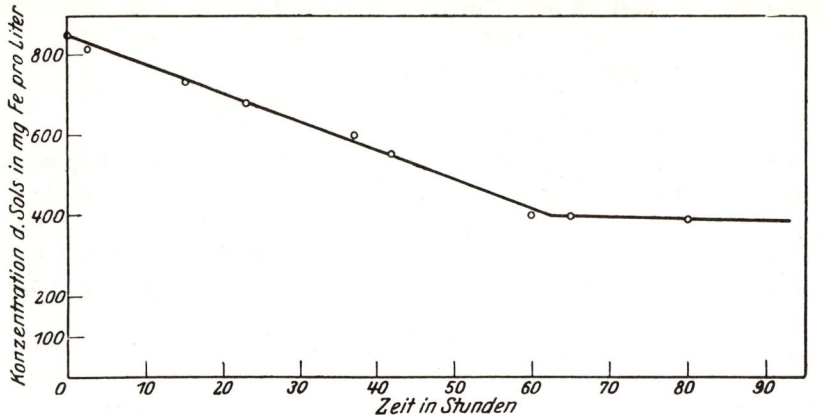

Kolloid Beihefte

Figure 3. Coagulation of an α-FeOOH sol by stirring (4). Ordinate: residual sol concentration in mg Fe/l. Abscissa: time of stirring in hrs. Rate of stirring: 1000 rpm.

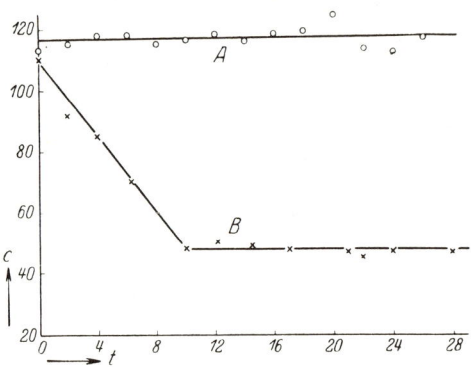

Elektrochemische Zeitung

Figure 4. Effect of stirring upon a CuO-sol in presence and absence of a liquid–air interface (5). Ordinate: residual sol concentration (mg Cu/l.). Abscissa: time of stirring in hrs at 1000 rpm. A: no liquid–air interface is present. B: air bubble present during first 10 hrs.

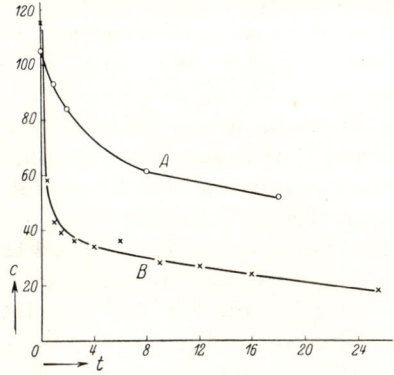

Figure 5. *Effect of stirring on a CuO sol in presence of a liquid–liquid interface and absence of a liquid–air interface (6). Ordinate: residual sol concentration (mg Cu/l.). Abscissa: time of stirring in hrs. Curve A: 600 rpm. Curve B: 1000 rpm.*

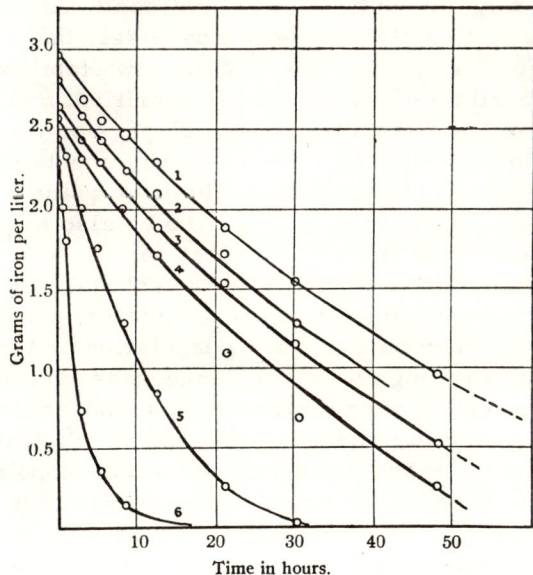

Figure 6. *Coagulation of iron oxide sol prepared according to Sorum [J. Am. Chem. Soc. (1928) 50, 1263] by bubbling air through it (7). Ordinate: residual colloid concentration. Abscissa: duration of bubbling. Curve 1: no electrolyte added to sol. Curves 2–6: increasing amounts of K_2SO_4 added to sol at time zero.*

absence of agitation. This led Heller (8) to suspect that the type of coagulation reviewed here may be produced in any colloidal solution provided its natural stability is reduced sufficiently. To check this, eight colloidal solutions were used whose "coagulation value" (9) was systematically reduced from 300 to 2.0 mmol of NaCl per liter. By using differing methods of preparation and differing degrees of dialysis it was found that a coagulation value of about 20 mmol/liter of a 1:1 electrolyte (such as NaCl) made the sols subject to "mechanical" coagulation. These results together with subsequent results on freeze coagulation (10) led to the following tentative explanation of "mechanical" coagulation (10):due to the difference in the dielectric constant in the interior of a colloidal solution and in the liquid-air interface, a critically reduced sol stability may lead to the situation where none of the particle collisions in the interior of the sol are successful while they may be in the interface. This importance of the dielectric constant for sol stability in general, was subsequently verified on a quantitative basis by Verwey and Overbeek (11). While therefore the explanation of the "mechanical" coagulation as a surface coagulation appeared reasonable also from the theoretical point of view, magneto-optical investigations (still unpublished at that time, see Section VIII) made it most likely that strong turbulence in the interior of a sol of critically low stability may lead to successful particle collisions due to a critical momentary predominance of the attractive London dispersion forces over the repulsive Coulomb forces as a result of a momentary deformation of the electrostatic double layer (10). The later quantitative theory (11) of colloid stability seemed to give substance also to this alternate possible cause of "mechanical" coagulation.

 The proposition that a critically low sol stability is necessary and sufficient to make "mechanical" coagulation possible no matter whether it may be a surface coagulation or true mechanical coagulation resulting from turbulence, was strengthened more recently by interesting experiments by Roe and Brass (12). These authors added to polystyrene latices which could be coagulated mechanically, more potassium palmitate or potassium laurate than had been necessary for the purpose of polymerization. They found that this addition increased the resistance of the system to mechanical coagulation appreciably, more the larger the soap concentration. This follows from Figure 7 in which these authors summarize their results. The ordinate gives a measure of the mechanical stability by indicating the time of stirring needed (at 14,000 rpm) in order to produce 100% coagulation. It is apparent that not only an increase in concentration of the long chain ions adsorbed at the surface of the polystyrene but also an increase in their chain length increased the mechanical stability.

In more recent years the interest in "mechanical" coagulation has been limited essentially to exploring the potential practical usefulness of the effect in order to remove colloids from aqueous systems. Comprehensive studies on this aspect were carried out by Mokrushin and collaborators (13-21) who concentrated on expanding the work of Stark (7) referred to above. The colloid coagulated by bubbling air through the systems was collected in a foam produced at the top of the liquid by addition of a suitable foaming agent. Further contributions to this practical aspect of "microflotation" were made subsequently by Matijevic and collaborators (22-24).

Objectives of the Studies Reviewed in the Following Sections of This Paper

There was little doubt in view of the literature results reviewed above that the liquid-gas and liquid-liquid interface, under certain conditions, might be capable of promoting coagulation (25). In absence of any quantitative theory, a decisive test of this reasonable hypothesis was not possible. It appeared most desirable to provide such a test. With the aid of such a test it could be hoped, in addition, to decide whether and when the "mechanical" coagulation may be a pure surface coagulation or a bulk coagulation, due for instance to turbulence, or may be a combination of the two possibilities.

A careful scrutiny of the literature data showed that a series of results reported appeared contradictory. A resolution of the respective problems also appeared mandatory. The first problem here concerns the kinetics of coagulation by stirring, shaking or by air bubbles. Figures 2, 3 and 4 show a linear decrease in colloid concentration as a function of time of agitation. This suggests that the coagulation is a _zero_ order reaction. On the other hand, Figures 1, 5 and 6 show more complicated rate curves. The fact that Figures 2, 3 and 4 were obtained in absence of electrolyte while Figure 1 and all curves except curve 1 of Figure 6 were obtained in presence of electrolyte, cannot explain the differences because the curves in Figure 5 also were obtained in absence of electrolyte. Moreover curve 1 of Figure 6 is not a straight line either. Stark believed that the experimental results given in Figure 6 indicate that the theory of Smoluchowski applies. In this theory the coagulation process is considered as a _second_ order reaction which differs, however, from the conventional second order reaction by the fact that the reaction products themselves are reactants. In other words, after one individual colloidal particle has aggregated with another, this secondary particle may in turn aggregate again with either another primary or secondary particle and so forth. On scrutinizing closely Stark's experimental results, it is seen that the rate constant derived on the basis of the Smoluchowski theory is not really a constant but varies considerably with the progress of coagulation both in presence and absence of additional electrolyte. That the Smoluchowski theory is not applicable to mechanical coagulation by stirring alone had in fact been found already by Freundlich and Kroch. An approach to applicability of the Smoluchowski equation was indicated only in presence of sufficient electrolyte. To complicate things even more, it may be mentioned that Mokrushin found that the colloid extraction by air bubbles followed the rate of a <u>first</u> order reaction (<u>15</u>) and that the Schulze-Hardy rule did not apply (<u>15</u>).

A second intriguing problem is the effect of electrolyte addition to the stability of a sol which can be coagulated mechanically. In the work by Freundlich and Basu (<u>2</u>) and by Freundlich and Kroch (<u>3</u>) the addition of electrolyte promoted coagulation. In the work by Freundlich and Loebmann (<u>4</u>) on the other hand, the rate of mechanical coagulation was found to be reduced at low electrolyte concentrations while it was increased at higher ones. In the experiments by Stark (<u>7</u>), finally, only promotion of coagulation was found on adding electrolyte.

Theory of Surface Coagulation

Theoretical equations for the reaction rate expected of the surface coagulation were developed by J. Peters and the writer (<u>26</u>). The equations arrived at are based on the following assumptions:

1) The coagulation proceeds exclusively in the liquid-air surface, and it is a bimolecular reaction.

2) The contribution of aggregates to the rate of coagulation (Smoluchowski rate increment) is neglected. In addition, the aggregates, by their return to the bulk, are assumed not to affect significantly the surface area available for occupancy by unreacted primary colloidal particles.

3) There is sufficient convection to exclude diffusion of particles to and from the surface as a rate-determining factor in surface coagulation.

4) There exists a definite steady state for the distribution of primary colloidal particles between bulk and surface. This distribution is assumed to follow a Langmuir adsorption isotherm.

5) The rate of adsorption is assumed to be large enough relative to the rate of mechanical surface renewal so that the adsorption equilibrium is not affected.

The resulting differential equation is

$$-\frac{dc}{dt} = \frac{S}{V} K_0 \frac{c^2}{(K_1 + K_2 c)^2} \tag{1}$$

Here, S is the surface area at constant rate of surface renewal; V is the volume of the solution; c is the concentration of the colloid in solution (moles/ml); t is time; and K_0, K_1 and K_2 are constants.

Equation (1) can be rearranged to

$$\frac{c}{(-dc/dt)^{1/2}} = \left(\frac{V}{K_0 S}\right)^{1/2} (K_1 + K_2 c) \tag{2}$$

for the purpose of convenient graphical checks of the validity of Equation (1).

Consequently, the plot of $[c/(-dc/dt)^{1/2}]$ vs c should be linear and the ratio of slope to intercept should represent K_2/K_1.

Equation (1) has the following properties which are of interest in connection with the reaction order of surface coagulation:

$$-\frac{dc}{dt} = k'\left(\frac{S}{V}\right) \quad \text{if} \quad K_2 c \gg K_1$$

$$-\frac{dc}{dt} = k''\left(\frac{S}{V}\right) c^2 \quad \text{if} \quad K_2 c \ll K_1$$

This shows that the equation derived for a bimolecular surface reaction may also satisfy an apparent zero-order reaction, an apparent first-order reaction being an intermediate possibility. The apparent reaction order depends therefore, (1) on the colloid concentration and (2) on the value of the constants K_1 and K_2, which depend on the distribution of the colloidal particles between bulk and surface. For a given value of K_1 and K_2, one would expect that the apparent order of reaction increases with increasing degree of coagulation, the initial period of an apparent zero-order reaction covering a larger portion of the coagulation process, the lower the initial colloid concentration, c_0. Moreover at equal c_0 and c, the probability of an apparent zero-order reaction would be higher, the higher the affinity of the colloidal particles is for the liquid-air surface (the more hydrophobic the particles are in case of a water-air surface).

On integrating Equation (1), one obtains on introducing f for c/c_0

$$\frac{1-f}{f} + 2c_0 \frac{K_2}{K_1} \ln \frac{1}{f} + \left(\frac{c_0 K_2}{K_1}\right)^2 (1-f) = K_s t \qquad (3)$$

where $K_s = c_0 K_0 S/K_1^2 V$. From this equation it follows readily that with an increase in K_2/K_1 the reaction rate at any stage of the coagulation changes gradually from the typical second-order reaction to that of a typical zero-order reaction. This explains again the puzzling fact mentioned in the preceding section that different authors found different orders for the reaction rate of coagulation induced in different systems by mechanical agitation or bubbling of gases.

Experimental Test of the Theory of Surface Coagulation

The first tests of the theory of surface coagulation were carried out by Peters (26) studying the coagulation produced by bubbling nitrogen gas through a colloidal solution of α-FeOOH prepared in the same manner as the sols used by Freundlich and Loebmann (4). The strongly anisometric crystals of α-FeOOH had an average length of 83 nm, a width of 10 nm and the standard deviation from these dimensions was sufficiently small to characterize the systems as nearly monodisperse, disregarding a relatively minor amount of nearly cubical crystals present at the same time. Figure 8 in which c_r represents the concentration of the uncoagulated (residual) colloid in percent of c_0, shows that equation (2) satisfies the coagulation process extremely well down to a residual concentration of merely 5% of the starting concentration. The slope of the straight line increases, as one expects, with the rate at which nitrogen gas is bubbled through the colloidal solution, the two cases considered being a flow rate of .65 and of 1.65 cm^3/sec. On taking into account the difference in the experimentally determined bubble sizes in cases I and II and the differences of the rate of bubbling in the two cases one arrives at a result that is the same within ±4%. This further supports the theoretical equation.

Although the probability of a "mechanically" effective turbulence is rather remote on bubbling gas through a solution, special tests were carried out (26) where even the possibility of minor

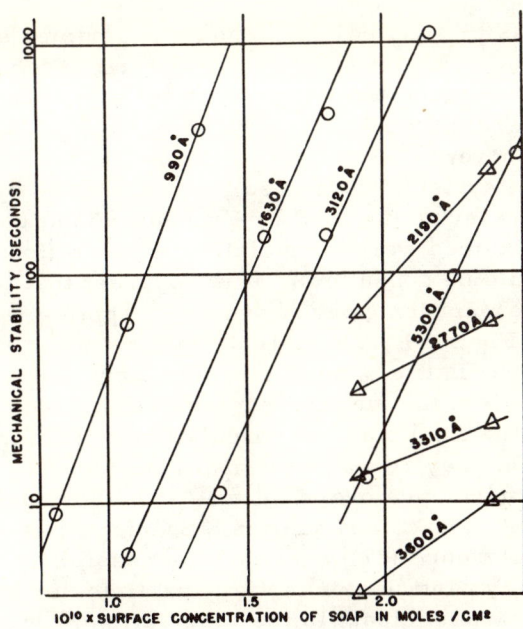

Figure 7. Effect of soaps upon the "mechanical stability" of polystyrene latex (12). Abscissa: concentration, in the water–air interface of K-palmitate (○) and K-laurate (△). Ordinate: number of seconds required to produce complete coagulation by high speed stirring (14000 rpm) of a 60 ml sample containing 24 volume percent of polymer. Parameter: particle "size" in Angstrom units.

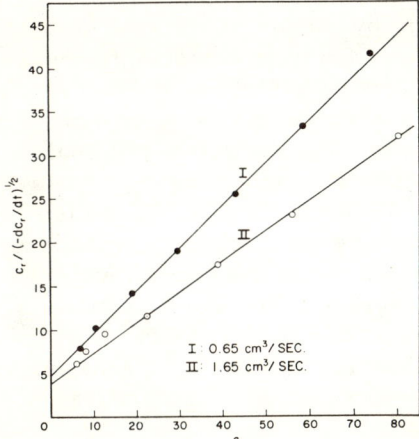

Figure 8. Coagulation of α-FeOOH sol by nitrogen bubbles. Test of equation (2) (26). c_r: colloid concentration of residual sol in percent of original colloid concentration, c_0 (370 mg Fe/l.). Average particle diameter: 83 nm. pH: 6.5. Parameter: rate of N_2-flow.

turbulence was eliminated by extending experiments over a period of 2000 hrs replacing bubbling of gas through a solution by gentle mechanical renewal of the liquid-air interface. These experiments also were fully satisfied by the equation of surface coagulation given above.

The same systems were used for coagulation by stirring (27) using the same technique used by Freundlich and Loebmann (4). The curves obtained were essentially the same (for details reference (27) should be consulted). It is apparent that the equation for surface coagulation given above again applies very well as is evident from Figure 9. The only deviations occur in the final stage of the coagulation when the residual concentration has fallen to less than 15% of the starting concentration. The cause of these deviations in the final stage of coagulation is still unresolved. This final stage may represent a manifestation of a coagulating effect due to turbulence overshadowed at higher concentrations by the effect of surface coagulation. Special experiments [(see Figure 4 in reference (27)] excluded the possibility that at this stage secondary particles or tertiary particles or still larger aggregates may participate in the coagulation process by providing an additional reaction locus in the interior of the colloidal solution.

The Role of Turbulence

In order to detect whether and to what extent turbulent flow may actually produce or participate in coagulation - a possible theoretical reason for it having been mentioned in Section II - experiments had to be carried out in which surface coagulation could be excluded as a factor. To this effect, the same type of colloidal solutions discussed above were circulated, by DeLauder (28), in a closed ring of Tygon tubing by means of a Sigma Motor Finger pump. Since Tygon tubing was found to contain impurities which have a small but distinct stabilizing effect on aqueous colloidal solutions, the tubing was subjected to a very rigorous cleansing process prior to use. On leaving the same colloidal solution, as used for the other experiments, for extended periods of time in the thoroughly cleaned and closed circular Tygon tubing ring, no coagulation could be detected by checking the turbidity of the system. On the other hand, on operating the pump - which had no contact whatever with the colloidal solution - a definite particle aggregation occurred although the effect was far too small to be detectable by any means other than by light scattering (turbidity) measurements, the aggregates formed being ultramicroscopic in size. Figure 10 shows that the turbidity determined at 6250 Å increased by a little more than 10% in 10 hrs of pumping. Since air bubbles, however small, had been rigorously excluded, the only possible causes of particle aggregation were either the turbulence created at the point where the squeezing action of the pump at the Tygon tubing took place, or the interface between the Tygon tubing and the colloidal solution. It could be shown that the solid-liquid interface in fact can be a locus of particle aggregation (28). It becomes detectable, however, only if sufficient turbulence is created in the system so that the fine skin of coagulum of ultramicroscopic thickness formed at the solid-liquid interface is removed so as to free the solid-liquid interface as a site for further aggregation. Since the colloidal solution in the Tygon tubing, if left at rest for days, did not show any change in turbidity in absence of pumping, the coagulation produced as evident from curve section I in Figure 10, was therefore due either to turbulence directly or to the turbulence induced removal of a colloidally-thin layer of coagulum from the liquid-solid interface. A differentiation between these two possibilities is not possible at the present time. No matter whether the contribution of turbulence was thus direct or indirect, it is evident that the effect is very small compared to the effect of a coagulation at the liquid-air interface. This is particularly well demonstrated by Section II of the curve in Figure 10. Here, at the time t_1, a small air bubble was introduced into the system and kept at the point of maximum turbulence (where the pumping action took place). Immediately a steep increase in turbidity was observed. Additional experiments (28) showed that placing and keeping the air bubble rather at the top of a vertical section of the loop where it remained quiescent and undeformed did not produce any increase in turbidity. This, therefore, proved that the increase in turbidity observed in Section II of Figure 10 was due to the constant removal, due to the turbulence induced continuous deformation of the air bubble, of an ultramicroscopic layer of aggregates.

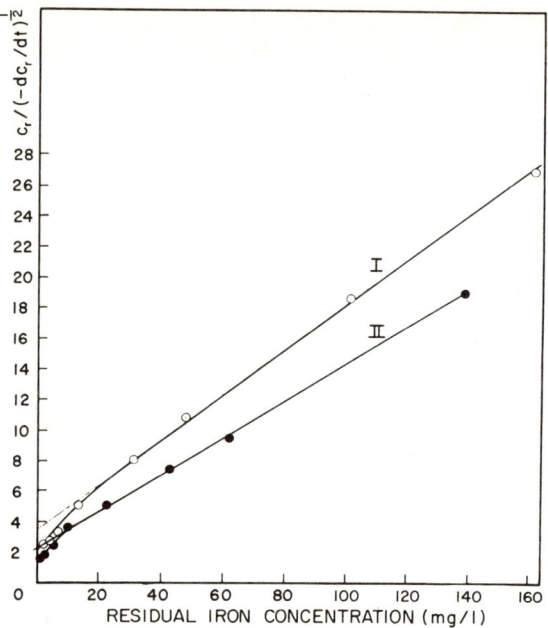

Figure 9. Coagulation of α-FeOOH sol by stirring. Test of equation (2) (27). I: 600 rpm. II: 900 rpm. c_0: 205 mg Fe/l.

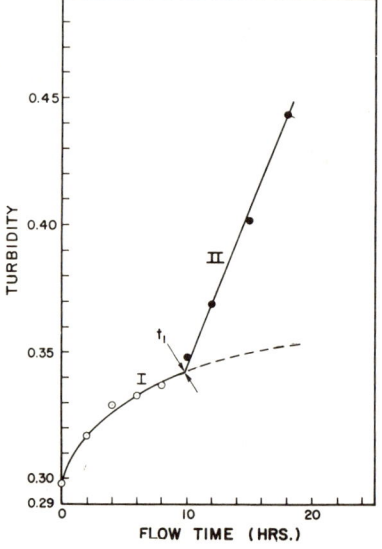

Figure 10. Coagulation of α-FeOOH sol during locally turbulent flow in loop of Tygon tubing (28). I. In absence of any liquid–gas interface. II. In presence of single air bubble in turbulent section of flow. t_1: time at which bubble is introduced. Flowrate: 150 ml/min.

Factors which Promote or Impede Surface Coagulation

Aggregation of two colloidal spheres in a dilute system becomes likely when the positive maximum of the potential energy curve is less than 15 kT (9 kcal/mol)(11). Investigation (27) of the temperature-dependence of the coagulation produced by stirring in the α-FeOOH sols discussed above led to the result that the apparent activation energy of surface coagulation is 8.93 kcal/mol, remarkably close to the theoretical value for sphere coagulation. The potential energy maximum was evidently lower in the water-air interface than in the interior of the sol stable at rest. Since a decrease in the height of the potential energy maximum occurs on raising the value of the Debye-Hückel constant, κ, it was expected in line with the earlier observations reported above (8, 10) that in very stable systems in which neither bubbling of gases through the solution nor stirring produces coagulation, one may bring it about by adding suitable electrolytes in a concentration sufficient to depress the potential energy maximum to the critical value in the water-air interface but not yet in the interior of the system. This was tested successfully on polychlorostyrene and polystyrene latices. Results obtained on the former are illustrated by Figure 11(28a). Up to a concentration of 0.15 mol KCl/liter latex stirring had little effect on the latex whose concentration was chosen at 185 mg/liter. At still higher KCl concentration stirring coagulation occurred, more the larger the electrolyte concentration. It is important to note that only from about 0.2 mol/liter KCl on significant bulk coagulation made its appearance, complicating the entire process. Completely analogous results were obtained on bubbling nitrogen through the latex rather than by employing stirring (28a). The situation is not quite this simple in the case of colloidal solutions of inorganic crystals. Thus, in the α-FeOOH systems discussed here primarily one observes at low KCl concentrations first an increase in the stability of the systems toward coagulation by stirring or bubbling until at a KCl concentration - which is still below that at which bulk coagulation becomes pronounced - the trend is reversed. An example of stabilization is given in Figure 12 (29). At more than 1.14 mmol KCl the trend is reversed, i.e. the electrolyte then promotes coagulation by stirring. The small increase to 1.17 mmol KCl was sufficient in this case to reverse the trend at the same rpm (900). Essentially the same results were obtained more recently (30) by DeLauder for other than 1,1 electrolytes, the only difference being that in the case of higher valent gegen-ions the concentra-

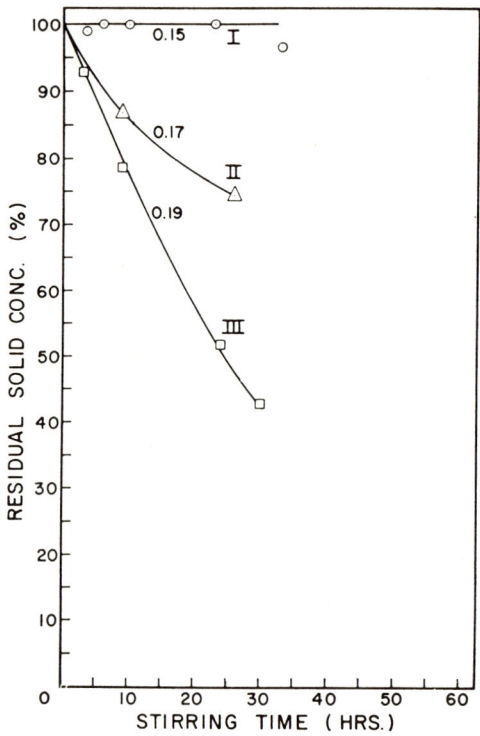

Figure 11. Effect of electrolyte upon the coagulation of Dow polychlorostyrene latex by stirring at 600 rpm (28a). Parameter: KCl concentration in mol/l. latex. c_0 of polymer: 135 mg/l.

tion range of the electrolyte within which stabilization occurs is considerably more limited and the stabilization itself is much less pronounced. Responsible for this stabilization in the interface may be not only a change in the interfacial electrostatic potential, but also competitive adsorption in the interface or hydrophylization of the colloidal particles. As regards the effect of competitive adsorption and/or hydrophylization, experiments by J. Peters on the effect of the addition of DDAO (N, N-dimethyl-n-dodecylamine oxide) proved the importance of the second and/or third factor mentioned. This compound, which at the low concentrations employed, did not affect the surface tension of the α-FeOOH sol nor lead to foam formation increased the stability of the sol toward mechanical action drastically, as is evident from Figure 13 (31). If the third factor was decisive, it would mean that the adsorption of the colloidal crystals in the interface had been reduced due to adsorption of DDAO at their surface making them thus more hydrophilic and therefore less inclined to go into the liquid-air interface (32). Drastic complications of this stabilizing effect occur if one employs additives which promote foam formation. In that case, the tremendous increase in liquid-air surface due to the foam may actually overcompensate the stabilizing effect of the additive. A good example is that of polyethyleneglycol which at extremely low concentrations (10^{-9} to 10^{-5} mol/liter) promotes the surface coagulation of α-FeOOH sols but at concentrations in excess of 10^{-4} mol/liter impedes it (31).

Morphology of the Aggregates Formed

A characteristic feature of colloidal systems coagulated by either bubbling gases through them or by mechanical agitation or shaking is that they exhibit, on slight agitation for the purpose of homogenizing the system, Schlieren of a strikingly silky character. The phenomenon at once indicates a strong anisometric character of the aggregates which to a large extent are too small to settle within less than a few hrs. The phenomenon, first observed on the classical sols subjected to mechanical coagulation, those of CuO and α-FeOOH, is common to all systems investigated more recently (33), e.g. those of TiO_2 and Cr_2O_3. In the case of the iron oxide systems, the anisotropy of the aggregates formed by coagulation by stirring could be investigated easily by means of magneto-optics because of the strongly paramagnetic character of the colloidal particles which leads to an appreciable

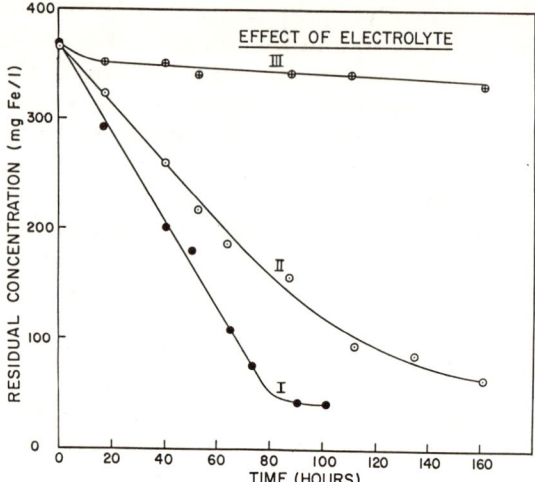

Figure 12. Effect of electrolyte upon the coagulation of α-FeOOH sol by stirring (29). Abscissa: time of stirring (same as time elapsed since addition, if any, of electrolyte). I: stirring at 900 rpm in absence of KCl. II: stirring at 900 rpm in presence of 1.14 mmol KCl. III: coagulating effect of 1.14 mmol KCl in absence of stirring.

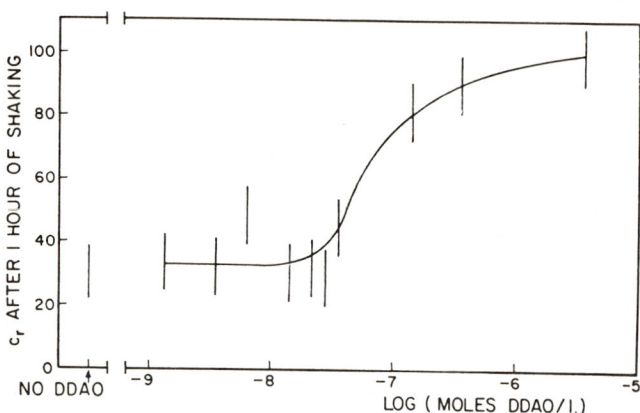

Figure 13. Effect of surface active agent upon coagulation, by 1-hr shaking, of α-FeOOH sol (31). DDAO: N,N-dimethyl-n-dodecylamine oxide. Frequency of vertical test tube oscillation: 6/sec. Amplitude of vertical test tube oscillation: 3.5 cm. Frequency of simultaneous horizontal test tube oscillation: 1/sec. Amplitude of horizontal test tube oscillation: 0.2 cm.

orientation in the magnetic field. A few examples are given in Table I (34) which pertains to an α-FeOOH sol of the type discussed above. The quantity, β, represents the phase difference between the extraordinary component of a linearly polarized beam traversing a cell containing a small amount of the sol placed into the magnetic field of a field strength of 28,000 Gauss, observation being perpendicular to the direction of the magnetic field. This phase difference is directly proportional to the birefringence ($n_e - n_0$). The quantity, δ, represents the rotation of the plane of polarization (strictly speaking, the rotation of the major axis of the elliptical vibration), an effect which could be due, in part, to linear dichroism and, in part, to dityndallism, i.e. to anisotropic light scattering (35). Contribution of the dichroism was eliminated, however, by carrying out the experiments at 620 nm where the absorption of α-FeOOH is negligible. The uncoagulated sol is seen to exhibit a significant degree of magnetic birefringence but no measurable anisotropic light scattering. On shaking the sol in a shaking machine for two hours, the oscillation period of the machine being 1/sec - which resulted in a 50% coagulation - a major increase in birefringence was observed and a small but now well measurable degree of anisotropic light scattering manifested itself. On producing a nearly 100% coagulation by increasing the period of the shaking machine to 6/sec and extending the treatment to six hours, a truly tremendous ellipticity and in addition, an unusually large rotation of the plane of polarization is observed. These very large effects which, in virtual absence of any primary particles, are due exclusively to aggregates, show that they are highly anisotropic. Part of this anisotropy is due to the flake-like character which also follows easily from dark-field microscopy but a considerable portion of the effect must also be attributed to intrinsic anisotropy, i.e. to non-random orientation of the primary particles aggregated within a flake, the simple reason being that the negative sign of the ellipticity, if due to birefringence, can in these systems, as explained elsewhere (36) be due only to predominance of intrinsic anisotropy. Part of the extremely large ellipticity may, of course, represent an additional effect that of pseudo-birefringence, resulting from ellipticity of the light scattered in the forward direction - an effect postulated many years ago (35) for anisometric particles not small compared to the wavelength.

Table I. Effect of Coagulation by Shaking upon Magnetic Birefringence and Magnetic Dityndallism (Anisotropic Light Scattering) of An Iron Oxide Sol
Coagulation Value (9): 2 mmol NaCl/liter
Wavelength Used: 620 nm
Fieldstrength: 28,000 Gauss
No Electrolyte Added

Treatment	Degree of Coagulation (%)	β	δ
Sol at rest	0	$-6°50'$	$0°\ 0'$
Shaken for 2 hours (1 period/sec)	50 ± 2	$-8°19'$	$<-0°10'$
Shaken for 6 hours (6 periods/sec)	95 ± 2	$-172°20'$	$-22°40'$

Why are the ultramicroscopic aggregates flake-like and therefore possess form anisotropy? The probable reason follows readily on recalling that on the liquid-air surface of many colloidal solutions whose natural stability is limited, a very thin microscopically or even visually visible skin forms on long standing. They are readily observable for instance, on V_2O_5 sols prepared according to the well known Biltz method. The anisotropic silky aggregates produced by coagulation by gas bubbles or by stirring may, therefore, represent fragments of the two dimensional coagulum formed at the interfaces at which the respective coagulation takes place. This explanation agrees with the concept of surface coagulation treated throughout this paper.

It is useful to add that there are two additional factors which may produce and contribute to an intrinsic anisotropy of the silky aggregates: one is stress orientation of the primary particles within the ultramicroscopic flakes under the influence of turbulence. This follows from results obtained on colloidal solutions of β-FeOOH prepared according to the well known Graham method. In one such case (37) the sol stability was so high that only in presence of 270 mmol NaCl visually perceptible aggregate formation occurred one day after electrolyte addition. Before the addition, vigorous shaking had no effect whatsoever. In presence of 1000 mmol NaCl, which led to complete coagulation by the time of the optical measurements, β had the value of $-2°4'$ in the sample coagulating without shaking and a value of $-111°20'$ in the sample which had been shaken vigorously while coagulating quantitatively (the measurements were made on the samples after suspending the aggregates temporarily). The second factor which also may be involved, is the tendency of anisometric colloidal particles to aggregate in a regular manner if the coagulation is sufficiently <u>slow</u>, i.e. when only collisions occurring under favorable mutual constellations are effective. This results in anisotropic aggregates which have a structure of either smectic or nematic character. This "ordered" coagulation is described elsewhere (38). It occurs also in ordinary bulk coagulation.

Acknowledgement

The work done by William B. DeLauder was supported by a grant from the Paint Research Institute.

Literature Cited

1. Von Smoluchowski, M., Physik. Z. (1916) 17, 587; Z. Phys. Chem (1917) 92, 129.
2. Freundlich, H. and Basu, S. K., Z. Phys. Chem. (1925) 115, 203.
3. Freundlich, H. and Kroch, H., ibid. (1926) 124, 115.
4. Freundlich, H. and Loebmann, S., Kolloid Beihefte (1929) 28, 391.
5. Zsigmondy, R., Z. Elektrochem. (1916) 22, 202.
6. Freundlich, H. and Recklinghausen, R. v., Z. Phys. Chem. (A) (1931) 157, 325.
7. Stark, H., J. Amer. Chem. Soc. (1930) 52, 2730.
8. Heller, W., Comp. rend. (1934) 198, 1776.
9. The "coagulation value" represents the concentration of 1:1 electrolyte (NaCl or KCl) needed to lead, 2 hours after its addition to the sol, to the formation of a boundary below the meniscus due to incipient settling of aggregates.
10. Heller, W., Comp. rend. (1934) 199, 354.
11. Verwey, E. J. W. and Overbeek, J. T. G., "Theory of the Stability of Lyophobic Colloids", Elsevier, New York (1948).
12. Roe, C. P. and Brass, P. D., J. Colloid Sci. (1955) 10, 194.
13. Mokrushin, S. G., Bull. Mendeleev All-Union Chem. Soc. (1953) 2, 26.
14. Mokrushin, S. G. and Potaskuev, K. G., Koll. Zh. S.S.S.R. (1956) 18, 215.
15. Schveikina, R. V. and Mokrushin, S. G., Colloid J. U.S.S.R. (1958) 20, 223.
16. Schveikina, R. V. and Mokrushin, S. G., Zhur. Priklad. Khim. (1958) 31, 943.
17. Schveikina, R. V. and Mokrushin, S. G., Zhur. Priklad. Khim (1958) 31, 1109.
18. Mokrushin, S. G., Koll. Zh. S.S.S.R. (1950) 12, 448.
19. Schveikina, R. V. and Mokrushin, S. G., Izv. Vyssh. Ucheb. Zaved. Khim. Khim. Tekhnol. (1958) 8.
20. Mokrushin, S. G. and Milyntina, M. I., Colloid J. U.S.S.R. (1953) 15, 217.
21. Mokrushin, S. G. and Sheina, Z. G., ibid. (1954) 16, 361.
22. Matijevic, E., Abramson, M. B., Schulz, K. S. and Kerker, M., J. Phys. Chem. (1960) 64, 1157.
23. Cassell, E. A., Matijevic, E., Mangravite Jr., F. J., Buzzell, T. M. and Blabac, S. B., AICHE J. (1971) 17, 1486.

24. Mangravite Jr., F. J., Cassell, E. A. and Matijevic, E., J. Colloid Interface Sci. (1972) 39, 357.
25. Stamberger, P. [J. Colloid Sci. (1962) 17, 146] opposed the concept of a surface coagulation by stirring.
26. Heller, W. and Peters, J., J. Colloid Sci. (1970) 32, 592.
27. Peters, J. and Heller, W., ibid. (1970) 33, 578.
28. DeLauder, W. J. and Heller, W., J. Colloid Sci. (1971) 35, 308.
28a. Heller, W. and DeLauder, W. B., J. Colloid Sci. (1971) 35, 60.
29. Peters, J. (1970) unpublished results.
30. DeLauder, W. B. (1971) unpublished results.
31. Heller, W. and Peters, J., J. Colloid Sci. (1971) 35, 300.
32. If the second factor were decisive, it would mean that preferential adsorption of DDAO impeded the adsorption of α-FeOOH. The result obviously would be the same in both cases.
33. Diop, S. and Heller, W. (1974) unpublished results.
34. Heller, W. (1937) unpublished results.
35. Heller, W., "Propriétés Magnéto-optiques des Solutions Colloïdales; Actualités Scientifiques et Industrielles," Vol. 806, Hermann and Co., Paris (1939). Rev. Mod. Phys. (1959) 31, 1072.
36. Heller, W., Quimfe, G. and Yeou Ta, Phys. Rev. (1942) 62, 479.
37. Heller, W. (1937) unpublished data.
38. Heller, W., J. Phys. Chem. (1937) 41, 1041. Parts of this paper are difficult to understand due to translation difficulties.

4

Kinetics of Ultracentrifugal Demulsification

ROBERT D. VOLD and ALICE ULHEE HAHN

Department of Chemistry, University of Southern California,
University Park, Los Angeles, Calif. 90007

Introduction

　　Since the first use of the ultracentrifuge in the study of emulsions in 1962 (1,2,3) there has been a considerable volume of work devoted to investigating what can be learned about emulsions by this technique (4,5,6), as also to discussing its limitations (7). Unfortunately there is great variability as to how different authors have treated the primary ultracentrifugal data, which precludes quantitative comparison of results obtained by different investigators and prevents interpretation in terms of generally accepted concepts of physical and colloid chemistry.
　　In their published work on Nujol-water-sodium dodecyl sulfate (SDS) emulsions Vold and Groot and Mittal reported a nearly constant rate of separation of oil after an initially more rapid separation, and used this constant per cent oil separated per minute as their criterion of stability (8,9), although alternative modes of expression were also attempted (10,11). In cases where the per cent oil separated per minute did not remain constant with time of centrifugation an empirical equation was proposed (12, 13) which represented the data well in many but not all cases. Rehfeld (2, 14) used the volume fraction of cream remaining after a certain time of ultracentrifugation, ϕ, as the criterion of stability, and developed an empirical equation, $\phi = \phi_0 t^{-n}$, where ϕ_0 is the volume fraction at zero time and n is an arbitrary constant. Garrett originally described stability in terms of resistance to flocculation (3), using the slope of the line relating distance of the cream-water boundary from the center of rotation to the time of ultracentrifugation as its measure. Later (5) he adopted resistance to separation of oil as the criterion of stability, using the slope of the linear relation found between log [volume % oil separated at 115,000 seconds) - (volume % oil at time t)] and time as its measure.
　　In the present paper earlier data obtained in this laboratory are reinterpreted, using either volume of oil separated, or volume per cent of cream remaining, as a measure of the extent of coales-

cence, rather than per cent oil separated, as in the earlier treatment of the data. When this is done, it is shown that the rate of separation of oil from all the emulsions hitherto studied can be represented by the usual rate equations of classical kinetics, either zero, first or second order. This makes possible the calculation of specific reaction rate constants, which are of great value for attempting to identify the rate determining step or steps in demulsification, since the values can be compared with theoretical values dependent on which of the many processes occurring before separation of bulk oil (6,8) actually determine the over-all observed rate. The order of reaction can also be used to investigate whether coalescence or flocculation is the more important process in determining the rate of demulsification of a given emulsion (15,16).

One of the more confusing aspects of the literature is the apparently equally successful treatment of the data on ultracentrifugal separation of oil from Nujol-water-SDS emulsions as either a zero order (1,8,11) or a first order (10,11) process. New data obtained with ultracentrifuge cell centerpieces of differing area are presented which make possible an unambiguous choice between these alternatives.

Experimental

Nujol from Plough, Inc., New York, was used without further purification. The manufacturer's specifications of its characteristics are: Saybolt viscosity, 360 to 390 at 100°F.; specific gravity 0.880 to 0.900 at 60°F.

The same sample of sodium dodecylsulfate (Eastman Kodak #5967) was used as in previous work (9). It was extracted with ethyl ether in a Soxhlet extractor for twenty hours to eliminate any traces of lauryl alcohol present as an impurity.

The methods used for preparation and characterization of Nujol-water-SDS (0.2 and 0.4% on the basis of the aqueous phase) emulsions were the same as described by Vold and Groot (1). Emulsions were ultracentrifuged at 39,460 r.p.m. at 20°C, and the volume of each layer (oil, cream, water) calculated from the positions of the boundaries on a photographic plate as described previously (1). Four different centerpieces, with sector angles of 2°, 2.5°, 4° and 6.5°, were used in order to determine the effect of different interfacial areas between clear oil and cream (residual emulsion) on the magnitude of the zero order rate constants.

The data for 50% Nujol-50% water-Triton X-100 and 50% Nujol-50% water-Tween 20 emulsions were taken from Mittal's results (17) at 25°C., as were also those for 50% olive oil-50% water-SDS emulsions. Using volume or per cent of cream remaining as a function of time as the variable, appropriate plots were constructed to test the conformity of the data to either first or second order rate laws.

Results

Nujol-Water-0.4% SDS Emulsions. In order to resolve the contradictory treatments of the rate of separation of oil from 50% Nujol -50% water-0.4% SDS emulsions as both first (10) and zero (1) order processes, such emulsions were ultracentrifuged in centerpieces with differing areas of interfacial contact between clear oil and residual emulsion (cream). The rate of a zero order reaction is independent of the concentration of the reacting material, and determined by the area of the surface at which the reaction occurs. Accordingly, if demulsification in this system is occurring according to a zero order rate law the values of the rate constants calculated from the slopes of the straight lines of volume of oil separated against time of ultracentrifugation should be in the same ratio as the sector angles of the centerpieces (which are directly proportional to the interfacial areas).

The results obtained are shown in Figure 1. The zero order reaction rate constants calculated from the slopes of the straight lines obtained are shown in Table I. It is clear from the data that the rate of separation of oil is directly proportional to the interfacial area between clear oil and emulsion, since a constant value for the rate constant per unit interfacial area is obtained irrespective of the sector angle or the volume of emulsion present in the cell initially. This result also supports the hypothesis (18) that with these emulsions coalescence is occurring only at the interface between creamed emulsion and bulk oil. That it is not also occurring throughout the emulsion phase but at a slower rate would seem also to be ruled out by the fact that if this were the case there would be an increase in the average drop size with time. This would have caused an increasing rate of separation of oil with time since it had been found (18) that the rate of separation of oil increased markedly with decreasing specific interfacial area of the emulsion. But here the rate remained constant over the period of one hour of ultracentrifugation.

Table 1. Zero Order Rate Constants for 50% Nujol-50% Water-0.4% SDS Emulsions

Sector Angle,°	Initial Volume of Cream, $mm.^3$	Rate of Separation of Oil $mm.^3/min$	Rate Constant $mm.^3\,min^{-1}mm.^{-2}$ x 10^2
		Emulsion HO43073	
2	88.5	0.42	1.76
4	220	0.82	1.71
6.5	318	1.38	1.78
		Emulsion AT22371	
2.5	120	0.53	1.78
4	250	0.85	1.78

Groot found (10), however, that a plot of the logarithm of the volume of cream remaining against time for a Nujol-water-0.4% SDS emulsion was also essentially linear except at very short times, which would be indicative of a first order process. The present data likewise give a nearly linear relation when plotted as log volume of cream remaining vs. time, much like the results for 0.4% SDS in his Figure 5.1. This result comes about because of the insensitivity of the log function when the amount of oil separated is relatively small, as is the case with 0.4% SDS emulsions. If y is the % oil separated at time t, and x is the volume fraction of creamed emulsion remaining, then $x = (100 - y)/100$. Since $\log x = (x-1) - 1/2(x-1)^2 + 1/3(x-1)^3 \cdots$, and x is close to unity and does not change greatly over the period of the experiment, a plot of log x against time will apparently be linear, as is also the plot of y against time. The direct dependence of the rate on the interfacial area between cream and bulk oil confirms that the process is actually zero order even though it would not be possible to reach a definitive conclusion on the basis of the kinetic plots alone.

Nujol-Water-0.2% SDS Emulsions. In previous papers from this laboratory (1, 18, 19) the slope of the linear portion of the plot of per cent oil separated from 50% Nujol-50% Water-0.2% SDS emulsions vs. time was used as a measure of their stability, which implies that demulsification is a zero order process. However, it was also recognized (1,6) that the relation was not truly linear at either short or at relatively long periods of ultracentrifugation. In order to establish whether the process is zero order or follows a different order, the rate of separation of oil was determined in ultracentrifuge cells of different interfacial area as was done with the 0.4% SDS emulsions.

The results obtained are shown in Figure 2. It is clear that the rate of separation of oil is varying with the time of ultracentrifugation, i.e., it is dependent on the volume fraction of emulsion remaining, rather than being constant as it would have to be for a zero order process. (Note, however, that over a limited time a rather good straight line could be drawn through the points obtained with the standard 4° sector, as was done in the earlier published work). When the logarithm of the volume of cream remaining is plotted against time, as in Figure 3, very good straight lines are obtained, the slope of which permits calculation of the first order rate constant. (Note: plots of log per cent cream remaining or log volume fraction of cream against time would have the same slope.) The values of the first order rate constant obtained for the same emulsion run in cells of sector angles 6.5°, 4°, and 2° are respectively 1.23, 1.23, and 1.15 x 10^{-2} min^{-1}. The constancy within experimental error independent of the difference in interfacial area between emulsion and bulk oil shows definitively that in the case of 50% Nujol-50% water-0.2% emulsions the rate-determining step in the ultracentrifugal

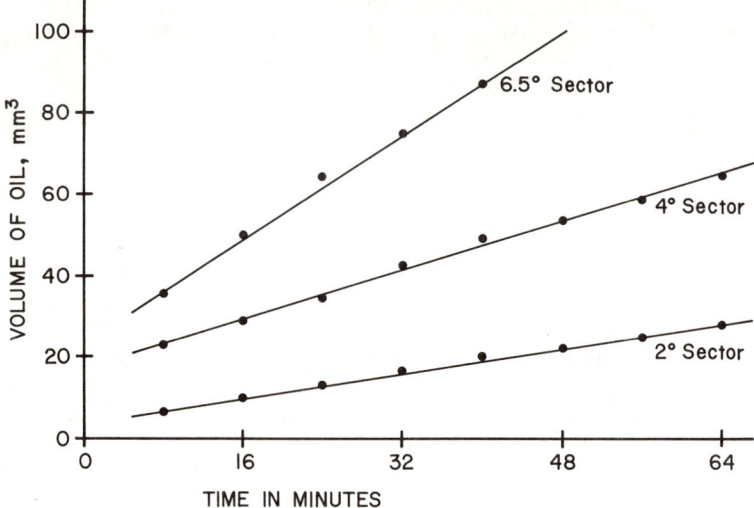

Figure 1. Volume of oil separated vs. ultracentrifugation time of 50% Nujol–50% water–0.4% SDS emulsion

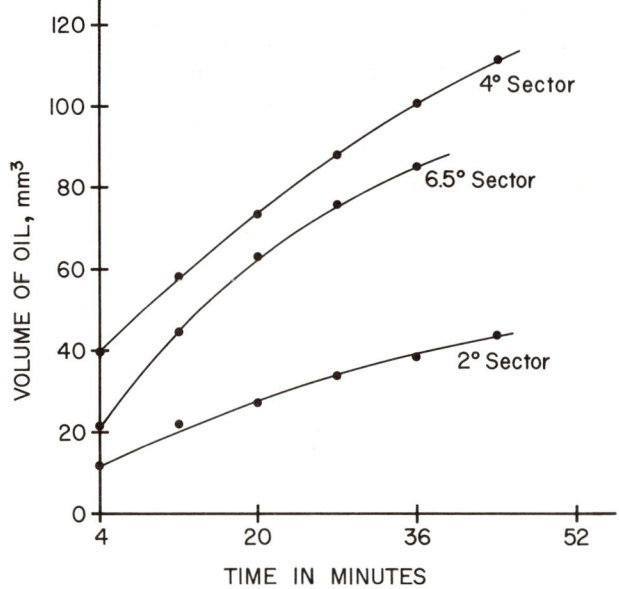

Figure 2. Volume of oil separated vs. ultracentrifugation time of 50% Nujol–50% water–0.2% SDS emulsion

separation of oil is a first order process.

It is interesting to compare the present rate constants with those calculated by Groot (20) and Mittal (11) for similar emulsions. Although the specific interfacial area of the particular emulsion on which the present rate data were obtained was not determined, it was made at nearly the same time and by an identical procedure (ten passes through the homogenizer) to other 0.2% SDS emulsions which were found to have specific interfacial areas of 1.65 and 1.80 x 10^4 cm^2 per ml. of emulsified oil. This is rather similar to Groot's emulsions of relatively low values of saturation adsorption, his emulsion 2020607 giving an interfacial area of 1.54 x 10^4 cm^2/ml oil when calculated on the same basis (54A^{o2} per adsorbed SDS molecule). On two different 0.2% SDS emulsions, Groot reported first order rate constants of 1.19 and 1.29 x 10^{-2} min^{-1}, at 25°C., which agree adequately with our value of 1.23 obtained in the standard sector cell at 20°C. Likewise Mittal's calculated first order rate constant becomes 1.15 x 10^{-2} min^{-1} for a Nujol-water-0.2% SDS emulsion of specific interfacial area 1.89 x 10^4 cm^2/ml oil at 25°C, when his result is expressed in terms of the natural logarithm rather than the logarithm to the base ten.

50% Olive Oil-50% Water-0.2% SDS Emulsions. The rate of separation of oil from this emulsion does not remain constant with time, but was found (6) to follow the empirical equation

t/% oil separated = 1/b·% oil max + t/% oil max,

where % oil separated is the quantity separated at time t, % oil max is the extrapolated limiting value at infinite time obtained from the Langmuir type equation fitting the data, and b is a constant. However, if Mittal's data (17) are treated according to classical kinetic theory it is found, as shown in Figure 4, that there is a linear relation between the reciprocal of the % cream remaining (volume of emulsion layer at time t x 100/initial volume of emulsion layer) and the period of ultracentrifugation. This is exactly the relation expected if the rate of separation of oil follows a second order rate expression. The specific reaction rate constant calculated from the slope of this line is 12.0 x 10^{-5} %$^{-1}$ min.$^{-1}$.

Figure 4 also shows a plot of the logarithm of the per cent cream remaining against time of ultracentrifugation. The distinct curvature of the plot rules out a first order mechanism for this emulsion.

50% Nujol-50% Water-Triton X-100 Emulsions. The data for these emulsions, in which the rate of separation of oil was neither constant, nor did it follow the empirical equation given above, were also taken from Mittal's dissertation (6,12,17). As is evident from Figure 5, the plot of the logarithm of the percent cream re-

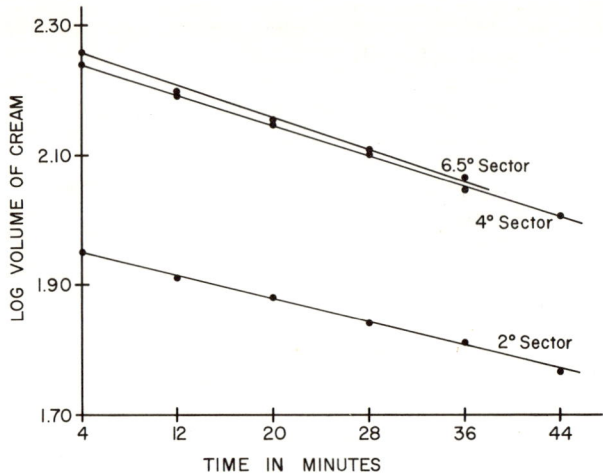

Figure 3. Log volume of creamed emulsion remaining vs. ultracentrifugation time of 50% Nujol–50% water–0.2% SDS emulsion

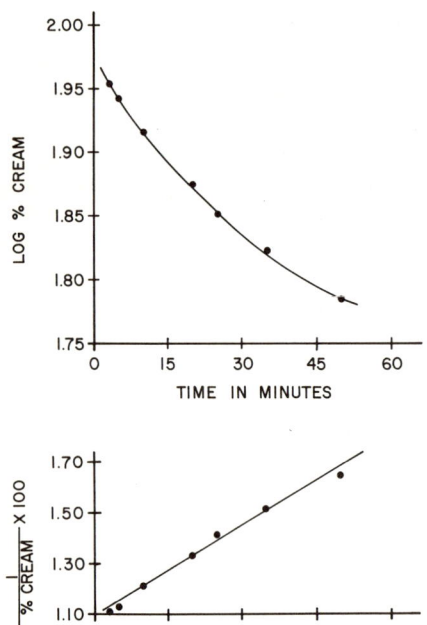

Figure 4. Demulsification of 50% olive oil–50% water–0.2% SDS emulsion treated by first and second order rate equations

maining against time of ultracentrifugation is linear, while a plot of the reciprocal of the per cent cream remaining against time shows a distinct curvature. Hence a second order mechanism is excluded, and it can be concluded that oil separates from Nujol-water-Triton X-100 emulsions in the ultracentrifuge according to a first order process. The rate constants calculated from the slopes of these lines are respectively 6.9×10^{-4} min^{-1} and 1.28×10^{-3} min^{-1} for 0.20% and 0.15% Triton-stabilized emulsions.

50% Nujol-50% Water-Tween 20 or Cetyl Pyridinium Chloride Emulsions. Proceeding similarly, the data for these emulsions (6,12, 17) were recalculated and plotted as either log % cream remaining or reciprocal of % cream remaining against the time of ultricentrifugation. In both cases the rate of separation of oil decreases steadily with time of ultracentrifugation, ruling out a zero order of reaction. The linearity of the plot of reciprocal of % cream remaining vs. time in Figure 6 establishes that oil separates from Tween 20 emulsions according to a second order rate law, while the non-linearity of the log % cream plot excludes a first order mechanism. A similar result was obtained with CPyCl-stabilized emulsions as shown in Figure 7, the separation of oil here too following a second order rate law.

The second order rate constants found for emulsions stabilized with 0.30%, or 0.15% Tween 20 are respectively 1.23×10^{-5} %$^{-1}$ min^{-1} and 1.79×10^{-5} %$^{-1}$ min^{-1}. The second order rate constant found for the Nujol-Water-0.1% CPyCl emulsion is 3.55×10^{-5} %$^{-1}$ min^{-1}.

Discussion

Van den Tempel (15,16) has analyzed very clearly the combined effect of flocculation and coalescence in the demulsification process of free-standing emulsions, and established the limiting conditions under which one or the other would be rate-determining. Where coalescence is rate-determining demulsification is expected to be a first order process, while if the rate of flocculation is rate-determining, demulsification is expected to follow a second order rate law. Sherman (21,22) emphasized that change in the number of globules with time is the only valid quantitative criterion of emulsion stability. He showed that in several cases plots of the logarithm of the total number of globules present varied linearly with time, corresponding to a first order mechanism, and indicating in terms of van den Tempel's analysis that the rate of coalescence was the rate-determining step in demulsification in these cases. Somewhat similar conclusions were reached by Srivastava and Haydon (23).

Before attempting to interpret the present results it is necessary to emphasize again the difference in physical state between a free-standing emulsion and an emulsion in an ultracentrifugal field. In the ultracentrifuge, because of the high centrifugal

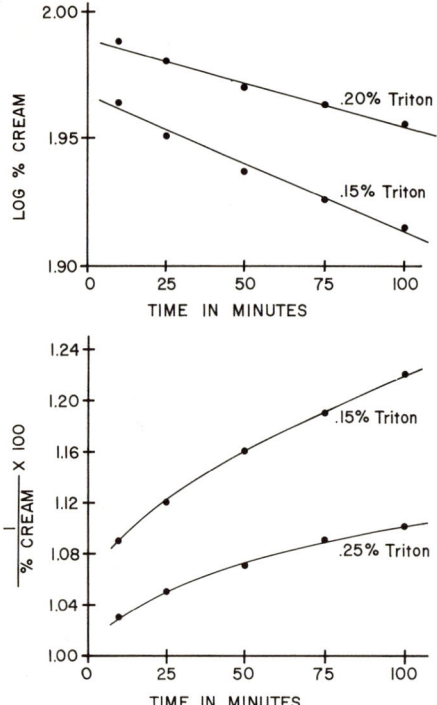

Figure 5. Demulsification of 50% Nujol–50% water–Triton X-100 emulsions treated by first and second order rate equations

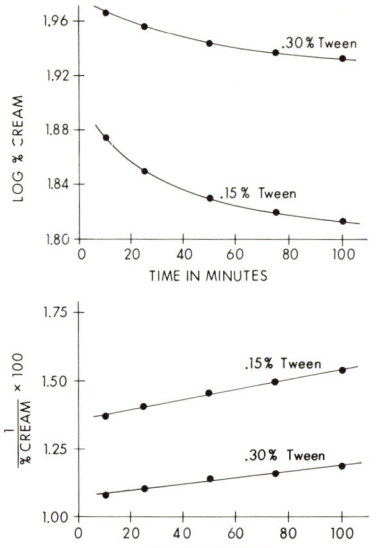

Figure 6. Demulsification of 50% Nujol–50% water–Tween 20 emulsions treated by first and second order rate equations

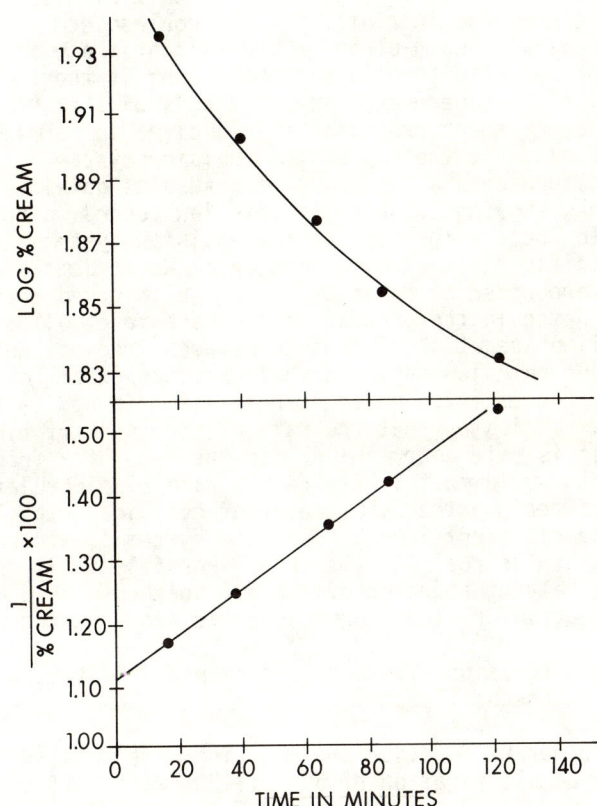

Figure 7. Demulsification of 50% Nujol–50% water–0.1% cetylpyridinium chloride emulsion treated by first and second order rate equations

force, the emulsified oil drops are very quickly flocculated and distorted into space-filling polyhedra (1,18). Hence the emulsion in the ultracentrifuge should be regarded as being already in a flocculated state at the beginning of the experiment, with the corollary that the rate of appearance of bulk oil is therefore a measure of either the intrinsic rate of coalescence--fusion of individual drops with each other in the bulk of the emulsion to form larger drops--or of oil-containing polyhedra at the top of the emulsion with the supernatant layer of bulk oil, or of one or another of the transport processes (6) involved in permitting free oil to accumulate at the top of the emulsion system.

Table II presents a summary of the results obtained in the present work, showing the kinetic rate law according to which demulsification occurs for each of the emulsion systems investigated and the specific reaction rate constant. Where demulsification follows a zero order of reaction it can be inferred that the rate-determining step in the process is coalescence of oil drops with the bulk oil phase at the interface between free oil and emulsion. A first order rate law suggests that coalescence of "drops" within the body of the emulsion phase is rate-determining. A second order rate law indicates that the rate of some sort of binary collision process is rate determining. In the case of a free standing emulsion this would most likely be the rate of flocculation, amenable to treatment by the DVLO theory of colloidal stability. However, in the ultracentrifuge, where the system is already flocculated, this can scarcely be the case. Possibly it is there a measure of the rate of approach of two oil polyhedra as a consequence of the drainage of residual aqueous phase from the interlammelar space.

Further discussion of these various possibilities will be presented in a subsequent paper.

Table II. Order of Reaction and Reaction Rate Constant for Ultracentrifugal Demulsification of 50% Oil-50% Water Oil-in-Water Emulsions.

Oil	Initial Conc. of Emulsifier wt %	Order	Rate Constant
Nujol[a]	0.4% SDS	0	0.82 mm^3 min^{-1}
Nujol[a]	0.2% SDS	1	1.23 x 10^{-2} min^{-1}
Olive[b]	0.2% SDS	2	12.0 x 10^{-5} $\%^{-1}$ min^{-1}
Nujol[b]	0.15% Triton X-100	1	1.28 x 10^{-3} min^{-1}
Nujol[b]	0.20% Triton X-100	1	6.91 x 10^{-4} min^{-1}
Nujol[b]	0.15% Tween 20	2	1.79 x 10^{-5} $\%^{-1}$ min^{-1}
Nujol[b]	0.20% Tween 20	2	1.43 x 10^{-5} $\%^{-1}$ min^{-1}
Nujol[b]	0.30% Tween 20	2	1.23 x 10^{-5} $\%^{-1}$ min^{-1}
Nujol[b]	0.10% CPyCl	2	3.55 x 10^{-5} $\%^{-1}$ min^{-1}

a. Data at 20°C from the present work
b. Data at 25°C from Mittal's investigations (17).

Literature Cited

1. Vold, R. D., and Groot, R. C., J. Phys. Chem. (1962) 66, 1969.
2. Rehfeld, S. J., J. Phys. Chem. (1962) 66, 1966.
3. Garrett, E. R., J. Pharm. Sci. (1962) 51, 35.
4. Garrett, E. R., J. Pharm. Sci. (1965) 54, 1557.
5. Garrett, E. R., J. Soc. Cosmet. Chem. (1970) 21, 393.
6. Vold, R. D. and Mittal, K. L., J. Soc. Cosmet. Chem. (1972) 23, 171.
7. Sherman, P., Soap Perfum. Cosmet. (1971) 44, 693.
8. Vold, R. D. and Groot, R. C., J. Colloid Sci. (1964) 19, 384.
9. Vold, R. D. and Mittal, K. L., J. Colloid Interface Sci. (1972) 38, 451.
10. Groot, R. C., "The Ultracentrifugation of Oil in Water Emulsions," pp. 36-37, 41-42, Ph.D. Dissertation, University of Utrecht, Utrecht, The Netherlands, 1965.
11. Mittal, K. L., "Factors Influencing the Ultracentrifugal Stability of Oil-in-Water Emulsions," pp. 93 and 106, Ph.D. Dissertation, University of Southern California, Los Angeles, California, 1970.
12. Mittal, K. L. and Vold, R. D., J. Am. Oil Chem. Soc. (1972) 49, 527.
13. Vold, R. D. and Mittal, K. L., J. Colloid Interface Sci. (1973) 42, 436.
14. Rehfeld, S. J., J. Colloid Interface Sci. (1974) 46, 448.
15. Van den Tempel, M. Rec. Trav. Chim Pays-Bas (1953) 172, 433.
16. Van den Tempel, M, "Proc. Second Int. Congr. Surface Activity" Vol. 1, p. 439, Butterworths, London, 1957.
17. Mittal, K. L., "Factors Influencing the Ultracentrifugal Stability of Oil-in-Water Emulsions," pp. 77, 80, 122, 132, Ph. D. Dissertation, University of Southern California, Los Angeles, California, 1970.
18. Vold, R. D., and Groot, R. C., J. Phys. Chem. (1964) 68, 3477.
19. Vold, R. D., and Mittal, K. L., J. Colloid Interface Sci. (1972) 38, 451.
20. Reference 10, p. 44, Table 5-1.
21. Sherman, P., "Proc. 4th Int. Congr. Surface Active Substances," Vol. 2, p. 1199, Gordon and Breach, London, 1967.
22. Boyd, J., Parkinson, C., and Sherman, P., J. Colloid Interface Sci. (1972) 41, 359.
23. Srivastava, S. N. and Haydon, D. A., Proc. 4th Int. Congr. Surface Active Substances, Vol. 2, p. 1221, Gordon and Breach, London, 1967.

5

Ultracentrifugal Technique in the Study of Emulsions

K. L. MITTAL

System Products Division, IBM Corp., Poughkeepsie, N.Y. 12602

Introduction

In view of the widespread applications of emulsion systems, it is very important to understand their making and breaking phenomena. Thermodynamically speaking, every emulsion is unstable and so it becomes increasingly important to comprehend the factors leading to its breakdown. The following four questions are extremely vital and are frequently asked in the realm of emulsion science and technology. (i) What constitutes emulsion instability? (ii) What are the suitable techniques to measure this instability? (iii) How can one test the emulsion instability on an accelerated basis? and (iv) Can accelerated tests predict the future behavior of emulsions stored under normal storage conditions, or in other words, what is the relationship between the accelerated instability and the shelf-like instability?

Instability of emulsions signifies a variety of connotations as expressed earlier (1). This can mean simple creaming, flocculation, coalescence, and the actual separation of the disperse phase. Creaming and flocculation do not represent emulsion instability as these are reversible processes, and a creamed or flocculated emulsion can be reverted to its original form. Coalescence and actual separation of the disperse phase are irreversible processes and do manifest emulsion instability.

In order to differentiate between these two phenomena, I had suggested (1) using the terms "microinstability" for coalescence, and "macroinstability" for the clear separation of the disperse phase.

The discussion above leads to a perennial question in emulsion science as to what should be the criterion for emulsion instability. Should it be the decrease in the number of particles with time or clear separation of the disperse phase. There is no consensus on this issue as there are proponents of both views. Sherman (2,3) emphasized that the change in the number of globules with time is the only valid quantitative criterion of emulsion instability, whereas Merrill (4) has stressed that the instability of emulsions should be expressed exclusively in terms of the actual separation of the disperse phase.

Apparently, recently, Merrill's views have been accepted by the investigators studying emulsion stability in the ultracentrifuge. It should be pointed out that maybe it is not wise to prescribe one general criterion of emulsion instability rather the intended use of the emulsion should dictate what it should be.

The literature is replete with the conventional methods of determining the instability of emulsions. Among the most widely used are the determination of the specific interfacial area, particle size distribution, mean particle size and the mean particle volume as a function of time (3-8). These methods are quite tedious as well as time consuming, although the application of Coulter counter (9) greatly increases their feasibility. Furthermore, these are not very meaningful with regards to the actual separation of the disperse phase. If the emulsion is too stable, these techniques may prove nugatory and furthermore these are not sensitive to the subtle but important changes taking place during storage of emulsions (10). These and other methods of studying emulsion deterioration have been reviewed (4,11-13) and their inherent limitations have been discussed.

On the other hand, emulsion instability has been determined by measuring the rate of creaming and the rate of separation of the internal phase, but has largely been restricted to normal gravitational conditions (14-18), which renders these processes very slow. Furthermore, there is uncertainty of complete phase separation, and these are only applicable to poor and unstable emulsions.

Ideally, one would like to predict the future behavior of an emulsion by conducting accelerated testing, and many such techniques have been described in the literature (<u>13</u>,<u>19-20</u>); these have been recently reviewed by Sherman (<u>20</u>). Unfortunately, all of these techniques are beset with undesirable traits. In order to draw meaningful results from accelerated testing it is imperative that the accelerated testing should merely accelerate the mechanism operative at normal storage or gravitational conditions without modifying it, and it should not introduce additional processes. None of the accelerated test methods fulfill all these requirements. Apropos, more recently, Petrowski (<u>21</u>) has determined emulsion instability by microwave irradiation and observed that after microwave heating, the emulsions exhibiting the highest surface temperature yielded the lowest amount of coalesced oil after centrifugation, but a correlation of microwave heating characteristics vs. long-term storage stability has yet to be established.

It is obvious from the discussion above that a rapid quantitative method of measuring emulsion instability is a desideratum and it is with this need in mind that the ultracentrifuge has been used.

Evolution of the Ultracentrifugal Technique

As pointed out earlier, Merrill (<u>4</u>) stressed that the stability of emulsions should be expressed exclusively in terms of the actual separation of the disperse phase and he developed a centrifugal method to study this process quantitatively. He studied a variety of emulsions (both o/w and w/o, and in the former both high and low density oils) and observed that the amount of oil separated (in o/w emulsions) was linear with time of centrifugation, and he expressed the slope of the straight line as a measure of the mechanical stability of emulsions. Furthermore, he found some agreement between the rate of oil separation by centrifugation with the results using the method of rate of decrease in the interfacial area. Cocktain, et al. (<u>22</u>) found a relation between log D (D is the root mean cube diameter of the particles) and the time of centrifugation. The early studies of Merrill were followed by a period of quiscence until 1962 when there was a great resurgence of interest and eventually commercially available untracentrifuge was exploited for the quantitative determination of emulsion instability.

Three papers (<u>10</u>,<u>23-24</u>) from three different groups of researchers appeared in 1962; the emulsion systems investigated were drastically different with the concomitant discordance in their findings. Garrett (23) studied highly stable emulsions which were amenable to only creaming or sedimentation; his results adhered to Svedberg equation. Furthermore, he attempted to correlate the rate of creaming in the ultracentrifuge to that under normal gravity. Mathematically, the calculations are valid but no experimental results are available to corroborate such correlations. In addition, he has implied that the shelf-life stability could be predicted from the ultracentrifugal results; this may not be so will be illuminated later.

Furthermore, creaming or sedimentation are not the true criteria for emulsion instability as discussed above.

On the other hand, Rehfeld (24), and Vold and Groot (10) used emulsion systems which were totally creamed during the initial period required for the ultracentrifuge to attain the pre-set speed. The emulsions showed breakdown, i.e., a clear separation of oil was discernible. Qualitatively, both groups are in agreement but the kinetics of separation of oil was found to be different. In a subsequent publication, Garrett (25) studied a variety of emulsions and determined the kinetics of the clear separation of oil. However, for some highly stable emulsions, he could study only the rate of creaming or sedimentation as was done earlier.

In a series of papers, Vold and Mittal (26-30) have studied a variety of emulsions and also the effect of operative variables on the kinetics of separation of oil for o/w emulsions. They have shown that depending upon the emulsion system, a variety of kinetic behaviors are possible (27), i.e., not only the quantity and rate of separation of oil but also the qualitative nature of the time dependence of the rate of separation of oil were found to change markedly with both the nature of the oil and the nature and also the concentration of the emulsifying agent.

Very recently, Rehfeld (31) has studied the ultracentrifugal stability of a variety of n-alkanes, benzene, and n-alkylbenzenes emulsions. For n-C_{17} to n-C_8 - in-water emulsions the results obeyed the empirical equation $\emptyset = \emptyset_0 t^{-n}$ where n varies from 0.14 to 0.17. $\emptyset$ is defined as the fraction of stable emulsion remaining uncoalesced.

Most recently, Vold and Hahn (32) have found that depending upon the emulsion system, the data for the ultracentrifugal kinetics of oil separation fit zero, first, or second order rate expression.

It is evident from the above discussion that the ultracentrifugal technique yields different results depending upon a host of factors-nature of the oil, nature of the emulsifier, concentration of the emulsifier, speed of rotation, etc. The main purpose in the present paper is to address the following questions: (i) What aspects of emulsion instability can be studied using ultracentrifuge? (ii) What is meant by ultracentrifugal stability? (iii) Is there a correlation between the ultracentrifugal instability and the shelf-life instability? Furthermore, merits and demerits of the ultracentrifugal technique are discussed.

Behavior of Emulsions in the Ultracentrifuge

Before discussing the different behaviors of emulsions, it is important to understand the various possible changes which take place in the emulsion once it is subjected to ultracentrifugation. The present author has used Beckman Model E Analytical ultracentrifuge and a similar equipment has been employed by other researchers. About 0.7 ml of emulsion is injected into the hole in the centerpiece which is contained in a cell housing. The light path is 12 mm and the cell is generally 4° sector shaped but it can be varied (32). The facilities are available for schlieren optics but in emulsion systems the transmission of visible light through various regions in the cell is generally used. Once the cell is put in the r4tor and the rotor is connected to the drive mechanism, the light passing through the emulsion contained in the cell presents the picture as shown in Figure 1a. The picture shown represents the true picture as seen in the view mirror (a facility available with the ultracentrifuge for visual inspection). The various regions are explained in the figure.

In the present communication, the discussion is primarily concerned with o/w emulsions because only these emulsions have been studied, but the treatment should be valid for w/o emulsions also. Moreover, in case of o/w emulsions, both when the oil is lighter or heavier than water should be amenable to ultracentrifugation, but for simplicity the behavior of emulsions containing lighter oil phase is described.

As the rotor is accelerated, there are certain discernible changes in Figure 1a depending upon the characteristics of the emulsion and the speed of ultracentrifugation. The changes in Figure 1a depend upon whether the emulsion is undergoing simple creaming or coalescence. These two cases are elaborated upon in the following paragraphs.

Case I. Simple Creaming. If the emulsion is too stable to break, i.e., to separate oil under the applied ultracentrifugal stress-as was the case with emulsion systems studied by Garrett (23)-then the following changes occur in region CG of Figure 1a as shown in Figure 1b: (i) a transparent region CD representing the light passing through the air above the emulsion in the cell will be visible and (ii) another transparent region will appear starting from the boundary G and extending towards the center of the rotor; this represents the water left as a result of the creaming of the emulsion. The relation between the length of the region FG and the time of ultracentrifugation depends upon the characteristics of the emulsion. The kinetics of creaming is discussed later.

Case II. Coalescence and the Eventual Separation of Clear Oil Phase. The Case I discussed above is an exception whereas Case II is generally observed. Rehfeld, Vold et al., and Garrett, in his recent work (25), have observed that their emulsion systems showed the behavior described under Case II. Once the rotor is accelerated and assuming that the emulsion is not so stable as to show creaming phenomenon exclusively, then a variety of behavior patterns are possible depending upon the nature and composition of the emulsion, temperature, and the ultracentrifugal stress.

Generally, two different regions appear invariably as shown in Figure 1c. Transparent region CD represents the light passing through the air above the emulsion and this is similar to what is observed in case of pure creaming (cf. Figure 1b). The length of the region CD depends upon how well the hole is filled with the emulsion, but there is always some air left resulting into this region. Second transparent region will start developing from the boundary G and extending towards the center of the rotor; this represents the creaming of the emulsion with the concomitant separation of the clear layer of water.

It has been observed repeatedly that the clearing of water is complete during the time it takes for the rotor to attain the preset speed. Rehfeld noted that the clearing of water was complete in 2 min. 50 sec. (time taken to attain a speed of 25,980 r.p.m.) and there was no further change in the length of region FG with time. Similarly, Vold et al. (10,26-30) observed the complete clearing of water in about 4 min. (time taken to attain a speed of 39,460 r.p.m.) for a variety of o/w emulsions (see Figure 1c).

Following the complete clearing of water, the emulsion droplets must assume distorted polyhedral shapes so as to occupy the total space. This space-filling structure of flocculated emulsion is described as foam-like or honey-comb structure.

Apropos, the complete clearing of water does not signify the absence of water in the creamed or flocculated emulsion; of course, there are water lamellae separating the droplets in the flocculated emulsion but this amount of water is immeasurably small.

Now the basic question is: What other changes are possible subsequent to the full appearance of regions CD and FG? Let it be added that the boundary D is a fine streak called miniscus; if the miniscus is broad, the boundary D is reckoned from the mid-point of the miniscus.

The answer to the proposed query is that depending upon the nature of the emulsion and other operative variables, the following changes are possible in region DF.

The emulsion may be too stable to separate any oil during the period the rotor is brought to speed and furthermore there is no separation of oil for some time even after the attainment of the speed. The zero time is reckoned from the time the rotor attains the pre-set speed. This situation dictates the absence of the transparent region DE representing the separation of oil, and the emulsion is said to have induction period. The induction period is the time elapsed before any separation of clear oil. (ii) The emulsion may not separate any oil during the initial period of bringing the rotor to speed but the separation ensues once the speed is attained. In such circumstances, the transparent region DE will not be present at zero time but will go on continuously growing after the zero time. (iii) The emulsion may be less stable resulting in the separation of oil even during the initial period of acceleration. This will culminate into the transparent region DE at zero time and its extent will increase with further time of ultracentrifugation (Figure 1d).

This comprehensive survey of the possible behavi4rs of emulsions leads to the discussion of the kinetics of separation of oil (once the oil starts separating), and this is presented in the later section.

Although in this paper, the two processes of creaming and the clear separation of oil are discussed separately, but it is possible that some emulsions might show a combination of both, as was the case with "bad" emulsions of Garrett (23).

Before leaving this subject, it is appropriate to point out that although the transparent region FG in both cases (see Figure 1b and 1c) result from the creaming of the oil droplets but in Figure 1c, the creaming or flocculation is complete in such a short duration as to render the situation unquantifiable with respect to kinetics of creaming.

What is Ultracentrifugal Stability?

After describing the various modes of behavior of emulsion in the ultracentrifuge, the question arises: What should be the criterion to represent and compare ultracentrifugal stability of various emulsions. The discussion here is primarily concerned with the emulsions conforming to Case II; the ultracentrifugal stability aspects of those emulsions showing behavior discussed under Case I, are presented later. The experimentally measurable quantities are the lengths of the various regions. As pointed out earlier, the length of region FG (representing water) remains constant during the experimentation with the result that the length of the total region DF (DE + EF) should also be constant. The regions DE and EF will change separately as a function of time of ultracentrifugation.

Vold and Groot (10,33-35) adopted the ratio DE/EF, i.e., percent oil separated as a measure of the breakdown of the emulsion and a plot of %oil separated (% oil sep) vs. time, t, was found invariably linear over large values of t. The slope of this plot was termed the rate of oil separation and the inverse of this was regarded ultracentrifugal stability. On the other hand, Rehfeld (24) took the lengths of the region EF at zero and 20 minutes as measures of emulsion breakdown.

His plots of EF/DF were linear with time and the slopes of these plots were measure of mechanical stability of emulsions, as he has termed it. Recently, Vold and Mittal (27) found that the relation between % oil separated and time is more complex and this gives rise to the difficulties in the proper choice of the criterion for ultracentrifugal stability. It should be pointed out that the lengths of various regions should be converted to corresponding volumes, V, by the well known formula for 4° sector cell.

$$V = \frac{\pi h}{90} (2 \Delta R_a + \Delta^2)$$

Where R_a is the distance from the center of rotation to the top of the cell and in the ultracentrifuge used it is 5.70 cm, h is the thickness of the cell which is 12 mm.

Figure 2 shows plots of %oil$_{sep}$ vs. time of ultracentrifugation for four different emulsions; and a great variation in the kinetic behavior of emulsion is evident from this figure. A summary of the kinetic studies of various emulsions is presented in Table I., and the various criteria for quantitative representation and comparison of the ultracentrifugal stability of emulsions are compiled in Table II. It is important to point out that different order of emulsion stability is possible depending not only upon the choice of the criterion but also upon the time of ultracentrifugation at which the chosen criterion is used to compare emulsions. This is shown in Tables IIIa and IIIb. Apparently, there is no single universally applicable criterion for representing ultracentrifugal stability quantitatively, rather the choice of the most meaningful criterion will be dictated by the particular application of an emulsion.

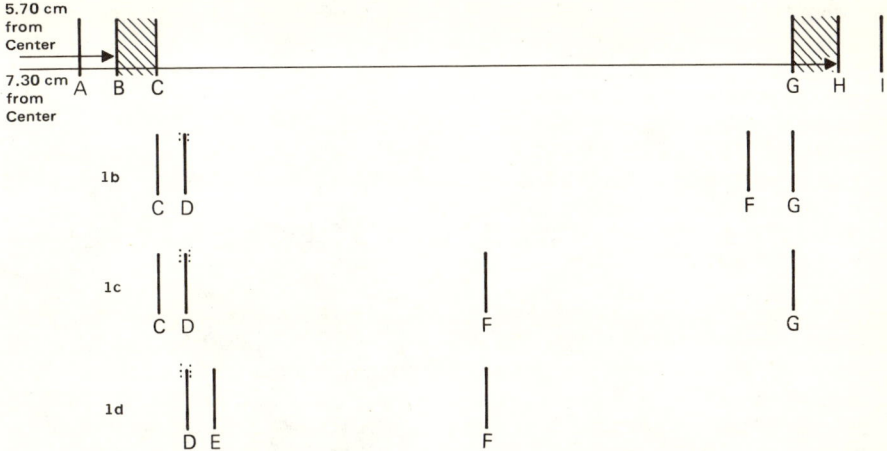

AB – Transparent region, represents light passing through the top reference hole.   BC – Dark region, is due to the metal of the cell holder.   CG – Opaque region, represents emulsion   GH – Dark region is due to metal of the cell holder.   HI – Transparent region, represents light passing through the bottom reference hole.   CD – Transparent region, represents air.   FG – Transparent region, due to the separated water.   DF – Opaque region, represents totally flocculated emulsion.   DE – Transparent Region, represents clear separated oil.   EF – Opaque region, due to the remaining flocculated emulsion.

Figure 1. Various possible behavior modes of an emulsion in the ultracentrifugal cell

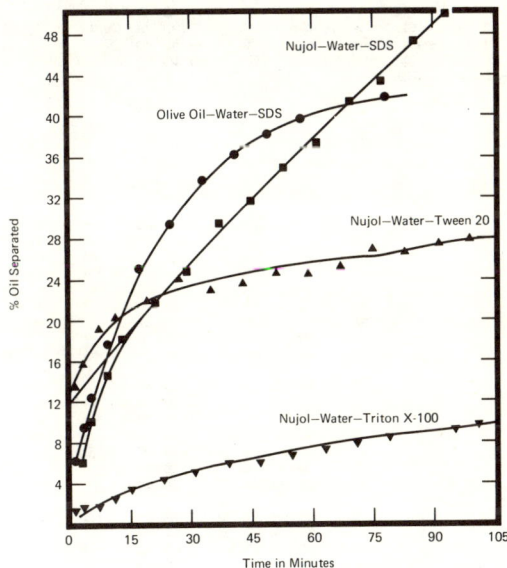

Journal of the Society of Cosmetic Chemists

Figure 2. Separation of oil at 39,460 rpm from 50% oil–50% water–0.2% surfactant emulsions. Nujol–SDS emulsions, M111968; olive oil–SDS emulsions, M060669; Nujol–Tween 20 emulsions, A 091069; Nujol–Triton X-100 emulsions, A 090969 (27).

Table I. Summary of the Kinetic Studies of Various Emulsions in the Ultracentrifuge

Emulsion	Kinetics	Reference
50% Nujol-50% water – 0.2% SDS	Zero order	10,29,33-35
"	First order	32,36
"	Rapid separation[a] zero order empirical equation plateau	27
50% Nujol-50% water – 0.4% SDS	First order	38
"	Zero order	32
50% Nujol-50% water-0.15 or 0.2% Triton X-100	First order	32
50% Nujol-50% water-0.15 or 0.2% Tween 20	Second order	32
"		
50% Nujol-50% water-0.1% CPC	Empirical equation[b]	27
50% Olive oil-50% water-0.2% SDS	Second order	32
"	Second order	32
50% Toluene-50% water-1% G-2151[c]	Empirical equation	27
50% Benzene-50% water-different Conc. of SDS	First order followed by zero order Sometimes zero order, sometimes no definite order	25 24
50% n-Alkanes-50% water-0.23% SDS	$\phi = \phi_0 t^{-n}$ (n=0.14 to 0.17)	31

Table I. (Continued)

Explanation: SDS   - Sodium dodecyl sulfate
CPC   - Cetylpyridinium chloride
G-2151 - Polyoxyethylene - 30 - Stearate (mol. wt. ca. 1600)

a. These different stages were observed when the emulsion was ultracentrifuged for seven hours.

b. The empirical equation reads

$$\frac{t}{\%oil_{sep}} = \frac{1}{b\%oil_{max}} + \frac{t}{\%oil_{max}}$$

where $\%oil_{sep}$ is the percent oil separated at time, t; $\%oil_{max}$ is the maximum amount of oil which can separate, and b is a composite constant.

c. If 7% G-2151 is used then the emulsions simply follow the sedimentation phenonenon, i.e., a plot of log (distance of creamed emulsion-water boundary) vs. time is linear.

Table II. Various Criteria for the Ultracentrifugal Stability of Emulsions

Criterion	Reference
Induction period[a]	37
Absolute amount of oil separated at time, t	27
% oil$_{sep}$ at time, t	27
Volume fraction of creamed emulsion remaining at time, t	24
Extrapolated % oil at zero time[b]	10
Rate of oil separation at time, t	27
Rate of oil separation after a given % oil$_{sep}$[c]	27
Zero order rate constant	10, 24
First order rate constant	32, 38, 25
Second order rate constant	32
% oil$_{max}$[d]	26, 27
Composite constant, b	26, 27
The empirical parameter, n	31

Note: In case of emulsions undergoing sedimentation or creaming, the sedimentation constant can be used for comparative purposes.

a. Induction period is the time elapsed before separation of any oil.

b. In many cases, there is an initial rapid separation of oil and the extrapolated % oil at zero time represents the extrapolation of the curve of % oil$_{sep}$ vs. time to zero time.

c. This is given by

$$\frac{d(\%oil_{sep})}{dt} = \frac{b(\%oil_{max} - \%oil_{sep})^2}{\%oil_{max}}$$

and for definitions of various terms, refer to the empirical equation described above. The above equation permits ready calculation of the instantaneous rate of oil separation after separation of any given fraction, much as 10% or 30%.

d. This represents the limiting amount of oil which will separate at infinite time, and this can be determined either experimentally or from the empirical equation (see Table I.).

Table IIIa. Order of Stability Based on Quantity of Oil Separated (27)

	% Oil Separated After			
	25 min	75 min	Oil_{max} (%)	
Most stable	Nujol-Triton X-100	Nujol-Triton X-100	Nujol-Tween 20	
	Nujol-Tween 20	Nujol-Tween 20	Olive oil-SDS	
	Nujol-SDS	Olive oil-SDS	Nujol-SDS	
Least stable	Olive oil-SDS	Nujol-SDS		

Table IIIb. Order of Stability Based on Instantaneous Rate of Separation of Oil (27)

Rate of Separation in % Oil per Minute After

	25 min	75 min	100 min
Most stable	Nujol-Triton X-100	Nujol-Tween 20	Nujol-Tween 20
	Nujol-Tween 20	Nujol-Triton X-100	Nujol-Triton X-100
	Nujol-SDS	Olive oil-SDS	Olive oil-SDS
Least stable	Olive oil-SDS	Nujol-SDS	Nujol-SDS

Creaming or Sedimentation

The classical expression relating the distance η of the particle from the axis of rotation with the time, t, of centrifugation is given by

$$\log \frac{x_2}{x_1} = \frac{V(\rho - \rho^1)(r.p.m.\, 2\pi)^2 (t_2 - t_1)}{k\eta} \qquad (1)$$

where x_2 and x_1 are the positions of the particle at time t_2 and t_1 respectively. V is the volume of the creaming particle, ρ and ρ^1 are the densities of the particle and the medium, k is the shape constant of the particle, η the co-efficient of viscosity of the dispersion medium, and r.p.m. represent the speed of the rotor.

Equation (1) can be simplified to Equation (2) for a given particle of a given density.

$$\log x_2 = Kt + \log x_1 \qquad (2)$$

where $K = \dfrac{V(\rho - \rho^1)\, 4\pi^2 (r.p.m.)^2}{k\eta} = S\,(r.p.m.)^2 \qquad (3)$

S is simply the sedimentation constant given in units of Svedberg, t is the time from the start of the centrifuge and x_2 is the position of the particle at time t.

Garrett (23,25) in his studies of creaming of emulsions found that both Equation (2) and (3) were obeyed as shown in Figures 3 and 4. On the other hand, Rehfeld, Vold et al., could not study the creaming according to Equations (2) and (3) as their emulsion systems were totally creamed in 3-4 minutes.

A few comments regarding the use of Equations (2) and (3) and the creaming of emulsions are in order: (i) Actually these equations are valid at constant r.p.m., and so in the case of changing speed this can cause some error, and if the time is reckoned after the speed is attained then there is uncertainty in estimating the extent of sedimentation during this initial period. (ii) The rate of creaming simply measures the rate of smallest particles and so it cannot profitably be used to differentiate between different emulsions. (iii) These equations are valid

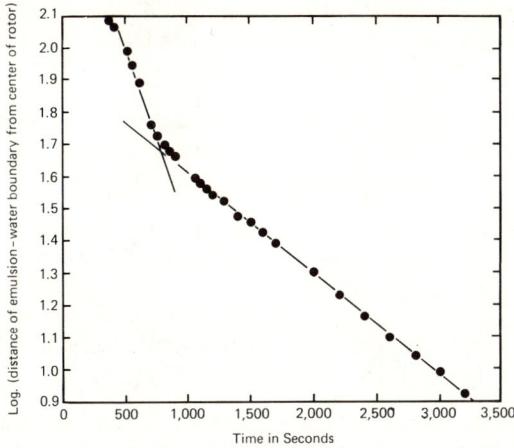

Journal of the Society of Cosmetic Chemists

Figure 3. Example of flotation of oil particles from the bottom of 50% toluene–water emulsion prepared with 7% G-2151 surfactant. The logarithm of the distance of the emulsion–water boundary from the center of the rotor is plotted against the time in seconds of ultracentrifugation at 24,630 rpm (25).

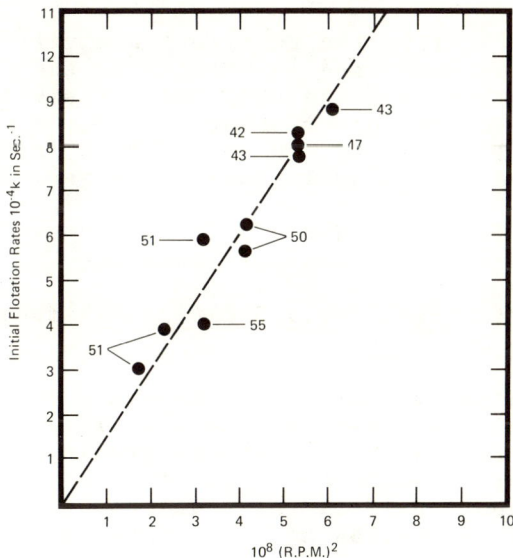

Journal of the Society of Cosmetic Chemists

Figure 4. Initial flotation rates in sec^{-1} as a function of ultracentrifugal $(rpm)^2$ for a 50% toluene-in-water emulsion stabilized with 7% G-2151. The numbers indicate age in days of the emulsions at the time of centrifugation (25).

below a certain r.p.m. (this will depend upon the emulsion characteristics) and at higher r.p.m.'s, clear separation of oil starts. (iv) Garrett found that the creaming phenomenon depends upon the concentration of the emulsifier used; if the concentration of the emulsifier is low, then the emulsion may not be amenable to quantification.

Garrett (23) has derived an expression relating the sedimentation in the ultracentrifuge to that under normal gravity

$$\frac{t_g}{t_c} = 4.856 \times 10^{-6} \, \ell \, (r.p.m)^2 / \log(\frac{\ell}{d} + 1)$$

where t_g is the time under normal gravitational conditions; t_c under ultracentrifugal conditions; l is the distance of sedimentation; d is the initial distance from the axis of rotation to the ultracentrifuge. Assuming a value of d of 7.08 cm (between 7.07 cm and 6.31 cm) and based upon the r.p.m. values ranging from 12,590 to 35,600, he calculated that it will take an average of 311 days (range of calculations from 271 to 376 day) for a sedimentation of 0.77 cm under normal gravitational conditions.

Mathematically, the relations are valid but caution should be exercised in predicting the future behavior of emulsions based upon the ultracentrifugal results. Furthermore, creaming is not a true criterion of emulsion instability and so not much insight into the breakdown of the emulsions can be gained through such studies. Moreover, it is obvious from the results compiled in Table I, that most of the emulsions studied are not amenable to creaming phenomenon.

Comparison of Shelf-life and Ultracentrifugal Stability of Emulsions

As pointed out earlier, there is a definite need for an accelerated test for emulsion instability and the results from such test should predict the shelf-life behavior of emulsions. Furthermore, the accelerated test should fulfill the requirements set forth earlier. In the following paragraphs, a comparison between the behavior of an emulsion under normal storage conditions and in the ultracentrifuge is made. (i) Under normal gravitational conditions, an emulsion consists of spherical droplets suspended in the dispersion medium, whereas in most cases, the drops are distorted to polyhedral or honey-comb structure in the ultracentrifuge. (ii) Under shelf-life, flocculation maybe rate determining in the separation of the disperse phase whereas in ultracentrifuge, the emulsion is already flocculated. (iii) Factors influencing the flocculation may be important under normal conditions whereas in the ultracentrifuge, these may play no important role. (iv) In normal storage, steps of flocculation and coalescence cannot be differentiated whereas in the ultracentrifuge coalescence can be separated from flocculation and can separately be studied. (v) There are subtle changes taking place in the emulsion during storage and before oil separation, but these changes cannot be studied in the ultracentrifuge. (vi) Ultracentrifugal stress may affect the drainage of water between the droplets, and the rupture of the emulsifier films, thereby modifying the mechanism of coalescence operative under normal gravitational conditions. (vii) Factors affecting the rate of oil separation, the kinetics of oil separation, and the effects of operative variables may be different under ultracentrifugal and shelf-life conditions.

Advantages and Disadvantages of the Ultracentrifugal Method

In spite of the difficulties and uncertainties associated with this technique, the method has numerous advantages over the conventional methods. (i) The technique is very time saving, the preparation, ultracentrifiguration, and the analysis of the results can be completed in two days. (ii) The results are reproducible(within 3 to 5%)and reliable.(iii) Excellent method to study the effect of operative variables such as concentration of emulsifiers,nature of oil phase, temperature etc. on the stability of emulsions. (iv) Excellent method for comparative purposes.(Emulsions of wide range of stability can be studied. (vi) The only reliable method to study the actual breakdown, i.e., clear separation of the disperse phase (vii) The conventional methods suffer from the drawback that the processes of flocculation and coalescence are difficult to be differentiated, but the ultracentrifugal technique can be used to study coalescence exclusively and without interference from flocculation. (viii) Valid for both o/w and w/o emulsions. (iv) Only a small quantity (0.7 ml) of emulsion is required, which is a special advantage in exploratory experiments. (x) The technique and the equipment can be sophisticated to determine the concentration of the emulsifier in the cleared aqueous phase; this can be very helpful in unravelling the mechanism of coalescence.

No method is perfect and the ultracentrifugal technique is no exception. The following drawbacks can be cataloged. (i) The equipment is expensive and the technique is exacting. (ii) The nature of the emulsion in the ultracentrifuge is quite different from that under normal gravity. (iii) The ultracentrifugal results may not be utilized directly to predict shelf-life stability because, as discussed above, the process of demulsification and the factors affecting this may be quite different from that under normal storage conditions.

Summary and Conclusions

1. Depending upon a variety of factors, inter alia, concentration of the emulsifier and the ultracentrifugal stress, an emulsion may undergo creaming or sedimentation; or it may coalesce with the eventual clear separation of the disperse phase. Most of the emulsions studied are prone to coalescence, but in some cases the emulsion composition and other experimental conditions are such that the emulsion is amenable to quantifiable creaming rate.

2. Caution should be exercised in predicting the rate of creaming under normal gravitational conditions, from the results obtained in the ultracentrifuge. Mathematically, the relations are valid but there are no experimental results available to corroborate such relations.

3. The physical make-up of the emulsion in the ultracentrifuge is different from that under normal storage; consequently, one should be careful in extrapolating results from the ultracentrifugal studies to the predictions of emulsion behavior under normal gravitational conditions. The effects of the operative variables may be different in two cases.

4. A host of criteria are possible for representing and comparing ultracentrifugal stability of emulsions. Different order of stability is possible (see Table III) depending upon the choice of the criterion and the time of ultracentrifugation at which the emulsions are compared.

5. The ultracentrifugal technique provides an ideal method to study the effects of operative variables on the process of coalescence and the kinetics of clear separation of the disperse phase.

6. All of the work reported has been confined to o/w emulsions, and the technique should be extended to w/o emulsions, and also to particle-stablized emulsions.

7. More work is needed before the utility of the ultracentrifugal technique as an accelerated test of emulsion stability is established. Emphasis should be in the direction of finding a correlation between the ultracentrifugal results and the shelf-life stability of emulsions.

Literature Cited

1. Mittal, K.L., J. Soc. Cosmet. Chem. (1971) 22, 815
2. Sherman, P., "Proc. 4th. Int. Congr. Surface Active Substances," Vol. 2, p. 1199, Gordon and Breach, London, 1967
3. Boyd, J., Parkinson, C., and Sherman, P., J. Colloid Interface Sci. (1972) 41, 359
4. Merrill, R.C., Ind. Eng. Chem. (Anal. Ed.) (1943) 15, 743
5. Groves, M.J. and Freshwater, D.C., J. Pharm. Sci. (1968) 57, 1273
6. Hallworth, G.W. and Carless, J.E., J. Pharm. Pharmacol. (1973) 25, Supp. 87
7. Jain, K.D. and Sharma, M.K., J. Indian Chem. Soc. (Feb. 1973) L, 122
8. Shinoda, K. and Saito, H., J. Colloid Interface Sci. (1969) 30, 258
9. Lien, T.R. and Phillips, C.R., Environ. Sci. Technol. (1974) 8, 558
10. Vold, R.D. and Groot, R.C., J. Phys. Chem. (1962) 66, 1969
11. Garrett, E.R., J. Pharm. Sci. (1965) 54, 1557
12. Becher, P., "Emulsions: Theory and Practice," IInd ed., Reinhold Publishing Corp., New York, 1965
13. Princen, L.H., in "Treatise on Coatings," R.R. Myers and J.J. Long, Eds., Vol. I, Part III, Chapter 2, Marcel Dekker , New York, 1972
14. King, A. and Mukergee, L.N., J. Soc. Chem. Ind. (1939) 58, 243

15. Cheesman, D.F. and King, A., Trans. Faraday Soc. (1940) **36**, 241
16. Cockbain, E.G. and McRoberts, T.S., J. Colloid Sci. (1953) **8**, 440
17. King, A. and Mukergee, L.N., J. Soc. Chem. Ind. (1940) **59**, 185
18. Lederer, E.L., Kolloid Z. (1935) **71**, 61
19. Parkinson, C. and Sherman, P., J. Colloid Interface Sci. (1972) **41**, 328
20. Sherman, P., Soap Perfumery Cosmet. (Nov. 1971) p. 693
21. Petrowski, G.E., J. Amer. Oil. Chem. Soc. (1974) **51**, 110
22. Cocktain, J.R. and Wynn J.B., J. Pharm. Pharmacol. (1952) **4**, 959
23. Garrett, E.R., J. Pharm. Sci. (1962) **51**, 35
24. Rehfeld, S.J., J. Phys. Chem. (1962) **66**, 1966
25. Garrett, E.R., J. Soc. Cosmet. Chem. (1970) **21**, 393
26. Mittal, K.L. and Vold, R.D., J. Amer. Oil Chem. Soc. (1972) **49**, 527
27. Vold, R.D. and Mittal, K.L., J. Soc. Cosmet. Chem. (1972) **23**, 171
28. Vold, R.D. and Mittal, K.L., J. Pharm. Sci. (1972) **61**, 869
29. Vold, R.D. and Mittal, K.L., J. Colloid Interface Sci. (1972) **38**, 451
30. Vold, R.D. and Mittal, K.L., J. Colloid Interface Sci. (1973) **42**, 436
31. Rehfeld, S.J., J. Colloid Interface Sci. (1974) **46**, 448
32. Vold R.D. and Hahn, A.U., this volume, p.64.
33. Vold, R.D. and Groot, R.C., J. Soc. Cosmet. Chem. (1963) **14**, 233
34. Vold, R.D. and Groot, R.C., J. Phys. Chem. (1964) **68**, 3477
35. Vold, R.D. and Groot, R.C., J. Colloid Sci. (1964) **19**, 384
36. Mittal, K.L., "Factors Influencing the Ultracentrifugal Stability of Oil-in-Water Emulsions," pp.93 and 106, Ph.D. Dissertation, University of Southern California, Los Angeles, California, 1970
37. ibid. p.80
38. Groot, R.C., "The Ultracentrifugation of Oil-in-Water Emulsions," pp. 36-37,41-42, Ph.D. Dissertation, University of Utrecht, The Netherlands, 1965

6

Electric Emulsification

AKIRA WATANABE and KEN HIGASHITSUJI
Faculty of Textile Science, Kyoto University of Industrial Arts and Textile Fibers, Matsugasaki, Kyoto 606, Japan
KAZUO NISHIZAWA
Nippon Filcon Co. Ltd., Nagaokakyo, Kyoto 617, Japan

Introduction

Although the dispersion of liquids by electric method has been known for a long time ([1], [2], [3], [4], [5]), it has rarely been used in practical emulsification processes. This is because of the fact that, except for the emulsification of mercury by Ilkovic ([3]), the liquid is dispersed in air at first and then introduced into another liquid phase containing stabilizing agent. In addition, the voltage used is very high, ranging from several to several ten thousand volts.

In connection with the studies on electrocapillary phenomena at oil/water interfaces ([6], [7], [8], [9]), the present authors have succeeded in producing w/o or o/w type emulsions directly by using much lower voltages as compared with the above method. The emulsions thus formed are in general highly monodisperse. Since physical properties of emulsions are strongly dependent on the size distribution, detailed studies on the present method appear to provide us with a useful means to clarify the mechanism of emulsification as well as of properties of emulsions in general.

Experimental

Materials. Except for the cases of applications, the materials used are as follows:
The inorganic (KCl) and organic (tetrabutylammonium chloride, TBAC) electrolytes were of Analytical Grade, which were used without further purification. The surfactants, sodium dodecylsulfate (SDS) and cetylpyridinium chloride (CPC), were of Extra Pure Grade, and Span 80 was of Commercial Grade. The organic solvent, methylisobutylketone (MIBK), of Extra Pure Grade, washed by sodium hydroxide aqueous solution and pure water, was distilled and then saturated with water before use. Ion exchange water was redistilled from an all Pyrex apparatus, which gave water of specific conductivity ca. 1 micromho/cm. This was used for all experiments.

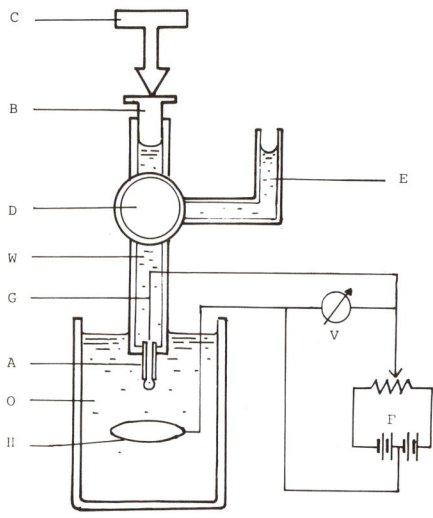

Figure 1. Schematic of the apparatus for electric emulsification. A: Teflon tube, B: Syringe, C: Driving motor, D: Three-way cock, E: Water phase reservoir, F: D.C. power supply, G, H: Pt electrodes, O: Oil phase, V: Voltmeter, W: Water phase.

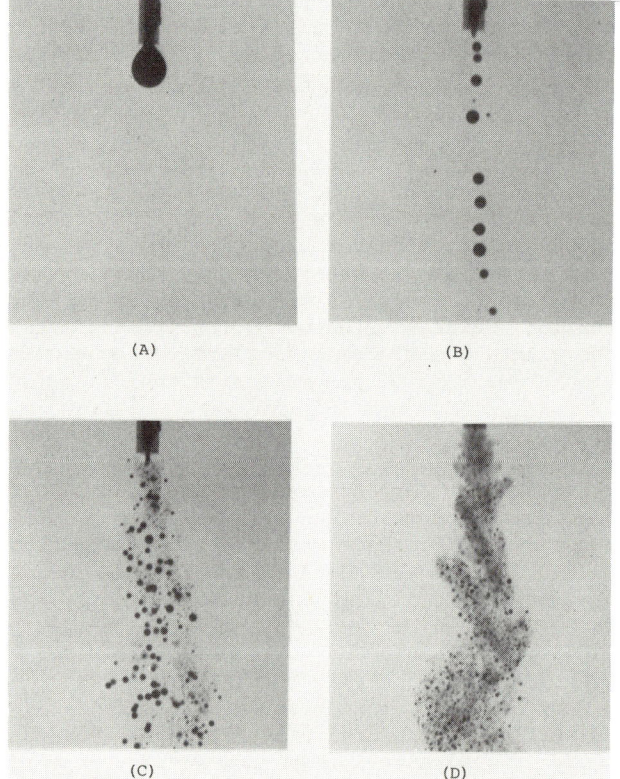

Figure 2. Photographs of electric emulsification of water in MIBK system. A: 0 volt, B: 100 volts, C: 250 volts, D: 500 volts.

All glassware was thoroughly cleaned and steamed before use.

Apparatus. The outline of the apparatus for electric emulsification is shown in Figure 1. The water phase in the syringe B is introduced into the oil phase through the Teflon tube A (inner dia., ca. 0.5 mm) at the tip of the glass syringe, by driving the motor C. The platinum electrodes G and H in water and oil phases, respectively, are connected to the variable D.C. supply F. When sufficiently high voltage is applied to the o/w interface, the dispersion of water phase takes place at the tip of the capillary and w/o type emulsion is produced.

In the case of formation of o/w type emulsions, the oil phase is introduced from the syringe into water phase.

In the present paper the sign of applied voltage E is always that of water phase with reference to that of oil phase.

Hycam 16 mm High Speed Camera is used to observe details of the phenomena which occur during the emulsification.

Results

Emulsification of 0.01 mol/dm^3 KCl in MIBK. When the applied voltage E is zero, large drops are successively formed at the tip as shown by the photograph in Figure 2, A. The size of drops decreases with increasing E (Figure 2, B), and at 250 volts a shower of fine drops is formed, although relatively large drops are also formed simultaneously (Figure 2, C). This is the minimum voltage necessary to give rise to continuous emulsification and is called "the critical voltage of emulsification", E(crit.), a measure of the difficulty of emulsification ($\underline{4}$). At 500 volts the emulsification takes place violently, presenting a fog-like appearance (Figure 2, D).

Photographs taken by the high speed camera are shown in Figure 3. It is noticed that the o/w interface is strongly deformed at high voltages, with a very sharp protuberance. At the critical voltage the dispersion takes place at this point. The protuberance moves around rapidly and irregularly at this stage.

The Effect of Composition on the Critical Voltage of Emulsification. The critical voltage of emulsification E(crit.) is strongly influenced by the composition of oil and water phases. As an example the effect of SDS concentration of oil phase on E(crit.) for w/o type emulsification is shown in Figure 4 ($\underline{9}$), where the water phase contains 0.0001 mol/dm^3 or 0.01 mol/dm^3 KCl. The oil phase is MIBK containing various concentrations of anionic surfactant SDS. It is noticed that, when the SDS concentration in MIBK is increased, E(crit.) decreases at first. This is ascribed to the adsorption of SDS at the w/o interface and also to the increase in electric conductivity, that is the decrease in ohmic drop in oil phase.

However, for SDS concentrations higher than 0.0001 mol/dm^3

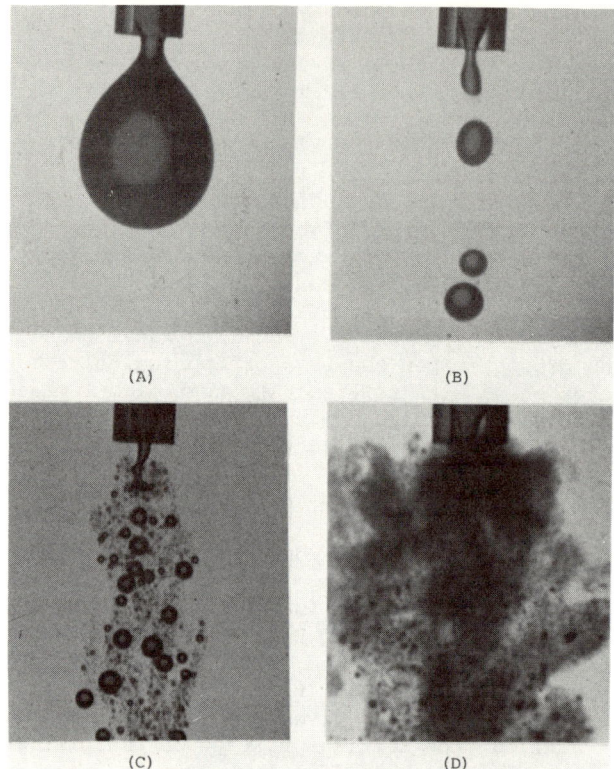

Figure 3. Photographs of electric emulsification taken by the high speed camera for water in MIBK system. A: 0 volt, 200 exposures/sec. B: 150 volts, 200 exposures/sec. C: 250 volts, 8,000 exposures/sec. D: 1,000 volts, 8,000 exposures/sec.

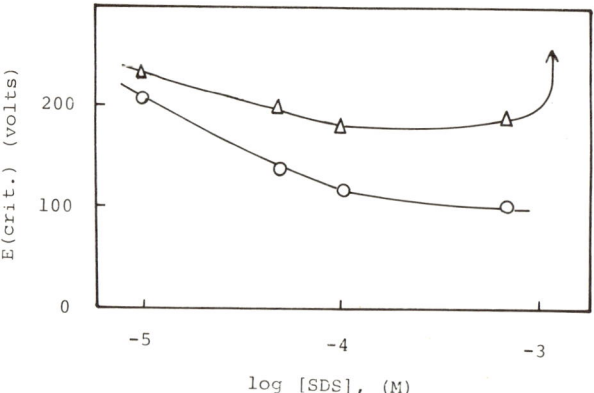

Figure 4. The effect of SDS concentration in oil phase on E(crit.). Water phase: KCl concentration, △: 0.0001 mol/dm³, ○: 0.01 mol/dm³. Oil phase: SDS in MIBK.

in the case of 0.0001 mol/dm^3 KCl, E(crit.) increases again, the emulsification becoming more difficult.

More or less the same behavior is found in the curve for 0.01 mol/dm^3 KCl.

It is interesting to find that this behavior is related to the change in the conductivity of oil phase. Figure 5 shows the relation between E(crit.) and the specific conductivity of oil phase $\kappa(o)$ for w/o type emulsification, the water phase being 0.0001 mol/dm^3 KCl aq. and the oil phase MIBK containing various concentrations of SDS. It is noticed that E(crit.) decreases with increasing $\kappa(o)$, the emulsification becoming easier. However, when $\kappa(o)$ approaches that of water phase $\kappa(w)$, i.e. 18.4 micromho/cm, no emulsification takes place under present experimental conditions, E(crit.) increasing indefinitely.

Figure 6 shows the relation between E(crit.) and $\kappa(w)$ for w/o type emulsification, the oil phase being 0.00005 mol/dm^3 CPC in MIBK and the water phase containing various concentrations of KCl (9). It is noticed that E(crit.) increases with decreasing $\kappa(w)$, and no emulsification takes place when the conductivity of water phase approaches that of oil phase $\kappa(o)$, i.e. 1.04 micromho/cm.

Opposite relations are found for o/w type emulsification as shown in Figure 7, where E(crit.) is plotted against $\kappa(w)$. The water phase contains various concentrations of KCl and the oil phase is 0.01 mol/dm^3 TBAC in MIBK. It is noticed that E(crit.) increases indefinitely as $\kappa(w)$ approaches $\kappa(o)$, i.e. 140 micromho/cm (Figure 7).

It is worth mentioning here that TBAC is electrocapillary inactive (1), and hence the electric emulsification takes place even in the absence of surface active materials.

These observations lead to the conclusion that the specific conductivity of the discontinuous phase must always be higher than that of continuous phase in order for the electric emulsification to take place.

<u>Particle Size Distributions</u>. The size distributions of emulsions thus formed electrically are much narrower than those of emulsions prepared mechanically (8). This is clear from Figure 8, in which the particle size distributions, measured microscopically, are shown for w/o type emulsions of the same composition, prepared by electric (A) and mechanical (B) methods. The water phase contains 0.01 mol/dm^3 SDS and the oil phase is 0.5 % Span 80 in toluene, the applied voltage being -300 volts in the case of electric emulsification.

Emulsions of much higher dispersity, with the average diameter lower than 0.1 µm, can be formed by choosing proper surfactants at proper concentrations. This can be proved by using the light-scattering technique (10).

<u>Stability</u>. The emulsions formed electrically are very

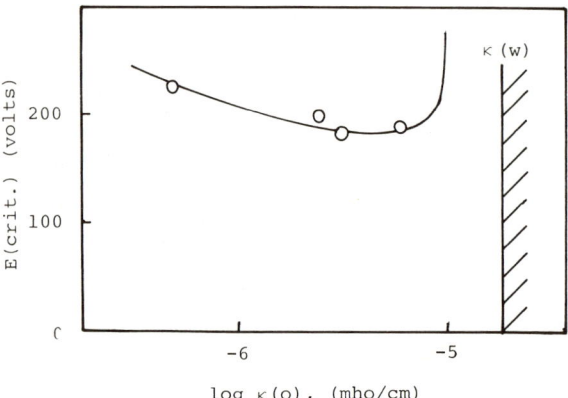

Figure 5. The influence of the specific conductivity of oil phase on E(crit.). Water phase: 0.0001 mol/dm³. Oil phase: SDS in MIBK.

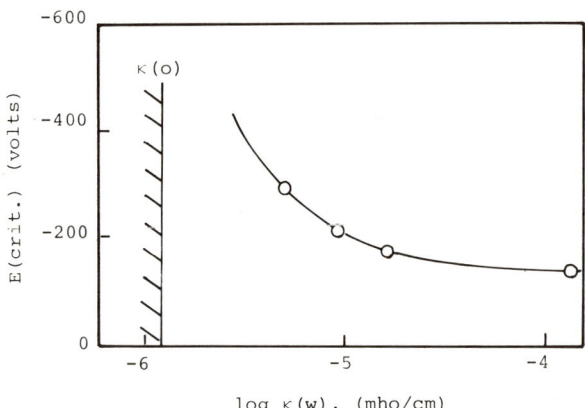

Figure 6. The influence of the specific conductivity of water phase on E(crit.). Water phase: KCl. Oil phase: 0.00005 mol/dm³ CPC in MIBK.

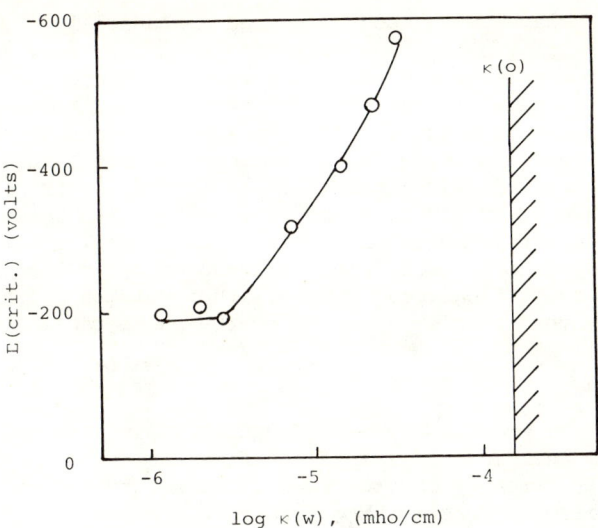

Figure 7. The influence of the specific electric conductivity of water phase on E(crit.). Water phase: KCl. Oil phase: 0.01 mol/dm³ TBAC in MIBK.

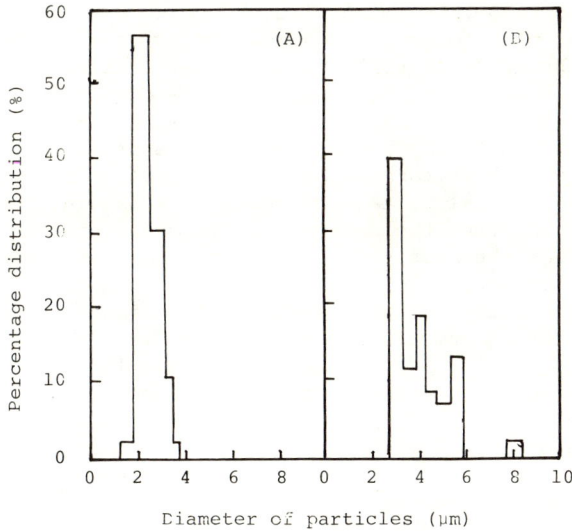

Figure 8. Particle size distributions measured by optical microscopy. A: Electric emulsification. B: Mechanical emulsification.

stable as compared with those prepared mechanically. This is due
to the small drop size, high degree of monodispersity and the
repulsive force between particle charges. The photographs of
emulsions, prepared by introducing fountainpen ink into MIBK, are
shown in Figure 9, both of which are taken 60 days after preparation. The left hand side is the emulsion formed by using the
homogenizer and the right hand side is formed electrically. Since
these emulsions contain no stabilizers other than those in the
original ink, drops begin to settle in a few min. after preparation and separate out completely after 3 days, in the case of
mechanical emulsification. However, the emulsion prepared electrically is very stable, as shown by the photograph.

Discussion

Since the process of emulsification is in principle a non-
equilibrium phenomenon, as is clear from the fact that the o/w
interface is strongly deformed and in ceaseless violent motion,
there are many factors which govern this phenomenon.

In the first place, we must consider the decrease in (macroscopic) interfacial tension by the applied potential difference,
the electrocapillary phenomena at o/w interfaces (6, 8). In
Figure 10 the interfacial tension γ is plotted against E, where
the water phase is 1 mol/dm^3 KCl and the oil phase 0.0001 mol/
dm^3 CPC or SDS in MIBK. It is noticed that γ decreases over the
negative polarization range in the case of cationic surfactant
CPC, becoming almost zero at E = -30 volts. For the anionic
surfactant SDS, on the other hand, γ decreases over the positive
polarization range, becoming almost zero at E = +50 volts.
These values of E for zero γ almost coincide with those of
E(crit.), the critical voltages of emulsification.

It has been found, however, that the emulsification also
takes place over the polarization range in which γ does not decrease in Figure 10 or even in the absence of surfactants, if a
sufficiently high voltage is applied. Since γ is higher than
zero at the critical voltage of emulsification in these cases,
we can conclude that the decrease in (macroscopic) interfacial
tension is not the necessary condition of electric emulsification.

From microscopic point of view, the fluctuation of interfacial tension always takes place by local changes in curvature
and concentration and also by the uneven distribution of current
density (11, 12, 13, 14). Hence, there are local regions in
which γ is lower than the average macroscopic value. Hence, it
is expected that a sharp protuberance can grow at such a point on
the surface of liquid drop and the local curvature increases.
Hence, the fluctuation is amplified there, thus increasing the
tendency to emulsification.

In the second place, the strong electric field at the interface must be taken into account, which helps to break the mechanical equilibrium at the interface locally.

6. WATANABE ET AL. *Electric Emulsification* 105

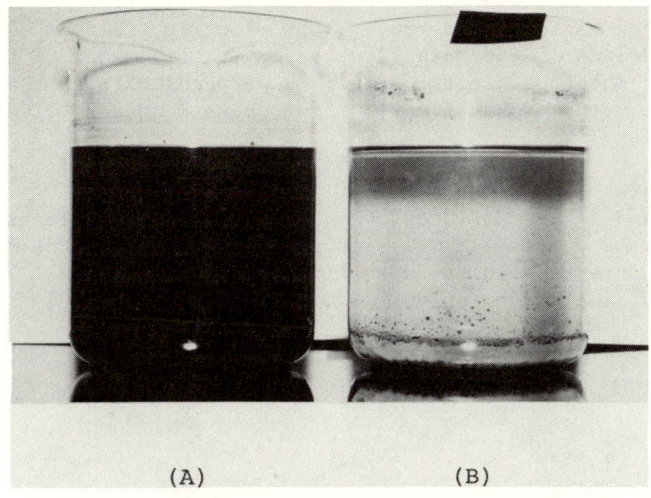

Figure 9. The stability tests of w/o type emulsions. A: Electric emulsification. B: Mechanical emulsification by using the homogenizer.

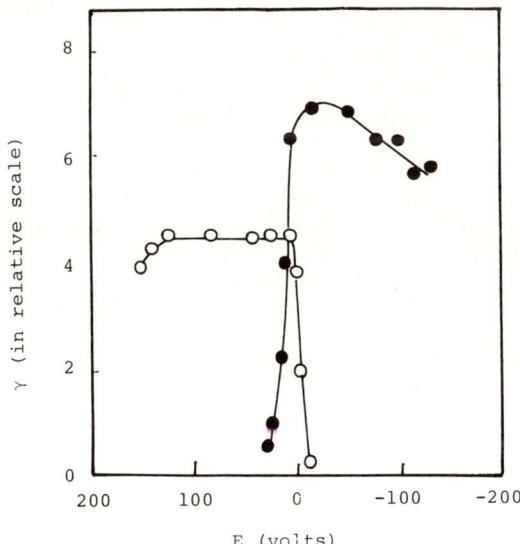

Figure 10. Electrocapillary curves at o/w interfaces. Water phase: 1 mol/dm³ KCl. Oil phase: 0.0001 mol/dm³ CPC (○) or SDS (●).

The mechanical force due to the electric field is equal to the difference in the Maxwell stresses at both sides of the interface. When the electric field is perpendicular to the interface, the normal force f acting from phase 1 to 2 is given by

$$f = (1/2)[\varepsilon'(d\psi'/dx)^2_{x=0} - \varepsilon(d\psi/dx)^2_{x=0}]$$

where ε and ε' are dielectric constants, ψ and ψ' electric potentials, of phases 1 and 2, respectively, and x is the distance from the interface in the direction from phase 1 to 2. In the absence of ionic adsorption, we have for x = 0

$$\psi = \psi'$$

and

$$\kappa(d\psi/dx) = \kappa'(d\psi'/dx)$$

where κ and κ' are specific conductivities of phases 1 and 2, respectively. From these three equations we obtain:

$$f \gtreqless 0 \quad \text{for} \quad \kappa/\sqrt{\varepsilon} \gtreqless \kappa'/\sqrt{\varepsilon'}$$

Thus, direction of the force f is governed by the values of $\kappa/\sqrt{\varepsilon}$ at both sides of the interface. The effect of ionic adsorption on the surface potential can be neglected, when a sufficiently high voltage is applied.

The above discussion is in accord with the experimental facts mentioned earlier. Since the change in specific conductivity dominates under the present experimental conditions, the square roots of dielectric constants of water and oil phases being almost equal to each other, the mechanical force is directed from the phase of higher to that of lower specific conductivities. This explains the relation between the type of emulsion formed and the specific conductivities of water and oil phases.

It is therefore concluded that the mechanical force due to the discontinuity of electric field at the interface, together with the local fluctuation of γ, acts to break the mechanical equilibrium at the interface and drops are thrown into the other phase with complicated motion of the liquid. Then, the deformation and the velocity of drop formation increase, and the continuous emulsification is accelerated.

Applications

In the case of ordinary emulsification processes, properties of emulsions, e.g. particle size distribution, stability, are usually controlled by selecting proper surfactants at high concentrations. This causes in many cases a considerable disadvantage, and in addition the use of synthetic surface active materials often gives rise to various difficulties. The present technique has in this sense a great advantage, since stable, nearly monodisperse, fine emulsions are formed in the absence of, or in the presence of very small amounts of, surface active materials.

This method has a vast field of applications, for example the dyeing, cosmetics, food and pharmaceutical industries (16). A few examples will be presented here briefly.

The emulsion for solvent dyeing can be prepared electrically by introducing the aqueous solution of dye into tetrachloroethylene in the presence of sorbitan mono-oleate. The leveling effect is good. Moreover, it is also possible in this method to control the particle charge according to the nature of the fiber to be dyed.

A cosmetic cream of high quality is formed electrically by introducing water into liquid paraffin containing 0.3 % sorbitan mono-oleate. The amount of surfactant is about one tenth of that in ordinary commercial cosmetic creams.

The electrocapillary spinning is a modification of the present procedure (15). For instance, we can spin threads out of gelatin by introducing 20 % aqueous solution of gelatin into absolute ethanol through a Teflon tube of ca. 0.5 mm in inner diameter (15). When the applied voltage is zero, the gelatin solution passes into the ethanol phase slowly and coagulates readily. At sufficiently high voltages, however, the gelatin solution forms a thread, which moves around violently at the front of the tip and is dehydrated rapidly. In Figure 11 photographs taken by the high speed camera at the speed of 200 exposures per sec. are shown. The gelatin powder thus formed has the specific area much larger than that of untreated one, and hence dissolves in water quickly. The initial dissolution velocity of the gelatin sample treated at 1,000 volts is about ten times larger than that of untreated one. This technique of electrocapillary spinning can be applied to any other soluble high polymers, e.g. polyvinylalcohol and gum arabic.

Acknowledgements

The authors wish to express their gratitude to Messrs. K. Kamada, M. Takayama and K. Ishizaki and to Misses K. Takubo and Y. Ebe for their assistance in the experimental work.

Literature Cited

1. Zeleny, J., Phys. Rev. (1914) 3, 69.
2. Vonnegut, B. and Neubauer, R., J. Colloid Sci. (1952) 7, 616.
3. Ilkovic, D., Coll. Trav. Chim. Tchécosl. (1932) 4, 480.
4. Nawab, M. and Mason, S., J. Colloid Sci. (1958) 13, 179.
5. Gopal, E.S.R., in "Emulsion Science", P. Sherman, Ed., p. 55, Academic Press, London, 1968.

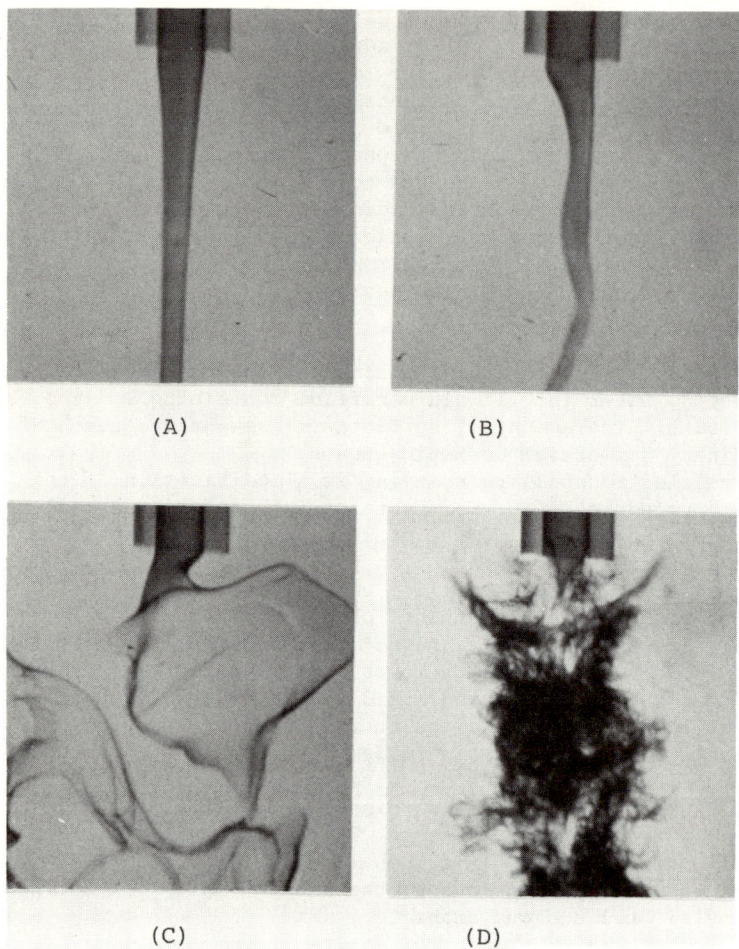

Figure 11. Photographs of electrocapillary spinning of gelatin taken by the high speed camera at 200 exposures/sec. A: 0 volt, B: 250 volts, C: 800 volts, D: 1,000 volts.

6. Watanabe, A., Matsumoto, M. and Gotoh, R., Nippon Kagaku Kaishi (1966) $\underline{87}$, 941.
7. Watanabe, A., Matsumoto, M., Tamai, H. and Gotoh, R., Kolloid-Z. Z. Polym. (1967) $\underline{220}$, 152.
8. Watanabe, A., Nippon Kagaku Kaishi (1971) $\underline{92}$, 575.
9. Higashitsuji, K., Nishizawa, K., Kamada, K. and Watanabe, A., Nippon Kagaku Kaishi (1974) $\underline{1974}$ (6),995.
10. Higashitsuji, K., Takayama, M., Nishizawa, K. and Watanabe, A., (1974), unpublished data.
11. Tolman, R.C., J. Chem. Phys. (1949) $\underline{17}$, 333.
12. Kirkwood, J.C. and Buff, F.P., ibid. (1949) $\underline{17}$, 338.
13. Buff, F.P., ibid. (1955) $\underline{23}$, 419.
14. Higashitsuji, K., Nishizawa, K. and Watanabe, A., unpublished data.
15. Ebe, Y., Takayama, M., Kamada, K., Nishizawa, K. and Watanabe, A., (1974), unpublished data.
16. Watanabe, A., Higashitsuji, K. and Nishizawa, K., (1974), unpublished data.

7

Macromolecular Emulgents in Emulsion Stability

S. N. SRIVASTAVA

Department of Chemistry, Agra College, Agra, India

Introduction

There has been always a lack of complete theoretical understanding of emulsion behaviour. Earlier workers (1-6) attempted physicohemical interpretation of emulsion characteristics in a rather qualitative manner. Later on some attempts were made in a semiquantitative way to account for emulsion stability in terms of the rate of decrease of specific interface with time (7). This was not, however, a very satisfactory theoretical approach and gave no clear insight into the mechanism of emulsion coagulation except that it did reflect a semiquantitative correlation between film strength and stabilizer concentration. Verwey (8) has considered the stability of emulsions in terms of the electrical double layer which exists at the liquid-liquid interface. The total potential drop was considered to occur in both the phases, the greater drop being in the oil phase. The D.L.V.O. theory of colloid stability should, therefore, be applicable to stablished emulsions. Recently this theory has been used by van den Tempel (9) and Albers and Overbeek (10) to explain the behaviour of both O/W and W/O systems using arbitrary values of the van der Waals constant.

In the present work by the author the stabilization of emulsion droplets by electrostatic forces and by the visco-elasticity of the adsorbed film has been examined. Emulsion globules covered by adsorbed macromolecular emulgent films exhibit appreciable stability to both flocculation and coalescence which together constitute the overall process of separtion of the disperse phase. The quantitative theory of these phenomena is examined on one hand to explain flocculation using the D.L.V.O. theory and controlling the double layer potential through the variation of pH and the bulk phase concentration. On the other hand, coalescence which follows flocculation is found to be governed by the rheology of the film. The functional van der Waals constants have also been assessed.

Present address: Unesco Science Centre, University of Alexandria (A. R. Egypt)

Theoretical

The flocculation of an emulsion is similar to that of a lyophobic colloid. It may be either irreversible occuring in the primary minimum or reversible occurring in the secondary minimum (if its depth is small). The quantitative aspects of both the possibilities have been dealt by the D.L.V.O. theory. However, for the macromolecular stabilized emulsion under investigation, having an average particle radius greater than one μm, flocculation occurs at secondary minima. The particles in this type of reversible flocculation are still separated by a distance of the order of 20 nm. This is diagrammatically depicted in Figures 1-2.

In the framework of the D.L.V.O theory (11), the interaction energy of the coated liquid droplets $V = V_R + V_A$ where V_R and V_A are respectively the repulsion and attraction energies. The repulsive energy was caluated using an approximate expression:

$$V_R = \frac{\epsilon a \psi_o^2}{2} \ln(1 + e^{-\kappa H}) \qquad (1)$$

where V_R is the energy of repulsion, ϵ, dielectric constant, a, the particle radius, κ, the Debye Huckel parameter and H the interparticle distance; ψ_o is the surface potential assumed to be equal to the zeta potential for the present system having $\kappa a \gg 1$.

The attraction energy V_A symbolizing the partially retarded van der Waals interaction was ascertained from the following equations:

$$V_A \approx \frac{-Aa}{\pi} \left(\frac{2.45\lambda}{120 H^2} - \frac{\lambda^2}{1045 H^3} + \frac{\lambda^3}{5.62 \times 10^4 H^4} \right) \qquad (2)$$

valid for H > 15 nm

$$\text{and} \quad V_A \approx \frac{-Aa}{12\pi} \left(\frac{\lambda}{\lambda + 3.54 \pi H} \right) \qquad (3)$$

valid for H < 15 nm

The interaction V can be calculated from the total number of particles associated with a given particle i.e. the degree of aggregation. D, can be computed from

$$D = 4\pi n_o a^3 \int s^2 \exp(-V/kT) \, ds \qquad (4)$$

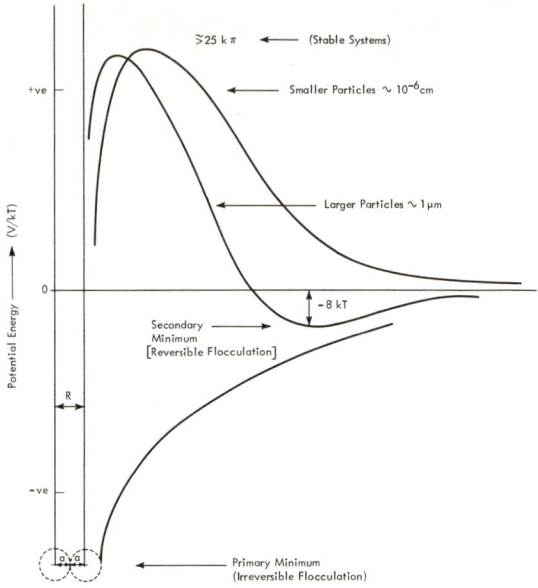

Figure 1. Characteristic interaction energy profiles in the light of the D.L.V.O. theory

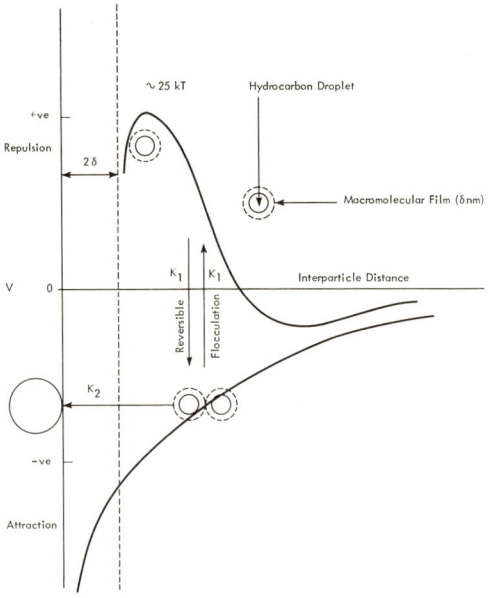

Figure 2. Diagrammatic representation of the coalescence of the coated emulsion droplets

where n_0 is the number of droplets per unit volume of the suspension and $s = 2 + H/a$. The intergrand is evaluated graphically.

In case of reversible flocculation under study, a singlet-doublet equilibrium temporarily exists in the system and the extent of flocculation can be determined from the formula

$$D = \frac{\text{number of doublets}}{\text{number of singlets + number of doublets}} \quad (5)$$

For the above calculations the value of the van der Waals constant, A is needed which was calculated from the equation.

$$A_{ii} = \pi^2 q_i^2 \cdot B_i \quad (6)$$

where q_i is the number of atoms of kind i contained in 1 c.c. of the matter of i th kind and B_i is the London constant relative to the same substance, the latter being calculated from the following three equations:

(a) The London equation, $B_L = 3/4 \, \alpha^2 h\nu_0$ \quad (7)

(b) The Slater-Kirkwood $B_{S-K} = 11.25 \times 10^{-24} n^{1/2} \alpha^{3/2}$ \quad (8)

(c) The Slater-Kirkwood-Moelwyn-Hughes $B_{S-K-M} = 3/4 \, (s_1 n)^{1/2} n \, \nu_e \, \alpha^2$ \quad (9)

where α is the polarizability, ν_0 is the characteristic London frequency n is the number of valency electrons, V_e the electronic vibration frequency and S_1 the effective number of dispersion electrons is given by the empirical relation $S_1 = 0.39n$. Because the emulstion droplets are immersed in an aqueous medium, the net constant is given by

$$A = \left\{ A_{\text{disperse phase}} + A_{H_2O} - 2A_{\text{disp.ph-}H_2O} \right\} \quad (10)$$

A is thus invariably positive and hence for two particles of any species embedded in a second medium there will always be a net attractive force.

Employing the above mathematical formulations the interaction of emulsion droplets stabilized by macromolecular substances such as bovine serum albumin (BSA), pepsin, sodium and gum arabic has been examined quantitatively in the light of the stability theory. The orthodox range of colloidal dimensions (1 - 100 nm) has been deliberately excluded and the range of the particle size chosen is such that the flocculation is reversible and occurs in the secondary minima of interaction energy profiles. Since such interactions may be altered by adsorbed macromolecules an attempt has also been made to study the influence of adsorption on van der Waals interaction using the following equation from Vold's mathematical treatment (12) described elswhere (13)

$$A_{MS}^{1/2} = -\left[(H_P-H_{PS})A_{PM}^{1/2} \pm \left\{(H_P-H_{PS})^2 A_{PM} - (H_S + H_P - 2H_{PS})(H_P \cdot A_{PM} + 12 V_A/f(P))\right\}\right] \div (H_S + H_P - 2H_{PS}) \quad (11)$$

where A_{MS} is the van der Waals constant of the adsorbed macromolecule in aqueous solvent. The H symbols are the so-called H functions.

Thus H_P is the H function of two bare spherical particles, H_S that of two coated spherical particles and H_{PS} is that of one bare particle separated by a distance of H+ δ (δ, being the film thickness) from coated particle. f(p) is the correction factor due to retardation. Here $δ \approx$ 1.385 nm taken from the work of Biswas (14) and Cockbain (15).

Materials and Methods

A crystallized sample of BSA was obtained from Armour Pharmaceutical Co., London. Pepsin, sodium alginate and gum arabic were British Drug House products. Analar petroleumether (B.D.H.) having a boiling range 120-160C° and a density of 0.756g cm^{-3} was used after purification through alumina column. Other materials were of A. R. grade and were used as such.

The emulsions were prepared by suspending 1.0% by volume of petroleum ether in 0.01% aqueous solution of the macromolecular emulgents and 0.01 MKCl. The mixture was then subjected to homogenization for 30 min. in a Fischer homogenizer with stainless steel stirrer and Pyrex glass container. The emulsions were then allowed to stand for half an hour and sampling was done under identical conditions.

The flocculation was observed haemocytometrically using an improved Neubauer model of haemocytomer and Fischer Tally counter under Leitz microscope with 15X95 magnification. The pH adjustments were done by a Cambridge pH meter. The electrophoresis measurements were carried out in a rectangular Northrupand Kunitz flat type microelectrophoretic cell which was mounted on the base of Carl Zeiss Jena microscope using reversible Ag-AgCl electrodes. The details are described elswhere (16).

For the large emulsion particles in the present study having $κ a \gg 1$ and charge densities $< 1.4 \times 10^4$ e.s.u. per cm^2 the following equation due to Helmholtz Smoluchowski was used.

$$\zeta = \frac{4\pi\eta}{\epsilon x} \cdot u \quad (12)$$

where the notations have the usual meaning.

The elasticities (shear moduli) of the macromolecular film were determined by the Oscillating bob method. For this a torsion pendulum based on the method of Biswas and Haydon (17, 18) was used.

Results and discussion

The original samples of emulsions prepared as described before had the following initial characteristics (Table I). The macromolecule coated emulsion drops were rather large having their radius greater than 1μm, a typical microphotograph along with size distribution curve being shown in Figure 3. Their stability was strongly pH dependent such that at the isoelectric point the flocculation was rapid but at pHs appreciably different they aggreated only slightly. This less degenerate state was a sort of quasi equilibrium state (Figure 4.) assumed to arise from the coexistence of monomers and dimers and is interpreted in terms of reversible flocculation occurring in the secondary minima of energy profiles as discussed below.

In the framework of the D.L.V.O. theory the interaction energy curves for the flocculating emulsions were constructed. They are all very similar having a high potential energy barrier 25 kT which obviates the possibility of the occurrence of the flocculation in the primary minimum. But they have secondary minima with a depth of 8 kT where particles are easily trapped leading to flocculation {cf. Overbeek et. al. (19) and Bagchi (20)}. Therefore, using Equation (4) one could calculate the degrees of aggregation D which are listed in Table II with corresponding zeta potentials and clearly show that aggregation is low and of the order of 20%.

In the aforesaid initial calculations of the energy profiles the working value of the van der Waals constant A used was 1.6×10^{-13} erg, a value for paraffinic hydrocarbon reported earlier (21). However, assuming the general validity of the D.L.V.O. theory to calculate the total interaction V (in Equation (4)) for a range of A values it is possible to find the effective constant which best fits the observed aggregation data. The procedure is illustrated in Figure 6 and the corresponding data are given in the last column of the Table II. Although these values are of the order of 10^{-13} erg they are appreciably higher than the value 1.6×10^{-13} erg for the paraffinic hydrocarbon. In fact the two values should not be the same because the latter is for the solid wax particles and those of the Table II are for the coated petroleum ether drops where the macromolecular adsorbate should apparently affect the constant. An attempt has been made to estimate a value for the material of the macromolecular film using Vold's treatment (12). The data are given in Table III. In view of the calculations presented in Table III,

Table I. Initial Characteristics of the Original Samples of Emulsions

Stabilizer	KCl content mol m^{-3}	Approximate mol. wt.	Initial droplet conc. ml^{-1}	Average radius nm	Normal pH	Zet pot. mV	Iso-electric point
BSA	0.01	70200	7.4X10^7	1.24	5.6	-32	5.0
Pepsin	"	35500	5.8X10^8	1.10	6.2	-112	2.8
Gum Arabic	"	1000,000	3.8X10^8	1.20	6.0	-82	—
Sodium Alignate	"	300,000	2.5X10^8	1.05	6.6	-95	4.8

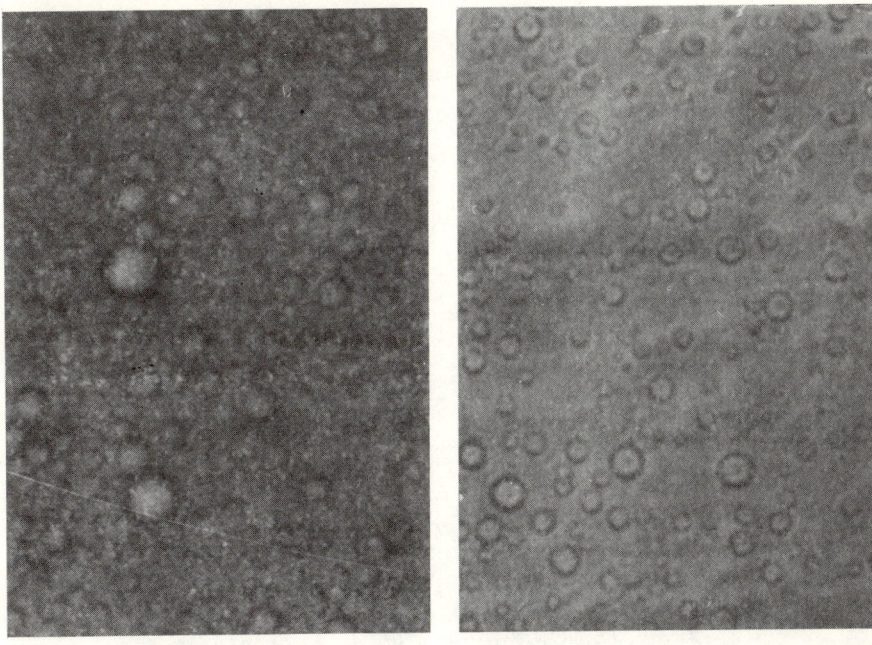

Figure 3a. Microphotograph of sodium alignate emulsion

Figure 3c. Microphotograph of gum arabic emulsion

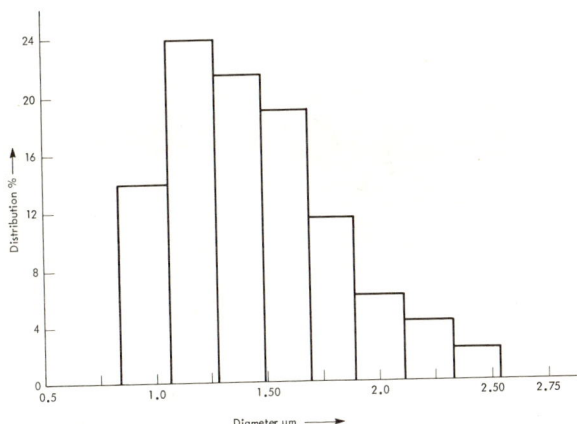

Figure 3b. Particle size distribution curve of gum arabic emulsion. Mean diameter: 1.2 μm.

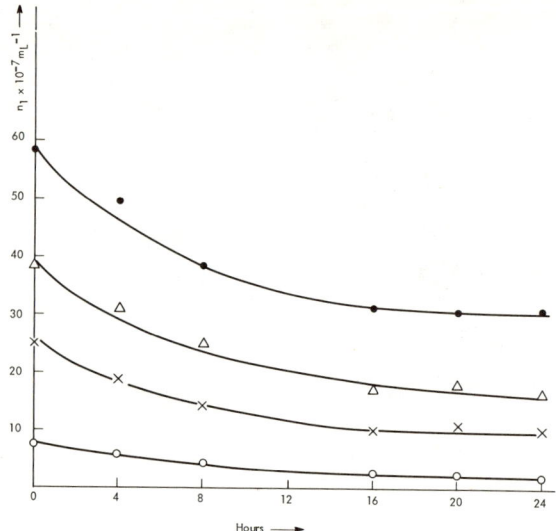

Figure 4. Plot of monomers vs. time. ⊙: BSA, $\zeta = -25$ mV, $Pb(NO_3)_2 = 1.75 \times 10^{-4}$M. ●: Pepsin $\zeta = -70$ mV, $UO_2(NO_3)_2 = 4 \times 10^{-5}$M. △: Gum arabic $\zeta = -57$ mV, $UO_2(NO_3)_2 = 5 \times 10^{-5}$M. ×: Sodium aliginate $\zeta = 64$ mV, $BaCl_2 = 1 \times 10^{-4}$M.

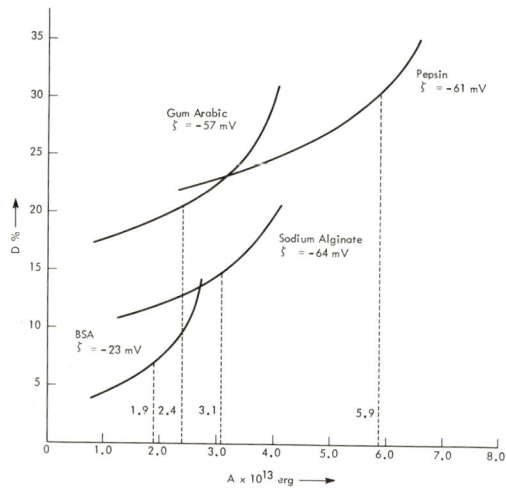

Figure 5. Degrees of aggregation of stabilized emulsions as a function of van der Waals constant

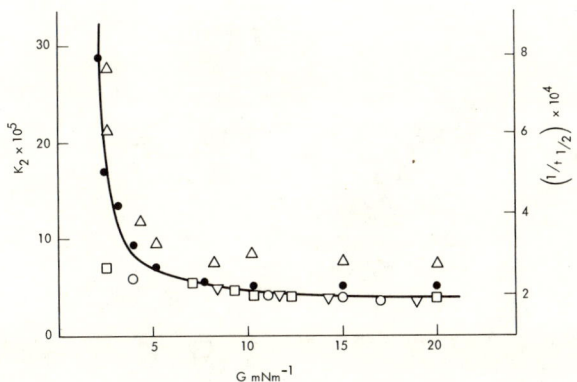

Figure 6. Comparison of the coalescence rates for the emulsions ($K_2 \times 10^5$) and the flat interface $\{(1/t\frac{1}{2}) \times 10^4\}$. ⊙: *phenol.* ▫: $C_{15}H_{29}SO_4^- + C_{15}H_{29}NMe_3^+$ *(1:1).* ▽: *n-decanol.* △: $KCl + H^+$. ●: *flat interface.*

Table II. D, ζ, and Effective van der Waals Constant

Stablizer	Electrolyte Content		pH	Observed D%	ζ mV	Effective $A \times 10^{13}$ erg
BSA	3.08×10^{-3}	M BaCl$_2$	5.7	7	-23	1.9
"	1.75×10^4	MPb(NO$_3$)$_2$	5.6	12	-25	1.8
Pepsin	4×10^{-5}	M UO$_2$(NO$_3$)$_2$	6.2	6	-70	5.2
"	1×10^{-6}	M Th (NO$_3$)$_4$	6.2	32	-61	5.9
Sodium Alginate	1×10^{-4}M	BaCl$_2$	6.6	14	-64	3.1
"	1×10^{-5}M	La (NO$_3$)$_3$	6.6	18	-18	3.0
Gum Arabic	5×10^{-5}M	UO$_2$(NO$_3$)$_2$	6.0	21	-57	2.4

Table III. The A for the Macromolecular Adsorbates

Adsorbate	in mV	Depth of Sec. Min kT	A 10^{13} MS.
BSA	-23	8	5.2
Pepsin	-61	6	3.4
Gum arabic	-57	9	4.5
Sodium alginate	-64	5.5	4.8

it will be speculative to comment on these values of A_{MS} except that they do seem to represent a correct order of magnitude.

Coalescence. Once the emulsion has flocculated irrespective of the mechanism, coalescence is bound to occur between the flocculated frops. Thus assuming that the aggregation has proceeded to the formation of dimers only the coalescence rate can be calculated from the equation (22)

$$-\frac{dn_2}{dt} = -k_2 \left(\frac{n_2 - n_1}{2}\right) \qquad (13)$$

where k_2 is the rate constant for coalescence and n_1 and n_2 are numbers of monomers and dimers respectively. The estimated values of k_2 are given in Table IV. It may be mentioned here that besides the dependence of coalescence on the intervening water film between the drops it also seems to depend on the elasticity G of the macromolecular coating of the film. However, it has to be recognized that in the process of coalescence one has also to overcome the steric barrier prior to coalescence although the measured elastic moduli may very well reflect a measure of the steric barrier. This is indicated from the following table where the rheological parameter G rather than the position of secondary minimum seems to govern the coalescence. This was further confirmed when the Visco-elastic films were presumably displaced or thinned by the addition of nonionic materials to the isoelectric emulsions. The resulting loss of G should give rise to increased coalescence which is indicated from the following table. It should, however, be noted that

Table IV. Dependence of Coalescence on G and the Position of the Secondary Minimum

	Elasticity G_{-1} mNm	KCl mol^{-3} m	pH	Position of the sec. min. nm	$K_2 \times 10^5$ sec	$kx6 \times 10^{10}$
BSA	20	0.01	5.5	15	5.6 ⎫ 5.7	
"	20	0.1	5.0	6	5.9 ⎭	1.2
"	8	0.01	4.3	18	6.9 ⎫ 7.5	
"	8	0.1	4.0	8	8.0 ⎭	1.4
Gum Arabic	22.5	0.01	6.2	17	5.2 ⎫ 5.3	
"	22.5	0.1	5.5	9	5.4 ⎭	1.1
"	12.0	0.01	4.5	19	6.2 ⎫ 6.3	
"	12.0	0.1	4.0	11	6.4 ⎭	1.3

Table V. Influence of Nonionic Detergents on Coalescence

Stabilizer	Isoelectric point	Detergent Conc.	Coalescence rate
BSA + 0.01 MKCl	5.1	0.05 M Phenol	6.2
" " "	"	2×10^{-4} M Decanol	6.8
" " "	5.1	1.5×10^{-6} $C_{14}H_{29}SO_4$ + $C_{14}H_{29}NMe_3$ (1:1)	7.6

the G values used here are for flat interfaces. But as pointed outed earlier (23) the coalescence rates of sperical coated emulsion drops and those of the order of 0.02 ml resting under flat interfaces seem to be similar (Figure 6.).

Lastly, in order to have an idea of the extent of flocculation relative to the conventional Smoluchowski constant, the effective rate constant of flocculation was calculated from the following equation (24, 25)

$$-\frac{dn_i}{dt} = K_1 n_1^2 - \beta(n_2 - n_1) - K_2(n_2 - n_1)/2 \qquad (14)$$

where β is the deflocculation rate constant and is assumed to be negligible when the coated droplets are in contact in relatively deep attraction energy minima. Hence neglecting $\beta(n_2 - n_1)$ and taking the appropriate measured values of other parameters of the above equation K, can be evaluated. Some values are given in the last column of the Table IV and are of the order of 6×10^{10} a much lower magnitude compared to the theoretical value of 6×10^{12}. This is, however, expected because of the slow aggregation of the system for reasons stated above.

In conclusion it may be remarked that although above considerations give a reasonably clear picture of the mechanism of the separation of disperse phase a detailed and more accurate study of the rheology of the macromolecular adsorbate along with theoretical calculation of coalescence rate is desirable.

Acknowledgement

The author is thankful to Dr. D. A. Haydon for his advice, Dr. C. Prakash for his help in the experiments.

Summary

Natural macromolecules such as bovine serum albumin, pepsin, gum arabic, sodium alginate, etc., have been used to promote petroleum ether-in-water emulsions. The choice of the emulgents was rather arbitrarily based on their biological significance. The stability of such coated emulsions drops was assessed in the light of the D.L.V.O. theory. Flocculation was studied both by the addition of electrolytes and some suitable surfactants, using haemocytometry. The corresponding electrophoretic measurements yielded zeta potentials. With the help of the latter and related parameters, interaction energy profiles of the systems under investigation could be constructed. These, when combined with the flocculation data, were found to be consistant with the D.L.V.O. theory. The functional Hamaker constants were also calculated.

This treatment explained the flocculation behaviour of the emulsions which along with the interpretation of coalescence gave an insight into the overall process of separation of phases of an emulsion.

Literature Cited

1. Selmi F., Nuovi Ann. Sci. Nat. Bologna, (1845), $\underline{2}$, 146.
2. Hatschek, E., "Foundations of Colloid Chemistry," MacMillan & Co., London, 32, 1925.
3. Hillyer, H.W. and Marshall, C.R., J. Am. Chem. Soc., (1903), 25, 513.
4. Donnan, F.G., Z. Physik, Chem, (1899), $\underline{31}$, 42.
5. Clayton, W., "Theories of Emulsions and Their Technical Treatment," p. 755, J. Churchill Ltd., London, 1954.
6. Hildebrand, H., J. Phys. Chem., (1941), $\underline{45}$, 303.
7. King, A., Trans Faraday Soc., (1941), $\underline{37}$, 168.
8. Verwey, R.J.W., Trans Faraday Soc., (1940), $\underline{36}$, 192.
9. Van den Tempel, M., J. Colloid Sci., (1958), $\underline{13}$, 125.
10. Albers, W., and Overbeek, J. Th. G., J. Colloid Sci., (1959), $\underline{14}$, 509.
11. Verwey, R.J.W. and Overbeek, J. Th. G., "Theory of the Stability of Lyophobic Colloids," Elsvier, Amsterdam, 1948.
12. Vold, M.J., J. Colloid Sci., (1961), $\underline{16}$, 1.
13. Srivastava, S.N., Indian J. Chem., (1968), $\underline{4}$, 211.
14. Biswas, B., Dissertation, Cambridge University, 1960.
15. Cockbain, E.G., J. Colloid Sci., (1956), $\underline{11}$, 575.
16. Srivastava, S.N. and Prakash, C., Bull. Chem. Soc. Japan, (1967), $\underline{40}$, 1754.
17. Biswas, B. and Haydon, D.A., Proc. Roy. Soc., (1963), A$\underline{271}$, 296.
18. Biswas, B. and Haydon, D.A., Kolloid Z.Z. Polym., (1962), $\underline{185}$, 31.
19. Overbeek, J. Th. G., et al., J. Phys. Chem, (1971), $\underline{75}$, 2094.
20. Bagchi, P., J. Colloid Interface Sci., (1974), $\underline{46}$, 86.

21. Scrivastava, S.N. and Haydon, D.A., Trans. Faraday Soc., (1964), 60, 971.
22. Srivastava, S.N., Dissertation, Cambridge University, 1962.
23. Srivastava, S.N., and Haydon, D.A., Proc. 4th Int. Cong. Surface Active Substances, Vol. 2, p. 1221, Gordon and Breach, New York, 1967.
24. Goodeve, C.F., Trans. Faraday Soc., (1939), 35, 342.
25. Gillespie, T., J. Colloid Sci., (1960), 15, 313.

8

Viscoelastic Properties of W/O Emulsions Containing Microcrystalline Wax

H. KOMATSU* and P. SHERMAN

Department of Food Science and Nutrition, Queen Elizabeth College, University of London, Campden Hill Road, London W8 7AH, England

Introduction

Dispersions of solid particles in fluid media exhibit viscoelasticity when subjected to low shear stress (1-6), provided the volume concentration of dispersed phase is not too small, because the three-dimensional structure resulting from particle flocculation and flocculate growth does not undergo substantial breakdown. Within the flocculated structure the particles are held together by weak attraction forces. Dispersions of fluid particles in fluid media i.e., emulsions, also exhibit viscoelasticity in a similar test situation (7, 8). Consequently, it was considered of interest to study the viscoelastic properties of systems in which dispersions of both solid and fluid particles are present simultaneously. Emulsions in which the continuous phase is a fluid containing dispersed solid particles can be regarded as representative of such systems. Because of their practical implications, the systems selected for study were water-in-liquid paraffin emulsions containing microcrystalline wax.

Experimental Methods

1. Preparation of Emulsions. Two series of emulsions were prepared in which the oil phase consisted of liquid paraffin (medicinal grade) containing microcrystalline wax (Croda Chemicals, Mobilwax No. 2305). In the first series all emulsions (samples E_1-E_5) contained 47.6% wt/wt distilled water (expressed as % wt/wt of the total emulsion), and the microcrystalline wax in the oil phase ranged from 20%-40% (expressed as % wt/wt of the oil phase). In the second series of emulsions (series E_6-E_{11} the water content ranges from 0-47.6% (expressed as % wt/wt of the total emulsion), and the weight fraction of microcrystalline wax in the oil phase was kept constant at 30%. The emulsifying agent employed was polyoxyethylene oleyl ether (Croda Chemicals, Volpo N3). In emulsion series 1 the concentration of emulsifying agent was kept constant, but in series 2 the concentration varied from double

the level employed in series 1 down to the same level. Table 1 shows the detailed composition of all emulsions in series 1 and 2.

The liquid paraffin was heated to 70°C, and it was maintained at this temperature while adding the microcrystalline wax. After all the wax had melted and dissolved in the oil the water phase, previously heated to 75°C, was slowly added while maintaining a continuous agitation with a mixer-homogeniser (Silverson Machines Ltd., London, England). When all the water had been added the emulsions were quickly cooled in a bath containing iced water at 0°C. The emulsions were then allowed to stand for 3-4 days at 25.0 ± 0.1°C before testing.

2. **Rheological evaluation of the emulsions.** The viscoelastic parameters of the emulsions, and also of the continuous phases (samples 0_1 - 0_5), were evaluated at low shear stress (1.26×10^2 - 2.27×10^3 dyne cm^{-2}) by applying a small air pressure to samples deposited in glass U-tubes and measuring the volume displacements. Details of the equipment design and of the experimental procedure will be published elsewhere (9).

The readings were converted into creep compliance-time data which were then analysed by conventional methods (10).

2. **Melting characteristics of microcrystalline wax in liquid paraffin.** Mixtures of microcrystalline wax were prepared in liquid paraffin at 75°C, and then cooled in iced water. The wax concentrations employed corresponded to those incorporated in the continuous phases of the emulsions.

Two series of solutions were prepared. One series contained the appropriate concentration of polyoxyethylene oleyl ether, while the other series did not contain any emulsifying agent. The melting characteristics of these mixtures were examined using a Du Pont 900 Thermal Analyzer with a heating rate of 1.0°C min^{-1}.

Results

1. **Rheological data.** Analysis of the creep compliance-time data for both the emulsions and the microcrystalline wax-liquid paraffin mixtures gave only four parameters in each case examined. These were an instantaneous elastic modulus (E_0), a retarded elastic modulus (E_1) and an associated viscosity (η_1) and a Newtonian viscosity (η_N).

In each case the creep compliance J(t) with time t could be represented by

$$J(t) = J_o + J_1 \left[1 - \exp(-t/\tau_1) \right] + \frac{t}{\eta_N} \quad (1)$$

Table I. Composition of the Emulsions and Continuous Phases Studied

a) Emulsions

Ingredients	Series 1					Series 2					
	E_1	E_2	E_3	E_4	E_5	E_6	E_7	E_8	E_9	E_{10}	E_{11}
Microcrystalline wax	10	12.5	15	17.5	20	30	27	24	21	18	15
Liquid paraffin	40	37.5	35	32.5	30	70	63	56	49	42	35
Polyoxyethylene oleyl ether	5	5	5	5	5	10	9	8	7	6	5
Water	50	50	50	50	50	0	10	20	30	40	50

b) Continuous phases

Ingredients	O_1	O_2	O_3	O_4	O_5
Microcrystalline wax	20	25	30	35	40
Liquid paraffin	80	75	70	65	60
Polyoxyethylene oleyl ether	10	10	10	10	10

where J_0 is the instantaneous elastic compliance, J_1 is the retarded elastic compliance and τ_1 is the retardation time.

The data in Table II show the following points: a) in the microcrystalline wax-liquid paraffin mixtures the values of all four rheological parameters increase as the concentration of microcrystalline wax increases. At the highest concentration of microcrystalline wax employed there is a disproportionate increase in the values of E_0 and E_1.
b) In emulsion series 1 the values of all four rheological parameters increase as the concentration of microcrystalline wax increases. The effect appears to be particularly noticeable for E_1 and η_1.
c) In emulsion series 2 the values of all four rheological parameters decrease as the concentration of water increases. The parameter particularly affected is E_0.
d) A comparison of the rheological data for emulsion series 1 and for the microcrystalline wax-liquid paraffin mixtures indicates that the values of the rheological parameters are lower for the emulsions when the wt/wt concentration of microcrystalline wax, expressed as a % total weight, is the same in both cases.

2. **D.T.A. data.** No significant difference was found between melting data obtained for microcrystalline wax-liquid paraffin mixtures with and without the addition of emulsifying agent.

Discussion

At the emulsification temperature of $70°C$ the microcrystalline wax is completely dissolved in liquid paraffin. Consequently the only process which needs to be considered at this stage is the flocculation ⇌ deflocculation of the water drops, resulting from the emulsification procedure, under the influence of the shear forces developed by the mixer-homogeniser.

Flocculation is a second order process in dispersed systems with a heterogeneous particle size distribution while deflocculation is a first order process (11), 12, 13). The rate constants for the two processes are influenced by the combined effects of Brownian motion and of shear due to a velocity gradient, so that the rate of change in the number (N) of flocs per unit volume of emulsion with time is given by

$$\frac{dN}{dt} = -(k_0 + k_1 \frac{du}{dz}) N^2 + (\beta_0 + \beta_1 |\frac{du}{dz}|) N \qquad (2)$$

where the constants k_0 and β_0 are shear-independent terms associated with flocculation and deflocculation respectively, k_1 and β_1 are the respective shear dependent terms and du/dz reflects the velocity gradient in the region immediately around the particles.

Table II. Rheological Data for Emulsions (Series 1 and Series 2) and Continuous Phases

		E_o ($Nm^2 \times 10^{-3}$)	E_1 ($Nm^2 \times 10^{-3}$)	η_1 ($mNms^{-1} \times 10^{-5}$)	η_N ($mNms^{-1} \times 10^{-6}$)
Emulsions Series 1	E_1	1.49	1.55	3.91	8.50
	E_2	4.53	2.01	2.74	7.04
	E_3	2.97	2.83	6.36	14.21
	E_4	5.92	6.40	19.40	27.8
	E_5	7.52	20.60	45.50	23.5
Emulsions Series 2	E_6	200.0	103.0	311.0	105.0
	E_7	33.6	83.4	211.0	53.3
	E_8	18.01	67.8	103.0	66.9
	E_9	7.93	41.0	71.3	33.6
	E_{10}	2.92	8.40	16.3	51.9
	E_{11}	2.97	2.83	6.36	14.2
Continuous phases	O_1	93.50	54.7	176.0	57.0
	O_2	238.0	139.0	295.0	58.4
	O_3	200.0	103.0	311.0	105.0
	O_4	281.0	175.0	536.0	130.0
	O_5	11,300	529.0	772.0	136.0

The shear dependent term β_1 contributing to deflocculation has also been defined in another way

$$\beta_1 = \frac{\lambda}{x_c} \qquad (3)$$

where λ is the distance over which disruptive forces operate and x_c is the distance over which the attraction forces between particles in the flocs extend (14). Hiemenz and Vold (11) suggest that λ is proportional to the radius (R) of the flocs and x_c is proportional to the radius (r) of the particles within the flocs, so that the second term on the right hand side of Equation (2) can be written as

$$\beta_0 + \rho \frac{R}{r} \left|\frac{du}{dz}\right| \qquad (4)$$

where ρ is a proportionality factor. The terms quoted in Equation (3) indicate that the rate of floc disruption should increase with increasing floc size, and also that the deflocculation rate should increase as the size of the particles within the flocs decreases. Nevertheless, studies with dispersions of carbon black-in-heptane indicated that the shear independent terms k_0 and β_0 in Equation (2) contribute significantly to flocculation $\rightleftarrows$ deflocculation when the particle size is well below 1μm.

In the present study both Brownian motion and shear contribute to flocculation $\rightleftarrows$ deflocculation because of the heterogeneous size distribution of the particles produced during emulsification, but the relative contribution of the two effects is less important in the present situation than the degree of flocculation achieved. The maximum size of floc that can be achieved under any shear condition decreases as the intensity of agitation increases, because the resistance of a floc to disruption increases as (floc diameter)^2 increases but the disruptive force arising from a shear field increases with (floc diameter)^3. The lifetime of flocs in a shear field does not depend on floc or particle concentration because the rate controlling factor is the magnitude of the shear field and not floc collision (13). At any intensity of agitation an equilibrium is established between flocculation and deflocculation. With increasing agitation intensity the rate at which the equilibrium is established increases and the equilibrium shifts more in the direction of deflocculation.

It would appear, therefore, that the shear conditions employed during mixing and emulsification in the present study will prevent the development of large size flocs at the emulsification temperature. Subsequently, following the discontinuation of mixing, and while the emulsion is cooling down, particle flocculation will predominate. In this phase flocculation $\rightleftarrows$ deflocculation of water drops will be controlled by two new effects. The first of these is flocculation in the absence of a shear field, and the rate of continued floc growth will decrease as the flocs increase in diameter. The second effect is a progressive retardation of floc growth due to increased viscosity of the oil phase as the emulsion cools down and the microcrystalline wax crystallises out. Overall, there is probably some growth of flocs as the emulsion cools down, but it is most unlikely that it proceeds to such an extent that the flocs link together to form a three-dimensioned structural network. This conclusion is of major importance in explaining the difference between the viscoelastic properties at low shear stress of the W/O emulsions in the presence and absence of microcrystalline wax.

During crystallisation from a liquid paraffin medium and in the absence of water drops, the microcrystalline wax particles will gradually form an interlinked, compact, network in three dimensions. Microcrystalline wax may be in the form of long needles, but in practice, other shapes are also present although their forms have not been identified (16), so crystalline regions are represented by dark areas. Crystallisation will be initiated, in this case, around randomly distributed nuclei, and eventually regions of densely packed cores of crystals will be linked together by narrow chains of crystals. The final distribution of crystals within the interlinked network will probably resemble that postulated for spherical particles (17), but the crystals will be able to pack more closely together in the cores than would be possible with spherical particles.

Crystallisation of microcrystalline wax from solution when flocs of water drops are already present cannot be regarded as a truly random process. It can occur only in the spaces between flocs or water drops. Particularly in the case of the W/O emulsions containing 50% wt/wt water drops the latter will approach a cubic packing geometry when the emulsifier-mixer is switched off, so that crystallisation of microcrystalline wax will proceed mainly in the voids between the water drops. As crystallization proceeds a pressure will be exerted on the water drops so that some distortion is possible and they will no longer retain their spherical shape. Under these conditions some crystals may penetrate into regions where the surfaces of water drops were originally in contact. The vital difference between the arrangements depicted in the absence and presence of water is that in the latter case the small flocs of water drops act as barriers which prevent the microcrystalline wax crystals from forming such a close and compact interlinked network. This latter structure (Figure 1) has greater distances between the densely packed core regions than would be developed for microcrystalline wax in liquid paraffin only where the wax/liquid paraffin ratio corresponds to that in the analagous emulsion. The emulsion has an internal structure similar to that of a foamed polystyrene in the sense that the microcrystalline wax network provides the solid framework and the water drops occupy the space which in the polystyrene foam is occupied by air cells.

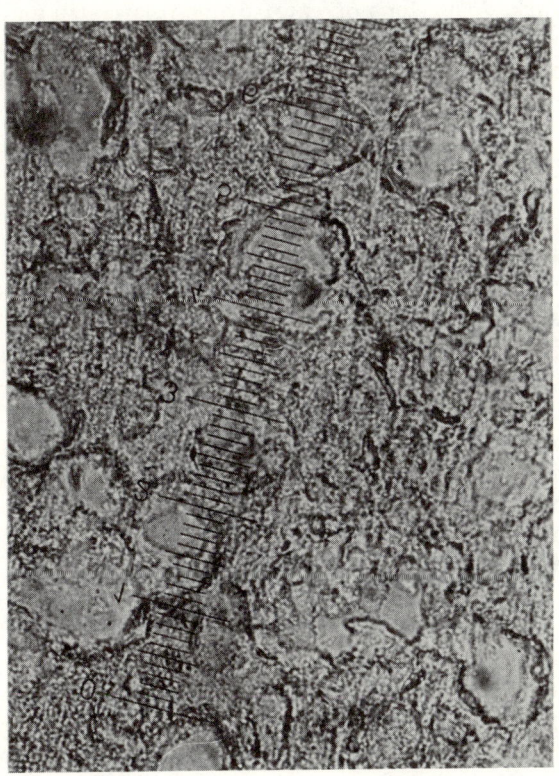

Figure 1. Structure of W/O emulsion containing microcrystalline wax as revealed by polarization microscopy (950×)

The attraction forces between the emulsified water drops in the flocs will be lower in magnitude than the attraction forces between the microcrystalline wax particles in narrow chains linking the densely packed cores of wax crystals together. Attraction forces between the wax crystals will be proportional to their size, but for the water drops the adsorbed emulsifier layer will also exert a substantial reduction (17). Thus, the viscoelastic properties of the emulsions depend primarily on the physical characteristics of the interlinked microcrystalline wax network. The flocs of water drops (their size and number) play a major part in the development of these characteristics since they determine the compactness of the crystal network and the distances between its constituent parts.

Literature Cited

1. Van den Tempel, M., J. Colloid Sci. (1961) 16, 284.
2. Nederveen, C.J., J. Colloid Sci., (1963) 18, 276.
3. Takano, M., Bull. Chem. Soc. Japan (1963), 36, 1418.
4. Payne, A.R., J. Colloid Sci., (1964) 19, 744.
5. Kambe, H. and Takano, M., Proc. 4th Intern. Congr. Rheol., (1965) 4, 537.
6. Papenhuizen, J.M.P., Rheol. Acta (1972) 11, 73.
7. Sherman, P., J. Colloid Interface Sci., (1967), 24, 107.
8. Sherman, P., Proc. 5th Intern. Congr. Rheol., (1970), 2, 327.
9. Komatsu, H. and Sherman P., J. Texture Studies, (1974) 5, 97.
10. Sherman, P., in "Emulsion Science," P. Sherman, Ed. Chap. 4, Academic Press, London, 1968.
11. Hiemenz, P.C. and Vold, R.D., J. Colloid Sci., (1965), 20, 635.
12. Swift, D.L. and Friedlander, S.K., J. Colloid Sci., (1964) 19, 621.
13. Reich, I. and Vold R.D., J. Phys. Chem. (1959), 63, 1497.
14. Gillespie, R., J. Colloid Sci., (1960), 15, 219.
15. Mason, S.G., Pulp and Paper Mag. Canada (1948), 49, 99.
16. Worth, A.H. "The Chemistry and Technology of Waxes," 2nd ed., p. 430, Reinhold, N.Y., 1956.
17. Vold, M.J., J. Colloid Sci., (1963) 18, 684.

*on leave of absence from the Shiseido Laboratory, Yokohama, Japan.

"True" Emulsion Polymerization of Acrylate Esters Using Mixed Emulsifiers

A. R. M. AZAD, R. M. FITCH* and J. UGELSTAD

Institute for Industrial Chemistry, Norwegian Institute of Technology,
7034 Trondheim—NTH, Norway

Introduction

The term "true" emulsion polymerization signifies the formation of a synthetic latex by polymerization of a monomer in the form of a fine emulsion. It is distinguished from "Suspension" or "bead" polymerization in that the monomer droplets are sufficiently small, so that the polymerization may follow classical Smith-Ewart kinetics (1) rather than bulk kinetics. Ugelstad and coworkers have shown that when vinyl chloride is agitated in water with a mixed emulsifier system of anionic surfactant and a normal long chain fatty alcohol a very fine emulsion is produced, and that the principal locus of the subsequent polymerization appears to be within the monomer droplets (2). This was further supported by similar studies on styrene (3). The evidence for the mechanism proposed has been indirect but compelling, based on two kinds of information: (a) particle size distributions of the product latexes and (b) the polymerization kinetics. Ideally one would like to compare the particle size distributions of the monomer emulsion with those of the polymer colloid formed, but the problem has been that no satisfactory method was previously available to measure the liquid monomer droplet sizes, as many are below the resolution of optical microscopes.

In this paper the work has been extended to acrylate esters. Furthermore, some preliminary results are given on a new technique for the electron microscopic determination of styrene monomer droplet sizes.

* Permanent address: Department of Chemistry and Institute of Materials Science, University of Connecticut, Storrs, Conn. 06268, USA

Experimental

Materials. Methyl methacrylate monomer was vacuum distilled twice before using. Fatty alcohols were also twice distilled *in vacuo*. Other materials have been described previously (4).

Procedures. Analytical methods and polymerization technique have been reported earlier (4).

Samples of the monomer emulsions were prepared by a staining technique in which a measured quantity of concentrated aqueous OsO_4 solution was added to the emulsion. After 10 and 15 minutes samples were dried on electron microscope grids and observed under the microscope.

Results and Discussion

Emulsifier Adsorption: Methyl methacrylate (MMA) is much more polar than vinyl chloride or styrene, so that emulsifier adsorption was expected to be less favorable. This was confirmed in experiments on the adsorption of sodium hexadecyl sulfate (SHS) in the presence of hexadencanol (HD) as shown in Figure 1. Only about 50% of the ionic emulsifier is adsorbed initially upon stirring with monomer, and this decreases with time. Vold and Groot showed that the addition of salt led to enhanced adsorption of sodium dodecyl sulfate onto Nujol emulsions along with greater stability (5). This is presumably due to the reduction in lateral electrostatic repulsions among adsorbate molecules as a result of the increase in ionic strength. With mixtures of ionic emulsifiers and straight chain fatty alcohols Goddard (6) showed that increased adsorption resulted solely from hydrophobic interactions of the long chains. Thus we expected that an increase in ionic strength in the mixed emulsifier system would act independently of the hydrophobic interaction to further enhance adsorption of the ionic emulsifier. The dramatic experimental confirmation of this is also shown in Figure 1. Thus the addition of sodium chloride to a concentration of 10^{-2}M was sufficient to cause 97% of the sodium hexadecyl sulfate to be adsorbed, although subsequent desorption upon continued stirring occurred rather rapidly. When the salt concentration was increased further to 10^{-1}M, 98% of the ionic surfactant was adsorbed, and this was accompanied by a marked retardation in the kinetics of both the adsorption and subsequent desorption (Figure 1). We should anticipate that a similar salt effect would be experienced upon the addition of an ionic initiator such as potassium persulfate.

The maximum size of the monomer emulsion droplets was reduced upon the addition of salt from $\sim$ 3 μm to $\sim$1 μm diameter, as determined by oil-immersion light microscopy. The minimum size could not be observed as it was below the resolution of these optics, and because we have been unable to develop an electron microscopic staining technique for acrylate esters.

If the amount of organic phase is increased while keeping the amount of mixed emulsifier and water constant, more emulsifier is adsorbed. Thus as the phase ratio of MMA: H_2O is increased from 1:3 to 1:1 to 2:1, a six-fold change, the maximum amount of emulsifier adsorbed increases from 75% to 99% in the absence of added salt. These results are for the system sodium hexadecyl sulfate (SHS): octadecanol (OD) (1:4 mole ratio) and are shown in Figure 2.

The results with MMA may be compared with those obtained earlier with styrene (3,4). The SHS/HD emulsifier/alcohol system is much more strongly adsorbed onto this more hydrophobic liquid, such that at a 1:3 mole ratio of SHS:HD, 98% of the ionic surfactant is adsorbed <u>without</u> added electrolyte. These and other results are shown in Figure 3. In the presence of ionic initiator we may expect even higher amounts of emulsifier adsorbed and slower rates of adsorption and desorption.

It must be stressed that the processes observed here are kinetically controlled. The emulsifier and fatty alcohol are both precent in the aqueous phase at the start of the emulsification process.

Small droplets formed in the region of high shear rate near the moving paddles are stabilized by the formation of a complex layer of emulsifier and fatty alcohol at the surface of the droplets. The ratio of fatty alcohol to emulsifier is thus initially relatively high and gives a high degree of stabilization towards coalescence. With time, the fatty alcohol is partly transferred to the interior of the droplets so that the amount remaining at the interface is apparently insufficient to maintain stability towards coalescence and consequent desorption of ionic emulsifier. The qualitative aspects of the interactions between ionic emulsifier and alcohol at the interface have been described by Vold and Mittal (7).

<u>"True" Emulsion Polymerization</u>: Ugelstad and coworkers have provided evidence that when free radicals are formed in the presence of these very fine particle size monomer emulsions, polymerization occurs almost exclusively in the emulsified monomer droplets (2,3,4). In favorable cases few if any new particles are nucleated in the aqueous phase. This behavior can be understood in terms of the theory of homogeneous nucleation

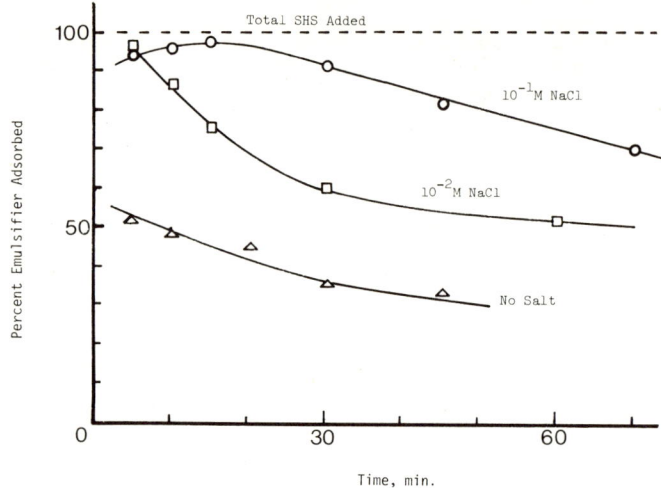

Figure 1. Kinetics of adsorption of sodium hexadecyl sulfate (SHS) (2g l^{-1} of H_2O) in the presence of twice as many moles of hexadecanol (HD) in methyl methacrylate emulsions containing various amounts of salt at 60°C; stirring rate, 600 rpm

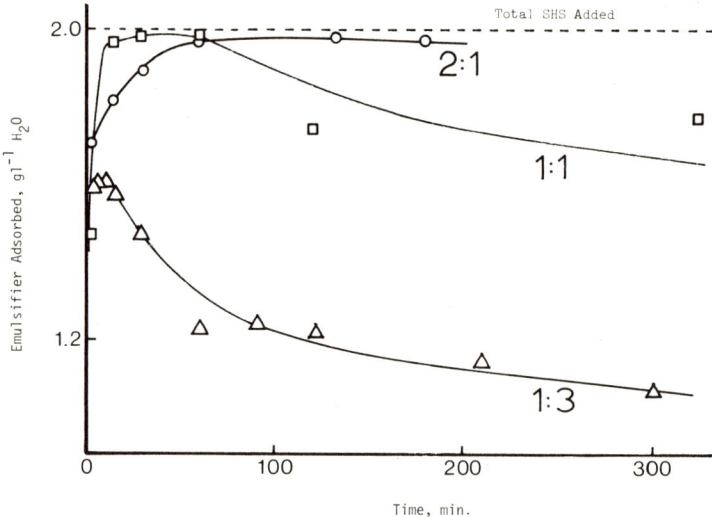

Figure 2. Kinetics of adsorption of sodium hexadecyl sulfate (SHS) (2g l^{-1} of H_2O) in the presence of four times as many moles of n-octadecanol (OD) in methyl methacrylate emulsions, at various phase ratios of MMA:H_2O as shown at 60°C; stirring rate, 600 rpm

of polymer colloids in which it is postulated that new polymer particles are formed by the self nucleation of growing primary oligomeric free radicals in the continuous phase. Particle nucleation only occurs where a primary radical is not previously captured by a monomer droplet. The probability of capture is proportional to the product of the number concentration and the radius (N·r) of the monomer droplets (8). In conventional emulsion polymerization the value of N·r is so low that the number of primary radicals captured by emulsified monomer droplets is negligible (9). In "true" emulsion polymerization, on the contrary, the value of N·r is so large that most, if not all, the primary oligomeric radicals are captured by monomer droplets and no or very little aqueous phase nucleation of new polymer particles occurs.

This picture is complicated by a consideration of the colloidal stability of the primary particles (10). Even if some new particles are formed, they may not be observed because they are unstable and coagulate into the larger particles or droplets (11). Thus the presence of even small amounts of emulsifier in the aqueous phase is important in stabilizing new polymer particles. Other factors, such as monomer solubility, the rate of initiation, ionic strength, and diffusion coefficients also affect the rate at which radicals are captured, and are discussed more fully elsewhere (8, 12).

The particle size distribution of the liquid styrene monomer emulsions has for the first time been obtained by OsO_4 staining prior to electron microscopy. An example is shown in Figure 4. The monomer is stained by a very rapid reaction which probably involves the formation of stable osmate monoesters and (polymeric) diesters (13,14):

4.

$$2 OsO_3 \longrightarrow OsO_2 + OsO_4 \quad + 2 OsO_3$$

(scheme showing formation of monoester and diester from alkene + OsO_4)

The reaction is over in a few minutes and is apparently quantitative. The mole ratio of OsO_4 to double bonds must be carefully controlled to ensure complete reaction whilst leaving a minimum of "excess" OsO_4 which tends to darken the background of the micrographs.

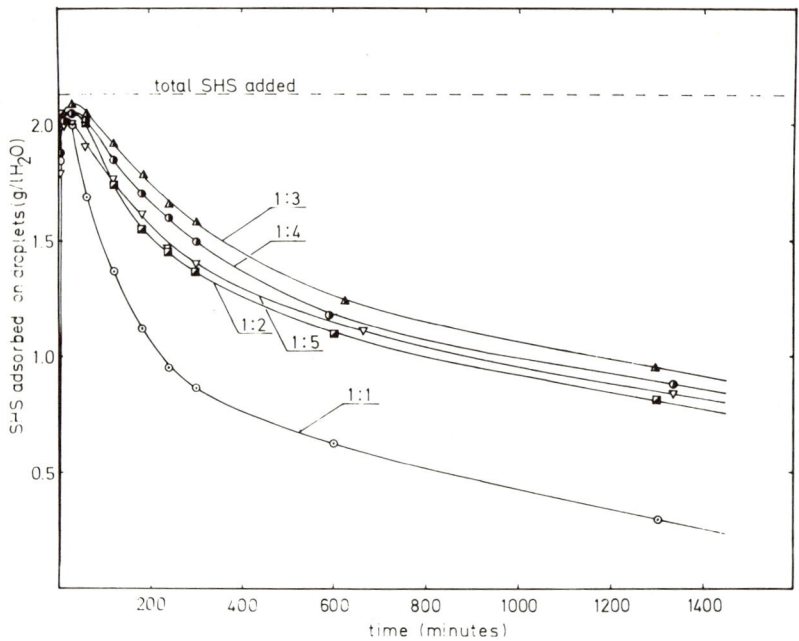

Figure 3. Kinetics of adsorption of sodium hexadecyl sulfate (SHS) at a concentration of 2.13 g l^{-1} of H_2O in the presence of varying amounts of hexadecanol (HD) in styrene emulsions, at 60°C; stirring rate, 600 rpm. Mole ratios of SHS:HD are shown for each curve.

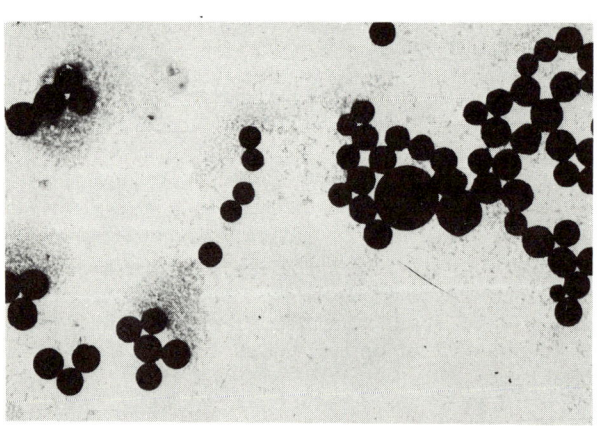

Figure 4. Osmium-stained styrene monomer emulsion. 213g l^{-1} H_2O SHS with 3 × HD (4640×)

If these osmate esters remain in contact with the aqueous environment long enough, hydrolysis presumably occurs to form among other decomposition products, OsO_2. Thus we have observed that if the fixed styrene droplets are allowed to stand for several hours at room temperature without separating the aqueous phase, subsequent electron microscopy shows no emulsion particles and only a dense gray background, presumably due to OsO_2.

In the case of acrylate and methacrylate esters, we have been unable to obtain satisfactorily stained EM specimens of the emulsified - monomer droplets, presumably because of a much more rapid hydrolysis of the more polar intermediate osmate esters.

Upon polymerization, the particle size distribution of the resultant polystyrene latex is almost identical to that of the original monomer emulsion. For instance the styrene monomer emulsion shown in Figure 4, when polymerized with potassium persulfate initiator, produced the latex shown in Figure 5. In addition to the larger particles, there is seen a number of new smaller particles formed we believe by aqueous phase nucleation and subsequently stabilized by small amounts of ionic emulsifier remaining in the aqueous phase.

In the case of MMA, in the absence of electron micrographs of the monomer emulsions, the evidence for "true" emulsion polymerization, although indirect, is convincing. For instance, the three emulsions shown in Figure 2, when polymerized, yielded the polymer colloids shown in Figures 6, 7 and 8. The particle sizes of the latexes formed from the 1:3 and 1:1 MMA:H_2O phase ratios are much smaller than the droplet sizes of the corresponding monomer emulsions as estimated by light microscopy. In contrast, the 2:1 phase ratio emulsion produced dramatically larger average size polymer particles. Once again evidence for some aqueous phase nucleation can be observed.

Summary and Conclusions

Mixed emulsifiers can be used to obtain with low shear rates extremely fine emulsions of polymerizable monomers. For the first time particle size distributions of styrene emulsions can be obtained by OsO_4 staining followed by electron microscopy. These systems exhibit "true" emulsion polymerization in that the emulsified monomer droplets serve as the principal loci of polymerization. Particle sizes of the resulting latexes are generally from about 250 nm to 2000 nm in diameter. The formation of new, smaller particles by aqueous phase ncuelation can be understood in terms of the theory of homogeneous nucleation of polymer colloids.

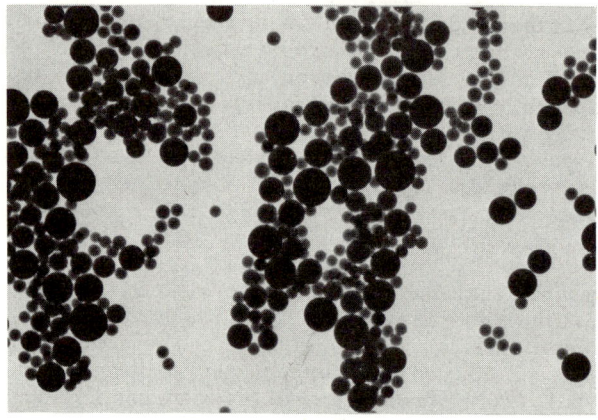

Figure 5. Polystyrene latex formed from the monomer emulsion shown in Figure 4 (4640×)

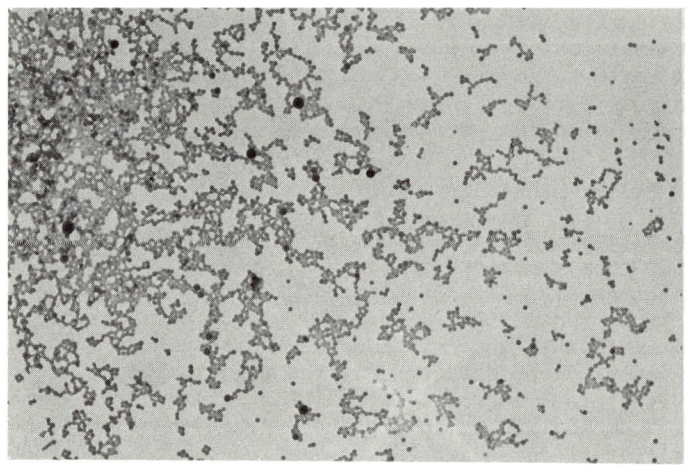

Figure 6. PMMA latex formed from emulsion at phase ratio of $MMA:H_2O = 1:3$. $2g\ l^{-1}\ H_2O$ SHS with $4 \times$ OD (5300×).

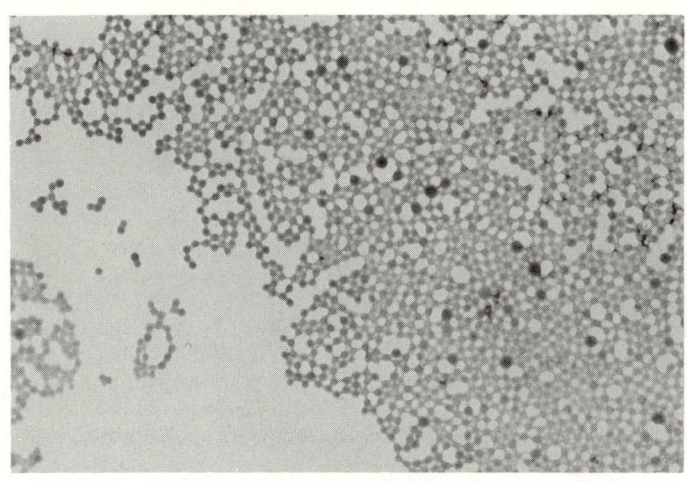

Figure 7. PMMA latex formed from emulsion at phase ratio of $MMA:H_2O = 1:1$. 2g l^{-1} H_2O SHS with $4 \times OD$ (5300×).

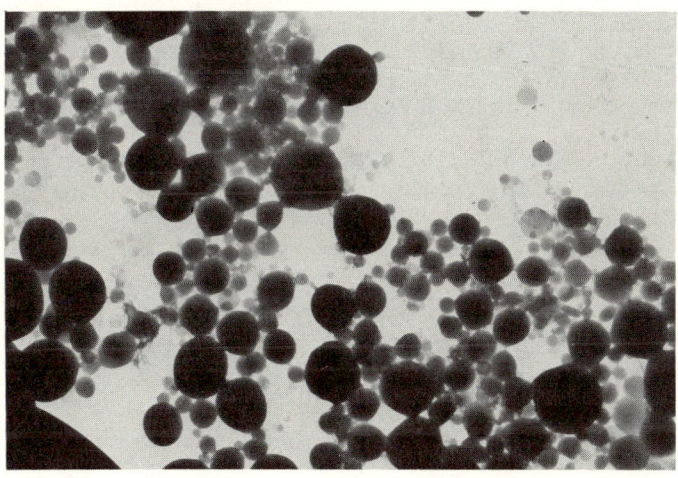

Figure 8. PMMA latex formed from emulsion at phase ratio of $MMA:H_2O = 2:1$. 2g l^{-1} H_2O SHS with $4 \times OD$ (5300×).

Acknowledgements

This work was supported in part by a Fellowship to A. R. M. Azad by the Norwegian Agency for International Development (NORAD), in part by a "Fulbright" Research Grant from the U.S. Educational Foundation in Norway (RMF) and in part by the Royal Norwegian Council for Scientific and Industrial Research (NTNF) for which the authors are most grateful.

Literature Cited

1. Smith, W. V. and Ewart, R. H., J. Chem. Phys., (1948), 16, 592.
2. Ugelstad, J., Flagstad, H., Hansen, F. K., and Ellingsen, J. Polym. Sci., Symposium No. 42, (1973), 473.
3. Ugelstad, J. El-Aaser, M. S., and Vanderhoff, J. W., J. Polym. Sci. Polym. Lett. Ed., (1973), 11, 503.
4. Ugelstad, J., Hansen, F. K., Lange, S., Makromol. Chem. (1974), 175, 507.
5. Vold, R. D., Groot, R. C., Proc. 4th. Int. Congr. Surface Active Substance, Vol. 2, p. 1237, Gordon and Breach New York, 1967.
6. Goddard, E. D., Kung, H. C. Chem. Specialties Mfr. Assoc., Proc. Ann. Meeting, (1965), 52, 124.
7. Vold, R. D., Mittal, K. L., J. Colloid Interface Sci. (1972), 38, 451.
8. Fitch, R. M., Brit. Polym. J., (1973), 5, 467.
9. Smith, W. V. and Ewart, R. H., J. Chem. Phys. (1948), 16, 592.
10. Fitch, R. M., Tsai, C. H., in "Polymer Colloids," R. M. Fitch, Ed p. 73 Plenum Press, New York, 1971.
11. Dunn, A. S., Chong, L. C.-H., Brit. Polym. J., (1970), 2, 49.
12. Fitch, R. M., Shih, L. B., Kolloid-Z. Z. Polym., in press.
13. Riemersa, J. C.,Biochim. Biophys. Acta, (1968), 152, 718.
14. Korn, E. D. J. Cell. Biol., (1967), 34, 627.

10

Kinetic Theory for the Slow Flocculation at the Secondary Minimum

PRANAB BAGCHI

Research Laboratories, Eastman Kodak Co., Rochester, N.Y. 14650

Introduction

The stability of lyophobic suspensions in the presence of ionizing molecules is adequately interpreted in terms of the balance of van der Waals attraction and the electrical double-layer repulsion by the Deryagin-Landau-Verwey-Overbeek (DLVO) theory (1). The total interaction potential as a function of the distances of separation of the surfaces of two particles is obtained by adding the electrical double-layer repulsion and the van der Waals (vw) attraction for various distances of separation. Usually it is observed that such total potential curves give rise to two potential minima. The extremely deep minimum at the point of contact between the two surfaces has been defined as the primary minimum and the shallow minimum at larger distances of separation as the secondary minimum (2).

However, the situation is somewhat different in the case of the stabilization of lyophobic suspensions by polymers or nonionic surfactants. It has been shown by various authors (3-20) that the nonionic interaction potentials between two particles with adsorbed nonionic polymers or surfactants owing to the osmotic (3-20) and the volume restriction (3-20) effects in good solvents give rise to large repulsion potentials. On the addition of the vw attraction, the total potentials in such systems show only a secondary minimum at large distances of separation between the particles but no deep primary minimum as in the case of ionic stabilization (except in the case of a poor solvent). The stability limits of such systems have been quantitatively interpreted by Bagchi in previous publications (19,21) in terms of the depths of such secondary minima. Doroszkowski and Lambourne (22), by measuring the repulsion energy between polymer coated particles using a film balance, have indeed shown that in the total interaction curves of such systems, there is no deep primary minimum.

Since the primary minimum is infinitely deep, the aggregation of suspensions at the primary minimum is irreversible. Such irreversible aggregation has been termed as coagulation by LaMer ([23]). However, the depth of the secondary minimum can be of any magnitude so both kinetically reversible or irreversible aggregation may take place at the secondary minimum. Such reversible aggregation at the secondary minimum has been termed by LaMer ([23]) as flocculation. Since aggregation rate is a measure of colloid stability, various investigators have devised methods of determining such aggregation (coagulation or flocculation) rate constants ([24-26]).

The rate constant, k_0, for fast bimolecular aggregation of monodisperse spherical particles of radius R in the absence of any repulsive or attractive potentials is given by the Smoluchowski ([27,28]) theory as

$$k_0 = 8\pi (2D) R , \qquad (1)$$

where D is the diffusion constant for monodisperse spherical particles of radius R. According to Einstein, $D = kT/6\pi\eta R$ for spherical particles, where k is the Boltzmann constant, T is the absolute temperature, and η is the viscosity of the medium. Thus k_0 may be rewritten as

$$k_0 = \frac{8kT}{3\eta} . \qquad (2)$$

It is important to notice that k_0 is independent of the particle radius as long as the latter is less than 0.1μm such that the condition for perikinetic aggregation is prevalent ([2,27,28]). The validity of Equation (2) has been established by various authors ([29-31]). One of the most important assumptions of the Smoluchowski theory is that all encounters between particles lead to irreversible contacts. Such an assumption, unlike in diffusion-controlled reaction kinetics where all collisions do not lead to reaction, is completely justified in the case of unprotected colloids because of the presence of the very deep primary minimum at the point of contact. This is why Equation (2) describes the rate constant of fast aggregation so well.

In the case of suspensions protected by ionic adsorption, however, the situation is quite different. In such a case, coagulation in the primary minimum involves the crossing of the stabilizing potential barrier, and Equation (2) is inapplicable. Fuchs (2,32) solved the diffusion equation in a potential field and showed that the coagulation rate constant k_0' in such cases may be written as

$$k_0' = \frac{k_0}{2R \int_0^\infty \exp\left[\frac{\Delta G_T(H_0)}{kT}\right] \frac{dH_0}{(2R+H_0)^2}}, \qquad (3)$$

where $\Delta G_T(H_0)$ is the summation of the vw attraction and the double-layer repulsion potential as a function of the closest distances of separation of the surfaces (H_0) of the two spherical particles. In the case of unprotected particles (absence of any surface charge), the interaction potential is not zero, as assumed by the Smoluchowski theory, owing to the presence of vw attraction. So the experimentally observed true fast flocculation rate constant, k_0', should be correctly written in the following manner as pointed out by McGown and Parfitt (33):

$$k_0' = \frac{k_0}{2R \int_0^\infty \exp\left[\frac{\Delta G_A(H_0)}{kT}\right] \frac{dH_0}{(2R+H_0)^2}}, \qquad (4)$$

where $\Delta G_A(H_0)$ is the vw attraction potential as a function of H_0. Since ΔG_A is negative, the value of

$$2R \int_0^\infty \exp\left[\frac{\Delta G_A(H_0)}{kT}\right] \frac{dH_0}{(2R+H_0)^2}$$

is usually between 1 and 1/2 as pointed out by McGown and Parfitt (33), which accounts for small differences between experimental fast coagulation rates and the Smoluchowski coagulation rate.

It is important to note that Equations (2), (3), and (4) correspond to aggregation at the primary minimum, i.e., the point of contact is at $H_0 = 0$. However, in the case of secondary minimum flocculation, the equilibrium distance of separation at contact is at a value of H_0 where the secondary minimum occurs. Thus the flocculation rate constant at the secondary minimum according to the Smoluchowski theory with Fuchs' correction should be given by k_s as

$$k_s' = \frac{k_0}{2R \int_{2R}^{\infty} \exp\left[\frac{\Delta G_A(H_0)}{kT}\right] \frac{dH_0}{(2R+H_0)^2}}, \quad (5)$$

where $2L$ is the value of H_0 at which the secondary minimum occurs. Hence, the experimentally observed stability ratio, W'_{EXP}, for secondary minimum flocculation, defined as the ratio of the flocculation rate constant of unprotected particles to that in the presence of protection, should be given by

$$W'_{EXP} = \frac{k_0'}{k_s'} = \frac{2R \int_{2L}^{\infty} \exp\left[\frac{\Delta G_A(H_0)}{kT}\right] \frac{dH_0}{(2R+H_0)^2}}{2R \int_0^{\infty} \exp\left[\frac{\Delta G_A(H_0)}{kT}\right] \frac{dH_0}{(2R+H_0)^2}}. \quad (6)$$

It has already been pointed out that the two integrals in Equation (6) are small because ΔG_A is negative. Also the contributions of these integrals for small values of H_0 where ΔG_A is substantially negative, are extremely small compared to the value of these integrals (34). In other words, the contributions to these integrals mainly come from ΔG_A at large distances of separation, where ΔG_A is still negative but small in magnitude. Thus the integrations of the above functions from 0 to ∞ or 2L to ∞ should make very insignificant difference. (This is easy to rationalize conceptionally in terms of the fact that when two particles are separated by very small distances, the probability of collision by diffusion in a given length of time is so high that the presence of a large attractive potential really does not enhance the process to any extent. Whereas, for particles separated by large distances the probability of collision by diffusion in the same length of time is much smaller than in the previous case and hence any attractive potential, even if small, enhances the process of bringing the two particles together.) Thus, according to the foregoing arguments,

$$2R \int_{2L}^{\infty} \exp\left[\frac{\Delta G_A(H_0)}{kT}\right] \frac{dH_0}{(2R+H_0)^2} \approx 2R \int_{0}^{\infty} \exp\left[\frac{\Delta G_A(H_0)}{kT}\right]$$

$$\cdot \frac{dH_0}{(2R+H_0)^2} \cdot \qquad (7)$$

Therefore, $W'_{EXP} = 1$ for secondary minimum flocculation according to the predictions of the Smoluchowski theory with Fuchs' correction. In terms of the uncorrected Smoluchowski Equation (2) such a prediction is easier to see. In the case of secondary minimum flocculation, the equilibrium distance of separation at contact is at $H_0 = 2L$. Therefore, one can say that the effective radii of the particles are R+L rather than R in such a case. According to Equation (2), k_0 is independent of R, which indicates that such secondary minimum flocculation rates should be independent of the position of the secondary minimum.

The above deduction using Smoluchowski theory is, however, not consistent with experimental observations of secondary minimum flocculation. Frens and Overbeek (35-37) have observed in colloidal silver sols that various degrees of reversible instability are produced at various electrolyte concentrations. They have concluded that such varied degrees of instability can be attributed to flocculation at the secondary minima of various depths. In nonionic stabilization, the stability of AgI dispersions in water and in the presence of poly(vinyl alcohol) [investigated by Fleer et al. (38,39)] and poly(styrene) (PS) latex dispersions stabilized by n-dodecylhexaoxyethylene monomer ($C_{12}E_6$) in water investigated by Ottewill and Walker (15) was interpreted by Bagchi (19,21) in terms of the depths of the secondary minima. The basic argument of Hesselink, Vrij, and Overbeek (18) in such an interpretation is that if the depth of the secondary minimum is 1 kT or less, suspensions would be infinitely stable as an energy of $\frac{1}{2}$ kT is available to a particle in diffusional motion.[2] In systems where the depth of the secondary minimum is 5 kT or larger, suspensions would be

---

[2] The average kinetic energy of a particle according to strict statistical theory is 3/2 kT rather than 1 kT, as frequently used in an approximate sense and as used in this paper. As the use of 3/2 kT in place of 1 kT would make only very minor differences in the computed results presented in this paper, 1 kT would be used as the average kinetic energy of a particle.

totally unstable as 5 kT is about the maximum energy that may be imparted to a particle owing to thermal motions or by shaking. Consequently, if the depth of the secondary minimum is larger than 5 kT, fast aggregation would take place giving rise to a situation which is kinetically similar to primary minimum fast coagulation. However, if the depth of the secondary minimum, $|\Delta G_W|$, is between 1 and 5 kT, slow flocculation would occur as observed earlier (4,16,17). It should always be remembered that the value of $|\Delta G_W|$ = 1 kT between stable and redispersible systems is a thermodynamic quantity whereas the value of $|\Delta G_W|$ = 5 kT for the boundary between redispersible and unstable systems, is an approximate physical estimate (18). The validity of the importance of the depth of the secondary minima in the interpretation of the stability of nonionic stabilization is again indicated by the recent work of Long et al. (40) in which they have shown qualitatively that the weak flocculation of latex particles is determined by the depths of such secondary minima.

The rather lengthy introduction of this paper was meant for pointing out the following two facts:

1. The importance of the secondary minimum in suspension stability, especially in the case of nonionic stabilization.

2. The Smoluchowski theory or Smoluchowski theory with Fuchs' correction is incapable of predicting the rate of slow flocculation at the secondary minimum.

Thus, in this paper a theory is developed for the prediction of the slow flocculation rate at the secondary minimum in light of the experimental work of Ottewill et al. (15).

Theory

The primary reason why Smoluchowski theory does not work in the case of secondary minimum flocculation is because of its basic assumption that all collisions lead to permanent contact. As the secondary minimum is not infinitely deep, such an assumption for the flocculation at the secondary minimum is incorrect. Also in the case of Fuchs' theory, it is considered that aggregation necessarily means contact between the particle surfaces, whereas secondary minimum flocculation implies trapping of two particles at the secondary minimum, which occurs at distances of separation larger than the point of particle-particle contact.

Figure 1 shows a typical curve of total potential as a function of H_0, the minimum distance of separation of the particle surfaces, with a secondary minimum at $H_0 = 2L$. In this figure the depth of the secondary minimum is designated as $|\Delta G_W|/kT$. The average energy of diffusion is 1 kT with a distribution of energies varying from zero to infinity according to the Boltzmann distribution of energies. Thus the energies of all bimolecular collisions are also given by a Boltzmann distribution of energies as follows (41):

$$df = \frac{1}{kT} \exp(-E/kT)\, dE, \qquad (8)$$

where df is the fraction of collisions with energy E. Thus, in a collision involving a secondary minimum of depth $|\Delta G_W|$, if the collision energy is larger than $(|\Delta G_W|/kT)-1$, the particles will bounce back and no attachment will take place. An energy of 1 kT is subtracted from $|\Delta G_W|/kT$ because an energy of 1 kT is available to a particle in translational or diffusional motion. Also if $|\Delta G_W| = 1$ kT and collisions take place with an energy E less than 1 kT, the particles will not stick together because rotational or translational diffusion energy available to such particles will pull them apart. If the collision energy, E, is less than $(|\Delta G_W|/kT)-1$, permanent contact will take place. Hence the fractions of collisions, Δf, with energy less than $(|\Delta G_W|/kT)-1$ leading to permanent contact will be given by the following integral,

$$\Delta f = \frac{1}{kT} \int_0^{\overline{E}} \exp(-E/kT)\, dE, \qquad (9)$$

where $\overline{E} = (|\Delta G_W|/kT)-1$ for $|\Delta G_W| \geqslant 1$ kT and

$\overline{E} = 0$ for $|\Delta G_W| \leqslant 1$ kT.

Hence,
$$\Delta f = 1 - \exp\left(1 - \frac{|\Delta G_W|}{kT}\right). \qquad (10)$$

It is important to note that in the case of diffusion-controlled bimolecular reaction rates one considers the overcoming of a potential barrier in order to make permanent contact. Consequently, one has to consider only those collisions with energy larger than the value of the potential maximum for reaction to take place. In such a case the integration of Equation (9) is carried out from 0 to the value of the potential barrier (or the activation energy). However, in the case of secondary minimum flocculation, the colliding particles are trapped in a potential trough, in which case integration of Equation (9) is carried out between 0 and $\bar{E}$.

The bimolecular flocculation rate is given by the general equation,

$$-\frac{dN}{dt} = k_s' (\Delta f) N^2 , \qquad (11)$$

where N is the number of particles per cm^3, t is time, and k_s' is the collision frequency given in the most rigorous form by the Smoluchowski equation with Fuchs' correction for the secondary minimum by Equation (5). Since ΔG_W is constant for a given situation, one can write the flocculation rate constant, k_s, at the secondary minimum as follows (42):

$$k_s = k_s' (\Delta f) = \frac{k_0}{2R \int_{2L}^{\infty} \exp\left[\frac{\Delta G_A(H_0)}{kT}\right] \frac{dH_0}{(2R+H_0)^2}}$$

$$\cdot \left\{ 1 - \exp(1 - \frac{|\Delta G_W|}{kT}) \right\}. \qquad (12)$$

The rate constant for the coagulation of the unprotected particles will be given by Equation (3) as in such a case aggregation takes place in the primary minima. So the experimental stability ratio, as defined previously, in the case of secondary minimum flocculation would be given by W_{EXP} as:

---

[3] Please note that in Equation (13) the effect of the hydrodynamic drag on the close approach of the two particles, formulated by Spielman (53) and Honig et al. (54) and recently experimentally substantiated by Hatton et al. (55), has been neglected. The reason behind this is that the correction factors for R = 30 nm and R = 30 + δ nm (where δ ~ 0 — 5 nm) are not too different from each other such that in the calculation of W_{EXP}, which involves the log of the ratios of the two correction factors, contribute insignificantly to the value of log W_{EXP}, calculated neglecting the hydrodynamic drag effect (56).

$$W_{EXP} = \frac{k_0'}{k_s} = \frac{2R \int_{2L}^{\infty} \exp\left[\frac{\Delta G_A(H_0)}{kt}\right] \frac{dH_0}{(2R+H_0)^2}}{2R \int_0^{\infty} \exp\left[\frac{\Delta G_A(H_0)}{kT}\right] \frac{dH_0}{(2R+H_0)^2}}$$

$$\cdot \frac{1}{1-\exp(1 - \frac{|\Delta G_W|}{kT})} \cdot \quad (13)$$

Since the approximation indicated in Equation (7) is still valid, one gets

$$W_{EXP} = \frac{1}{\left\{1 - \exp(1 - \frac{|\Delta G_W|}{kT})\right\}} \cdot \quad (14)$$

Equation (14) indicates that the rate of flocculation is solely determined by the depth of the secondary minimum, $|\Delta G_W|$. It is observed by inspection of Equation (14) that as $|\Delta G_W|$ gets larger W_{EXP} approaches 1; that is, for a deep secondary minimum flocculation rate approaches the Smoluchowski rate or the fast flocculation situation.

Calculations and Results

Figure 2 shows a plot of W_{EXP} as a function of ΔG_W calculated using Equation (14). It is observed that for $\Delta G_W = -1$ kT, $W_{EXP} = \infty$. As ΔG_W decreases, W_{EXP} decreases and becomes one at a value of ΔG_W of about -5.5 kT, in comparison with the total instability limit of 5 kT as estimated by Hesselink, Vrij, and Overbeek (18). It has been pointed out previously that the stability limits of PS latex dispersions in water and in the presence of $C_{12}E_6$ as investigated by Ottewill and Walker (16) can be reasonably well interpreted in terms of the depths of the secondary minima (21). Total stability in the scale of time means a stability ratio of infinity and total instability in the scale of time means a stability ratio of unity. For conditions intermediate between these two limits, slow flocculation was observed for the PS-LatexC$_{12}$E$_6$-water system by Ottewill and Walker (15). The stability ratios as

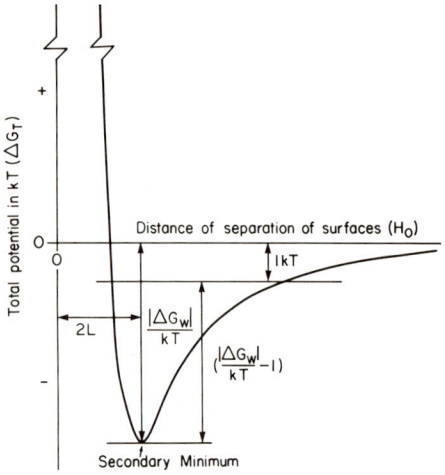

Figure 1. Schematic diagram of a secondary minimum

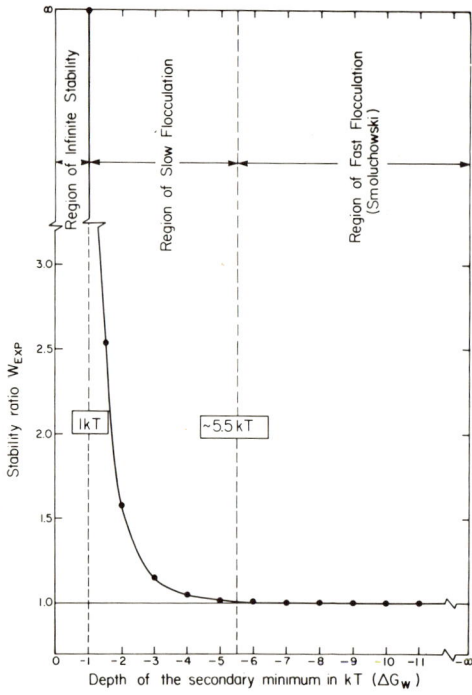

Figure 2. Plot of W_{EXP} as a function of ΔG_W according to Equation (14)

measured by them for this system in the presence of electrolyte beyond the critical flocculation concentration was previously (21) termed as the limiting stability ratio. These experimental stability ratios are directly comparable to the theoretical values predicted by Equation (14) since such experimental values correspond to the stability due to the nonionic stabilizer, which are known to give rise to only secondary minima. It is advised that the reader examine reference (21) in order to understand the applicability of Equation (14) to the data derived in reference (21) from the paper of Ottewill and Walker (15). Table I lists the values of the adsorption layer thicknesses of $C_{12}E_6$ on PS Latex (radius = 30 nm), the depths of the secondary minima as calculated by Bagchi (21), the experimental limiting stability ratios as determined by Ottewill and Walker (15), and the theoretically calculated values of the stability ratios from the values of $|\Delta G_W|$ using Equation (14) at various log molar equilibrium concentrations of $C_{12}E_6$ for the PS Latex-$C_{12}E_6$-water system investigated by Ottewill and Walker (15). In this calculation the Hamaker constant used for the PS Latex-water system was 5×10^{-14} erg as used by Ottewill and Walker (15). Figure 3 shows a direct comparison of the experimental W_{EXP} to the theoretically calculated W_{EXP} as a function of the log molar equilibrium concentrations of $C_{12}E_6$. The error limits in Table I and Figure 3 correspond to an error of +1 nm in the determination of the adsorption layer thickness as indicated by Ottewill and Walker (15).

Figure 5 shows the particle size dependence of the stability ratio of PS latex dispersions as a function of the equilibrium concentration of $C_{12}E_6$. The values of W_{EXP} in Figure 5 were calculated for particle radii ranging between 10 and 480 nm with a quite legitimate assumption that the adsorption layer thicknesses of $C_{12}E_6$ as a function of its equilibrium concentration are identical for PS latices of all sizes.

Discussions

The agreement observed in Figure 3 between the experimental limiting stability ratio and theoretically calculated stability ratio using Equation (14) can be considered to be rather good considering the uncertainties involved in the estimation of the Hamaker constant used in reference (21) to calculate the values of $|\Delta G_W|$. As pointed out previously, the theoretical value of W_{EXP} is solely dependent upon the

Table 1. Values of $2L$, ΔG_W, and Stability Ratios for the PS Latex-$C_{12}E_6$-Water System at Various Log Molar Concentrations of $C_{12}E_6$

Log molar equilibrium concentrations of $C_{12}E_6$	Twice the adsorption layer thicknesses, $2L$ in nm[b]	Depths of the secondary minima, ΔG_W in kT[b]	Experimental log W_{EXP}[a] Flocculated by La(NO$_3$)$_3$	Flocculated by HNO$_3$	Theoretical log W_{EXP} using the corresponding values of $|\Delta G_W|$[c]
-5.3	0 (→ 1.0)	-∞ (-→ -2.42)	---	---	0 (0 → 0.12)
-4.8	0.8 (→ 1.8)	-3.13 (-→ -1.18)	---	---	0.06 (0 → 0.80)
-4.4	2.8 (1.8 → 3.8)	-0.66 (-1.18 → -0.43)	---	---	∞ (0.77 → ∞)
-3.9	8.0 (7.0 → 9.0)	-0.13 (-0.16 → -0.10)	---	---	∞
-3.5	8.6 (7.6 → 9.6)	-0.11 (-0.14 → -0.09)	---	---	∞
-3.1	9.2 (8.2 → 10.2)	-0.10 (-0.12 → -0.08)	---	---	∞
-5.7	---	---	0	---	---
-5.5	---	---	0	---	---
-5.2	---	---	1.4	1.2	---
-4.6	---	---	∞	∞	---

[a] Taken from reference 15.
[b] Calculated in reference 21. Hamaker constant used for the PS Latex-water system 5×10^{-13} erg.
[c] Using Equation (14) and taking $(1 - |\Delta G_W/kT|) = 0$ for $G_W/kT \leq 1$.

value of $|\Delta G_W|$ [Equation (14)]. Hence, the accuracy with which predictions can be made using Equation (14) will depend entirely on the accuracy with which the depths of the secondary minima can be calculated. Since the objective of this paper is basically to develop the theory and to predict the observed trend, no extensive effort was made on an accurate estimation of the Hamaker constant. Nor was the Vold effect (43), whose importance in estimating the long-range interactions has been shown recently (44-46), considered in the calculation of the vw attraction for the PS Latex-$C_{12}E_6$-water system. However, it was shown previously (21) that the inclusion of the Vold effect to calculate the vw attraction is well within the error of estimation of the Hamaker constant for the PS Latex-water system. It is extremely interesting to note that the limits of stability ($W_{EXP} = 1$ for unstable-redispersible and $W_{EXP} = \infty$ for redispersible-stable) are in excellent agreement with the theoretical predictions as observed in Figure 3. This is true for the following reasons. In the stability limit of $W_{EXP} = \infty$ for stable dispersions the equilibrium value of $H_0 (=2L)$ is large. It is well known (21) that the vw attractions at large distances are much better represented by conventional theories than at smaller distances of separation. Also at the limit of instability and redispersibility ($W_{EXP} = 1$) where $|\Delta G_W|$ is about 5 kT, the rate of decrease of the vw attraction is so sharp compared to that for larger distances of separation (H_0) that the error in the value of 2L has very little effect on W_{EXP}; as for $|\Delta G_W|$ larger than 5 kT, W_{EXP} is essentially equal to unity (Figure 2). Thus for a redispersible system where $|\Delta G_W|$ is between 5 kT and 1 kT, a more quantitative agreement between Equation (14) and experimental results may be observed only if a more accurate estimation of the vw attraction can be made. The Parsegian and Ninham (48) type of calculation should be particularly useful for such estimations. It is interesting to note that if a value of 2.3×10^{-14} erg is used for the Hamaker constant of the PS Latex-water system (shown in Figure 4) a much better fit of the theory with the experimental results is observed. However, this does not, in any way, mean that the value of 2.3×10^{-14} for the Hamaker constant is any better than the value of 5×10^{-14} erg as used by Ottewill and Walker (15).

In Figure 5, it is interesting to note the theoretical predictions of the particle size dependence of the stability ratio of the PS latex dispersions as a function of the equilibrium concentration of $C_{12}E_6$. It

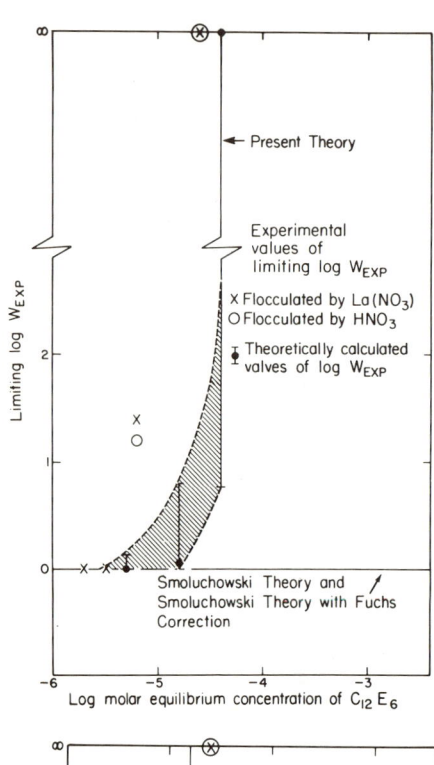

Figure 3. Comparison of the experimental limiting stability ratios for the PS Latex–$C_{12}E_6$–water system of Ottewill and Walker with the theoretical values of the stability ratio calculated using Equation (14) and the values of $|\Delta G_W|$ from Table I as a function of the log molar equilibrium concentration of $C_{12}E_6$. The error bars correspond to an error of ± 1 nm in the values of the adsorption layer thicknesses. Hamaker constant used for the calculation of $|\Delta G_W|$ is 5×10^{-14} erg as used by Ottewill and Walker. Particle radius = 30 nm.

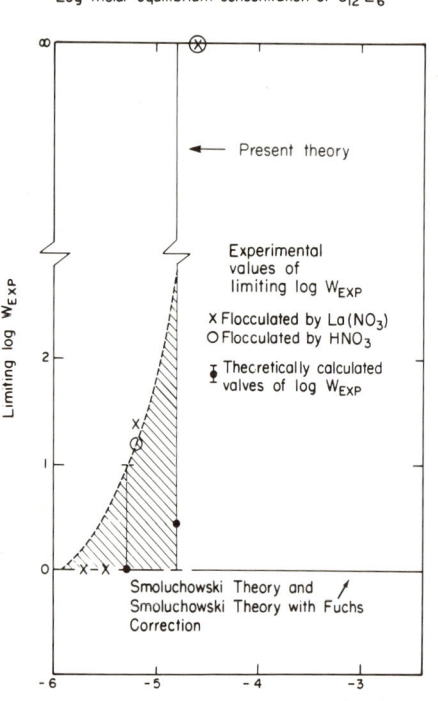

Figure 4. Same plot as Figure 3 except the Hamaker constant used for these calculations of $|\Delta G_W|$ is 2.3×10^{-14} erg

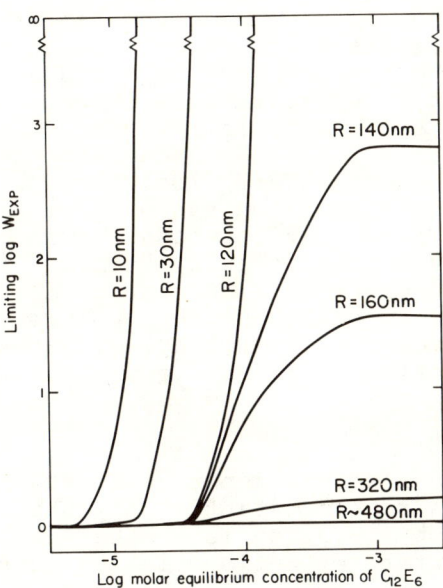

Figure 5. Size dependence of the stability ratio of dispersions of PS latex in water as a function of the equilibrium concentration of $C_{12}E_6$

is observed that for particles of radii less than
120 nm one is able to vary the stability ratio between
0 and ∞ as a function of $C_{12}E_6$ adsorption such that
$|\Delta G_w|$ ranges between 5.5 and 1 kT. However, for
particles of radii larger than 120 nm, the depths of
the secondary minima at saturation are between 5.5 and
1 kT. Consequently in such cases, after saturation,
one has dispersions which flocculate at constant rates
slower than the Smoluchowski rate and never reach complete stability. For particles larger than 480 nm the
depths of the secondary minima, even at saturation, are
larger than 5.5 kT. Hence, under all conditions such
systems would undergo fast flocculation.

So far the discussion has been on the interpretation of the perikinetic secondary minimum flocculation using the depths of such minima. Hiemenz and
Vold (49) have observed simultaneous flocculation - deflocculation in carbon black - poly(styrene)-toluene
and carbon black - poly(styrene)-cyclohexane systems
obeying the following kinetic equation:

$$-\frac{dN}{dt} = kN^2 - \beta N, \qquad (15)$$

where k is the flocculation rate constant and β is the
deflocculation rate constant. In light of the present
discussions, the question is whether one can equate k
with k_s and interpret deflocculation in terms of the
unsuccessful collisions. In the work of Hiemenz and
Vold (10,23) the kinetic units that underwent flocculation and deflocculation were large flocs of radius
above 200 to 300 nm. For such large floc sizes, as has
been pointed out previously, conditions for orthokinetic flocculation prevails. Thus the Smoluchowski
collision frequency as used in the derivation of
Equation (12) is inapplicable. Consequently, for such
large flocs the flocculation rate constant k may not be
compared with k_s. The flocs in such systems are of
random structure as theoretically deduced by Vold (50)
and Medallia (51,52) and electron microscopically
demonstrated for graphon suspensions in heptane by
Bagchi and Vold (10). The deflocculation of such
loosely bound and randomly structured flocs in this
system has been rightly interpreted by Hiemenz and
Vold (24,49) to be due to shearing in thermal gradients. Thus such a deflocculation process cannot be
rationalized in terms of unsuccessful collisions. However, the observation that such flocs deflocculate in
thermal gradients perhaps indicates that the binding
energy of such flocs comes from aggregation at secondary minima of depths less than 5.5 kT.

Acknowledgments

The author would like to gratefully acknowledge the thoughtful comments of Professor Egon Matijevic' of Clarkson College of Technology, Potsdam, N. Y., on this work. Thanks are also given to Dr. G. J. Robersen of the van't Hoff Laboratorium, Utrecht, for very kindly providing the hydrodynamic drag corrections mentioned in connection with Equation (13) and pointing out an error in Equation (2). The author would also like to very gratefully acknowledge the thoughtful suggestions of Professors Robert D. and Marjorie J. Vold during the preparation of this paper.

Summary

It is well known that the instability of a suspension can be due to slow flocculation at a secondary minimum in the particle-particle potential profile, for both ionic and nonionic stabilization. In most cases of ionic stabilization, the potential barrier to the primary minimum is not high enough and consequently one observes fast aggregation at the primary minimum. However, in the nonionic case, where the dispersion medium is a good solvent for the stabilizer, calculations indicate that unstable systems always exhibit a secondary minimum, which is the primary cause of instability. The measure of such instability is the aggregation rate. The Smoluchowski theory of aggregation is not generally valid for secondary minimum flocculation, since it makes the implicit assumption that every collision leads to permanent contact. Such an assumption is correct only in the case of primary minimum coagulation where the attraction potential, at the point of contact, goes to infinity. Fuchs' theory is inapplicable to secondary minimum flocculation because it considers only those collisions with energies high enough to overcome the potential barrier to make contact at the primary minimum as successful collisions. In the case of secondary minimum flocculations, only those collisions with energies less than the depth of the secondary minimum will lead to permanent contact or, in other words, successful flocculation. On this assumption, a kinetic theory for the secondary minimum flocculation is developed in this paper and applied to the poly(styrene (PS) latex-$\underline{n}$-dodecylhexaoxyethylene monomer ($C_{12}E_6$)water system as investigated by Ottwill and Walker. The theoretically predicted dependence of the limiting stability ratio due to the nonionic surfactant as a function of the equilibrium concentration of $C_{12}E_6$ seems to agree fairly well with the experimental observations of Ottewill and Walker.

Literature Cited

1. Verwey, E.J.W., and Overbeek, J. Th. G. "Theory of the Stability of Lyophobic Colloids," Elsevier, Amsterdam 1948.
2. Overbeek, J. Th. G., in "Colloid Science," H. R. Kruyt, Ed., Vol. II, Elsevier, Amsterdam, 1952.
3. Fischer, E. W., Kolloid Z., (1958) $\underline{160}$, 120.
4. Jackel, K., Kolloid-Z. Z. Polym., (1964) $\underline{197}$, 143.
5. Mackor, E. L., J. Colloid Sci., (1951) $\underline{6}$, 942.
6. Mackor, E. L., and van der Waals, T. H., J. Colloid Sci., (1952) $\underline{7}$, 535.
7. Clayfield, E. J., and Lumb, E. C., J. Colloid Interface Sci., (1966) $\underline{22}$, 269, 285.
8. Clayfield, E. J., and Lumb, E. C., Macromolecules, (1968) $\underline{1}$, 133.
9. Meier, D. J., J. Phys. Chem., (1969) $\underline{73}$, 3488.
10. Bagchi, P., and Vold, R. D., J. Colloid Interface Sci., (1970) $\underline{33}$, 405.
11. Bagchi, P., and Vold, R. D., J. Colloid Interface Sci., (1972) $\underline{38}$, 652.
12. Bagchi, P., J. Colloid Interface Sci., (1972) $\underline{41}$, 380.
13. Ash, S. G., and Findenegg, G. H., Trans. Faraday Soc., (1971) $\underline{67}$, 2122.
14. Evans, R., and Napper, D. H., Kolloid-Z. Z. Polym., (1973) $\underline{251}$, 329, 409.
15. Ottewill, R. H., and Walker, T., Kolloid-Z. Z. Polym., (1968) $\underline{227}$, 108.
16. Hesselink, F. Th., J. Phys. Chem., (1969) $\underline{73}$, 3488.
17. Hesselink, F. Th., J. Phys. Chem., (1971) $\underline{75}$, 65.
18. Hesselink, F. Th., Vrij, A., and Overbeek, J. Th. G., J. Phys. Chem., (1971) $\underline{75}$, 2094.
19. Bagchi, P., J. Colloid Interface Sci., (1974) $\underline{47}$, 86.
20. Bagchi, P., J. Colloid Interface Sci., (1974) $\underline{47}$, 100.
21. Bagchi, P., J. Colloid Interface Sci., in press.
22. Doroszkowski, A., and Lambourne, R., J. Colloid Interface Sci., (1973) $\underline{43}$, 97.
23. LaMer, V. K., J. Colloid Sci., (1964) $\underline{19}$, 291.
24. Hiemenz, P. C. and Vold, R. D., J. Colloid Sci., (1965) $\underline{20}$, 635.
25. Lewis, K. E., and Parfitt, G. D., Trans. Faraday Soc., (1967) $\underline{62}$, 1652.

26. Ottewill, R. H., and Shaw, J. N., Disc. Faraday Soc., (1966) 42, 154.
27. Smoluchowski, N. Von, Physik. Z., (1916) 17, 557, 585.
28. Smoluchowski, N. Von, Z. Physik. Chem., (1917) 92, 129.
29. van Arkel, A. E., and Kruyt, H. R., Rec. Trav. Chim., (1920) 39, 656.
30. Ottewill, R. H., and Wilkins, D. J., Trans. Faraday Soc., (1966) 62, 1652.
31. Lawrence, S. A., and Parfitt, G. D., J. Colloid Interface Sci., (1971) 35, 675.
32. Fuchs, N., Z. Physik., (1934) 89, 436.
33. McGown, D. N. L., and Parfitt, G. D., J. Phys. Chem., (1967) 71, 449.
34. Bagchi, P., Ph. D. Dissertation, University of Southern California, Los Angeles, California, 1970.
35. Frens, G., Ph. D. Dissertation, University of Utrecht, The Netherlands, 1968.
36. Frens, G. and Overbeek, J. Th. G., Kolloid-Z. Z. Polym., (1969) 233, 922.
37. Frens, G., and Overbeek, J. Th. G., J. Colloid Interface Sci., (1972) 38, 376.
38. Fleer, G. J., Ph. D. Dissertation, Laboratory for Physical and Colloid Chemistry, Agricultural University, Wageningen, The Netherlands, 1971.
39. Fleer, G. J., Koopal, L. K., and Lyklema, J., Kolloid-Z. Z. Polym., (1972) 250, 689.
40. Long, J. A., Osmond, D. W. J. and Vincent, B., J. Colloid Interface Sci., (1973) 42, 545.
41. Adamson, A. W., "A Textbook of Physical Chemistry," Academic Press, New York, 1973.
42. Smith, A. L., During his visit to the Kodak Research Laboratories (March 1974), Prof. A. L. Smith of the Liverpool College of Technology, England, has disclosed to me that a similar expression has been independently derived by him and his coworkers, F. W. McDowell (Ph.D. Thesis 1971) and W. Hatton (Ph.D. Thesis 1973). They might be publishing their derivation in the future.
43. Vold, M. J., J. Colloid Sci., (1961) 16, 1.
44. Osmond, D.W.J., Vincent, B., and Waite, F. A., J. Colloid Interface Sci., (1973) 42, 262.
45. Vincent, B. J. Colloid Interface Sci., (1973) 42, 270.
46. Becher, P., J. Colloid Interface Sci., (1973) 42, 645.

47. Padday, J. F., in "Thin Liquid Films and Boundary Layers," Spec. Disc. Faraday Soc., Vol. I., p. 64, Academic Press London, 1970.
48. Parsegian, V. A., and Ninham, B. W., J. Colloid Interface Sci., (1966) $\underline{21}$, 470.
49. Hiemenz, P. C., and Vold, R. D., J. Colloid Interface Sci., (1966) $\underline{21}$, 470.
50. Vold, M. J., J. Colloid Sci., (1963) $\underline{18}$, 684.
51. Medallia, A. I., J. Colloid Interface Sci., (1967) $\underline{24}$, 393.
52. Medallia, A. I., J. Colloid Interface Sci., (1970) $\underline{36}$, 115.
53. Spielman, L. A., J. Colloid Interface Sci., (1970) $\underline{33}$, 562.
54. Honig, E. P., Robersen, G. J., and Wiersema, P. H., J. Colloid Interface Sci., (1971) $\underline{36}$, 97.
55. Hatton, W., McFadyen, P., and Smith, A. L., J. Chem. Soc., Farad. Trans. I, (1974) $\underline{70}$, 655.
56. Robersen, G. J. personal communications (1974).

11

Stability of Sterically Stabilized Dispersions at High Polymer Concentrations

F. K. R. LI-IN-ON and B. VINCENT
School of Chemistry, University of Bristol, Bristol BS8 1TS, England

F. A. WAITE
I.C.I. (Paints Div.) Ltd., Slough SL2 5DS, England

Introduction

The use of polymer molecules, both natural and synthetic, to stabilise colloidal dispersions against coagulation has been widely practised in industry, and is of fundamental importance in many natural processes, both biological and environmental[1]. At present there is a need for a wide range of experimental work in this field, on well characterised systems, in order to test the several models and theories of steric stabilisation. One can distinguish at least two classes of experimental approach. Firstly, one may compare the stability of the dispersion in the presence and absence of polymer (at a given concentration). Stability is perhaps best defined in this context in terms of the ratio of the rate constants for aggregation in the two cases. Secondly, one may observe under what conditions a sterically stabilised dispersion will aggregate. There are perhaps four basic requirements for a dispersion to be effectively stabilised by adsorbed or anchored† polymer chains:

a) high surface coverage, i.e. no "bare" patches at the surface
b) strong adsorption
c) a "thick" adsorbed layer, to prevent the particles coming within a separation where the van der Waals interaction becomes effective.
d) the stabilising chains should be in "good" solvent environment.

Such requirements may be met by using AB block and AB_n comb structures (1-3). Here the (insoluble) A chains are the adsorbing

† by "adsorbed" we refer to the case where the stabiliser is physically adsorbed to the particle surface; by "anchored", where the stabiliser is chemically bonded to the surface, or even (partially) incorporated into the bulk structure of the particle.

(anchoring) component and serve to fulfil conditions (a) and (b) above, whilst the (soluble) B chains provide the actual steric barrier, fulfilling conditions (c) and (d). In principle, a sterically stabilised dispersion can be aggregated by adjusting conditions such that one or more of the pre-requisites (a) - (d) is no longer met. Examples for each one are as follows:

a) at low surface coverages bridging, flocculation may occur. (In some cases polymer may be displaced laterally to create bare spots).

b) weakly adsorbed polymers, e.g. homopolymers, may be displaced entirely from the particle surface by smaller molecules, e.g. ionic surfactants, which are themselves more strongly adsorbed but which do not themselves provide adequate steric stabilisation.

c) the thickness of the adsorbed layer is critical in determining the depth of the potential energy minimum in the particle-particle interaction curve. It has been shown by Long, Osmond and Vincent (4) that a critical depth for this potential energy minimum exists, <u>for a given particle number concentration</u>, at and beyond which aggregation begins to occur. Thus subtle changes in the thickness of the steric barrier, e.g. by changes in molecular weight or solvency, can change a system from being stable to being (weakly) aggregated.

d) in a series of papers (5-6) Napper has shown that by changing the solvency conditions for the stabilising B chains from a good solvent environment to a θ- or near θ-solvent, latex dispersions stabilised by AB block copolymers may be aggregated.

In all the systems studied to date, as far as we are aware, the effect of the concentration of polymer in the continuous phase on dispersion stability has not been considered. Indeed in most systems, the bulk polymer concentration is negligibly small. However, this is a significant point to consider in many practical dispersions, e.g. in drying latex paint films where the concentration of polymer in the continuous phase obviously must reach a high value as solvent evaporates.

In this work we set out to investigate whether high concentrations of homopolymer B in the continuous phase (approaching 100% polymer in fact) led to changes in the stability of dispersions stabilised by AB comb polymers. The specific system investigated was an aqueous polystyrene (PS) latex, plus a comb stabiliser with A = PS, B = poly(ethylene oxide) (PEO), molecular weight 750, to which was added homopolymer PEO, of various, relatively low molecular weights (200-4000), over a wide concentration range. Above a certain critical PEO concentration, for a given molecular weight, it was found that aggregation occurred. We present a preliminary report of the results obtained. It is intended that a more detailed report and discussion of other results will be published shortly.

Experimental

Materials. All solvents and monomers were redistilled before use. The polyethylene oxides used were the polyethylene glycol series 200, 400, 600, 1000, 1500, 4000, ex British Drug House, plus polyethylene glycol 800 ex Shell Chemicals Ltd. Carbowax 750 was obtained from Union Carbide Ltd. These materials were used without further purification.

Latex Dispersions. The polystyrene latex used was prepared by a method which is currently the subject of a patent application by I.C.I. (Paints Division) Ltd. Nevertheless it can be stated here that a free radical polymerisation route was used, and this gave a polystyrene latex dispersion, the particles of which were stabilised by a comb stabiliser, in which the polystyrene backbone is physically incorporated into the structure of the polystyrene particles. The methoxy-terminated poly(ethylene oxide) (M.W. 750) chains are therefore terminally anchored at the surface.

Characterisation of the Latex. The mean particle diameter of the latex from electron microscopy was 140 nm. Although not as monodisperse as some of the charge-stabilised polystyrene latices that have been prepared in the absence of added surfactants or polymers ([7](7)), we do not consider this to be a significant feature in interpreting the results of the experiments we report.

Microelectrophoresis and conductimetric titration indicated that the particles carried only a very small surface charge. The calculated zeta potential in 10^{-3} KCl, using the tables of Ottewill and Shaw ([8](8)), was less than 15 mV, i.e. not sufficient to play any significant role in determining the stability of the particles.

Flocculation. Flocculation kinetics were followed using a conventional turbidimetric technique ([9-10](9-10)), in which the optical density (OD) of the latex was recorded continuously as a function of time on the addition of a known amount of polymer solution. A Unicam SP 1800 spectrophotometer linked to a Unicam AR 25 pen recorder was used for this purpose. A wavelength of 540 nm was used, and the cell-housing of the spectrophotometer was thermostatted at $25°C$. The initial number concentration of the particles N_o, was kept fixed at 3.5×10^{10} particles per ml (corresponding to a particle volume fraction of 4×10^{-4}) for all the runs.

Oster ([11](11)) has shown that for an aggregation dispersion in which the particles are Rayleigh scatterers, the turbidity, τ, as a function of time, t, is given by,

$$\tau = AN_o V^2(1 + 2N_o kt) \tag{1}$$

where A is an optical constant related to the refractive index of the particles and the medium, V is the volume of a particle, and k the rate constant for the aggregation process.

As indicated in the Introduction, one may define the stability, W, of a dispersion as follows

$$W = k_o/k \tag{2}$$

where k_o is the rate constant under conditions of rapid coagulation (maximum rate) in the absence of added stabiliser, and is given by the Smoluchowski theory as follows,

$$k_o = \frac{4k'T}{3\eta} \tag{3}$$

where k' and T are the Boltzmann constant and absolute temperature and η is the viscosity of the continuous phase.

Combining Equations (1), (2) and (3) leads to

$$\tau = \tau_o(1 + \frac{8N_o k'T}{3W} \cdot t/\eta) \tag{4}$$

where τ_o is the initial turbidity (t = 0) of the dispersion. or

$$\frac{dOD_r}{d(t/\eta)} = \frac{8N_o k'T}{3W} \tag{5}$$

where $OD_r = OD/OD_o = \tau/\tau_o$. Thus the slope of a plot of OD_r versus t/η should directly reflect changes in W, effects due to changes in A and η, on changing the bulk polymer concentration, having been eliminated. Since the conditions for Rayleigh scattering are normally only fulfilled during the initial stages of aggregation, and thus non-linear plots are generally obtained, it is normal practice to take the limiting slope as t→0, i.e. in this case

$$\left[dOD_r/d(t/\eta)\right]_{t\to 0} .$$

<u>Viscosity</u>. The viscosities of the various PEO-water mixtures were determined at 25°C using a Cannon-Fenske capillary viscometer. The data obtained are shown in Figure 1.

Results

A typical plot of OD_r versus t/η is shown in Figure 2 for the latex in the presence of PEO 200. Below about 70% (by weight) PEO the system is stable. Between 70 and 80% there is a rapid increase in the initial aggregation rate. Thereafter, the rate of aggregation decreases again. As discussed above these effects cannot be attributed to viscosity or refractive index changes in the continuous phase, since these are eliminated in the method of plotting the data. Figure 3 shows the effects described more clearly. Here $(dOD_r/d(t/\eta))_{t \to 0}$ is plotted versus PEO concentration for various molecular weights. For each M.W. there is a critical minimum PEO concentration and only above this is aggregation observed. This critical concentration is plotted as a function of PEO molecular weight in Figure 4. From Figure 3 it can also be seen that the fall off in aggregation rate at higher polymer concentrations occurs for all the molecular weight samples studied.

Preliminary results have been obtained with a polystyrene latex dispersed in n-hexane, for which the stabiliser was a polystyrene-polyisoprene (PIP) block copolymer. The PIP chains in the block had a molecular weight of the order of 1000, and homopolymer PIP of the same molecular weight was added to the continuous phase at increasing concentrations. Again it was observed that above a critical minimum PIP concentration, aggregation occurs.

Discussion

Of the four mechanisms, outlined in the Introduction, concerning the various ways in which a sterically stabilised dispersion may be de-stabilised, one can probably only rule out with any confidence at this stage mechanism (b), i.e. desorption of the stabiliser on particle contact. It would seem (3) that the stabiliser is well anchored at the surface. Any of the other three mechanisms, either singly or in combination, or indeed some other mechanism not so far considered, could be responsible for the behaviour observed. However at this stage much more work needs to be done on the effects of e.g. molecular weight of the stabilising group, number concentration of the particles in dispersion, particle size and temperature, before any firm conclusions can be reached. It would also seem essential that other systems be studied. Work in all these directions is currently underway, and it is hoped to report these results in the literature in the near future.

Acknowledgements

The authors would like to acknowledge the many useful and

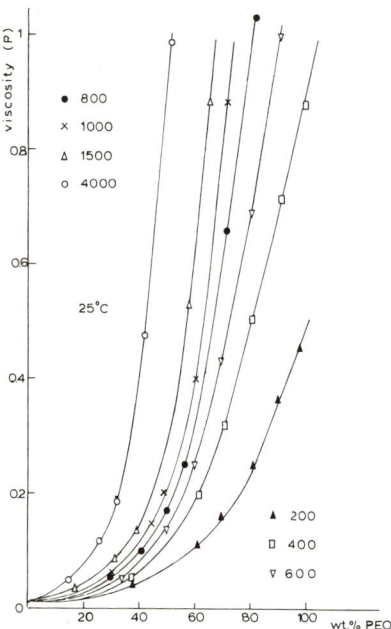

Figure 1. Viscosity of aqueous poly(ethylene oxide) solutions

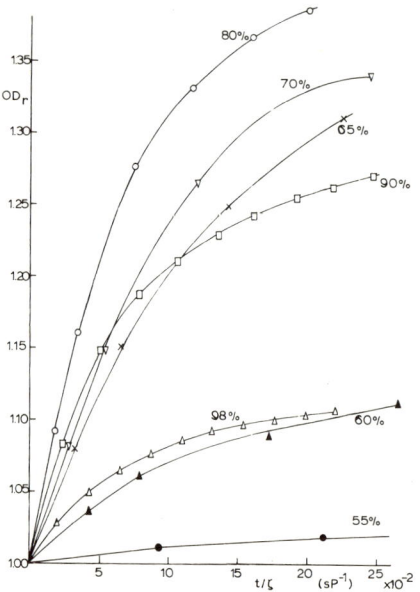

Figure 2. Corrected OD vs. time plots for the latex in the presence of various concentrations of PEO 200

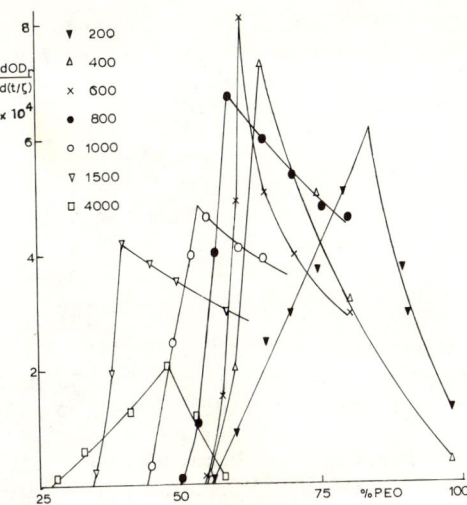

Figure 3. Stability curves for the latex as a function of poly(ethylene oxide) concentration

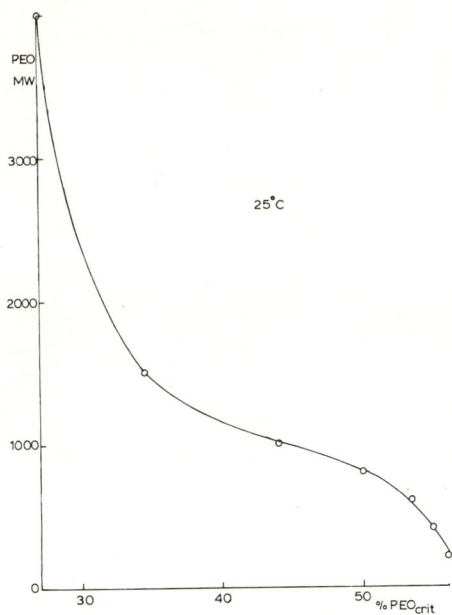

Figure 4. Critical poly(ethylene oxide) concentrations for flocculation as a function of molecular weight

stimulating discussions with Mr. D.W.J. Osmond (I.C.I., Paints Division) concerning this work. One of us (F.K.R.L.) would like to thank the University of Bristol for the award of a University Scholarship to help finance this work.

Literature Cited

1. Vincent, B., Adv.Colloid Interface Sci., (1974) 4, 193
2. Waite, F.A., J.Oil Colour Chem.Ass., (1971) 54, 342.
3. Barrett, K.E.J., Br.Polymer J., (1973) 5, 259.
4. Long, J.A., Osmond, D.W.J. and Vincent, B., J.Colloid Interface Sci. (1973) 42, 545.
5. Napper, D.H. and Hunter, R.J. in "M.T.P. International Review of Science, Physical Chemistry, Surface Chemistry and Colloids, Ser.1, Vol.7", M. Kerker, Ed., p. 289, Butterworths, London, 1972.
6. Napper, D.H., Ind.Eng.Chem.Prod.Res.Dev., (1970) 9, 467.
7. Goodwin, J.W., Hearn, J., Ho, C.C. and Ottewill, R.H., Br.Polymer J (1973) 5, 347.
8. Ottewill, R.H. and Shaw, J.N., J.Electroanal.Chem. Interfacial Electrochem. (1972) 37, 133.
9. Reerink, H. and Overbeek, J.Th.G., Disc.Faraday Soc. (1954) 18, 74.
10. Ottewill, R.H. and Shaw, J.N. Disc.Faraday Soc., (1966) 42, 154.
11. Oster, G., J.Colloid Sci., (1960) 15, 512.

12

Stability of Aqueous Dispersions in Polymer-Surfactant Complexes. Ethirimol Dispersions in Mixtures of Poly-(Vinyl Alcohol) with Cetyltrimethylammonium Bromide or Sodium Dodecyl Benzene Sulfonate

TH. F. TADROS

Plant Protection Limited, Imperial Chemical Industries Ltd., Jealott's Hill Research Station, Bracknell, Berkshire RG12 6EY, England

In a previous paper (1), the interaction of cetyltrimethylammonium bromide (CTABr) and sodium dodecylbenzene sulfonate (NaDBS) with poly(vinyl alcohol) (PVA) (12% acetate groups, $\bar{M}_v$ = 42,000) has been investigated in aqueous solution. A polymer-surfactant 'complex' or polymer nucleated micelle was postulated to form by oriented adsorption of the surfactant ions on the nonionic polymer chain, through some sort of 'hydrophobic' bonding between the hydrophobic end of the surfactant ion and the hydrophobic part of the polymer molecule. Adsorption isotherms on a hydrophilic surface such as silica (1), showed that at high pH, preadsorbed CTA^+ ions increased the adsorption of PVA possibly by 'anchoring' the dissociated -SiO- groups with the PVA chains. Provided there were sufficient CTA^+ ions to create enough hydrophobic sites on the silica surface, the PVA could increase to an amount beyond that required for monolayer coverage (assuming flat orientation) and PVA 'loops' with a proportion of PVA segments attached to the hydrophobised silica surface, result at the silica solution interface. At low pH, a silica surface with preadsorbed PVA also increased the adsorption of CTA^+ above that obtained in absence of PVA. The same increased adsorption of PVA was also observed on addition of NaDBS.

This paper extends previous work on the adsorption of these complexes to organic surfaces, an area of interest in the dyestuffs, pharmaceuticals, pesticides and other fields. Organic materials do not necessarily behave in a similar way to classical model systems such as silica and silver halides but adsorption behaviour is of great importance in these systems of industrial interest. We have therefore investigated the adsorption of these polymer-surfactant complexes on ethirimol (2), an organic pesticide having essentially hydrophobic characteristics but containing polar grouping and their effect on stability against flocculation. One would expect the surfaces of such a material to be heterogeneous with possibly varying degrees of adsorption on the various surfaces. It has a solubility of ~ 160 ppm in water. Due to the polyelectrolyte nature of the polymer-surfactant complex, one would expect a strong double layer repulsion force besides the steric force. Moreover, the chain-solvent interaction of the resulting polymer-surfactant complex is expected to be stronger than that of the nonionic polymer. This was proved before (1) by an increase of the cloud point of PVA on addition of NaDBS or CTABr. Therefore, one would expect a greater osmotic repulsive force for the polymer-surfactant stabilised system relative to that of the nonionic polymer.

Experimental

1. **Materials.** Ethirimol was twice recrystallised from ethanol and then dry ground in a coffee mixer. The surface area of the resulting powder, which was homogenised by mixing in a large container over rollers for 24 hours, was measured by BET Krypton adsorption, after degassing at room temperature in a helium atmosphere for 12 hours. The BET surface area was found to be 0.29 m^2/g.

CTABr, NaDBS, sodium lauryl sulfate, bromophenol blue were the same materials described before (1).

Analar sodium chloride, which was roasted in a silica crucible to constant weight, was used to control the ionic strength of the solutions.

Doubly distilled water from all glass apparatus was used for preparation of all solutions, which were always freshly made (kept for no longer than two weeks).

2. **Preparation of Aqueous Ethirimol Dispersions.** For the mobility experiments dispersions were prepared either by directly dispersing 0.1 g of ethirimol powder in 50 ml of the appropriate solution using ultrasonic irradiation, followed by filtration through a No. 2 sintered glass crucible to remove large particles or by dilution from a stock concentrated dispersion previously prepared by a wet milling process. In the latter case 7 g of ethirimol powder were dispersed in 400 ml doubly distilled water using ultrasonic irradiation followed by milling for 30 minutes to reduce the particle size. The resulting dispersion was stored in a constant temperature room at $20 \pm 1^\circ C$ to reduce crystal growth and kept continuously stirred using a magnetic stirrer. The resulting dispersion was heterodisperse and microscopic investigation showed that a significant proportion of the particles were asymmetric. An average particle diameter was estimated by sizing about 400 particles from a picture taken under the optical microscope. In the case of the asymmetric particles, the minor axis length was used to represent the diameter of the particle. The arithmatic mean particle diameter was found to be 1.65 μm.

3. **Determination of the Hamaker Constant of Ethirimol.** The Hamaker constant was calculated from refractive index measurements at various wavelengths. The refractive indices were measured at 25°C for four different wavelengths by the Becke line method ($\underline{3}$) using characteristic immersion liquids supplied by R.P. Cargille laboratories Inc. Since the refractive indices observed depend on the orientation of the crystal fragments it was only possible to determine the two extreme values. The refractive index values are shown in Table 1. The figures are accurate to ± 0.001. By convention n_α is the lower and n_γ the higher. The third index n_β (immeasurable) has an intermediate value. α, β and γ represent the various crystallographic planes of the ethirimol crystal. From the above data, the London dispersion constant B_{11} and Hamaker xonstant A_{11} were calculated using the method described by Gregory ($\underline{4}$). The effective Hamaker constant for interaction across an aqueous medium was calculated from the equation,

$$A_{eff} = (A_{11}^{\frac{1}{2}} - A_{33}^{\frac{1}{2}})^2$$

where A_{33} is the effective Hamaker constant for water ($\underline{4}$), taken to be 4×10^{-20}J. The calculated constants are shown in Table 2.

Table I. Refractive Indices of Ethirimol at Various Wavelengths

λ(nm)	$n\alpha$	$n\gamma$
486	1.557	1.723
546	1.554	1.705
589	1.549	1.696
656	1.544	1.677

Table II. London and Hamaker constants

Orientation	α	γ
B_{11}	5.43×10^{-76} J m^{-6}	6.47×10^{-76} J m^{-6}
A_{11}	6.52×10^{-20} J	7.76×10^{-20} J
A_{eff}	3.1×10^{-21} J	6.2×10^{-21} J
Average A_{eff}	4.6×10^{-21} J	

4. **Adsorption Isotherms.** 100 ml of CTABr (concentration range 5×10^{-5} - 4×10^{-3} mole/l) or NaDBS (10^{-4} - 2.5×10^{-3} mole/l) solutions with ionic strength controlled to 10^{-2} with NaCl were added to 2 or 4 g of ground ethirimol. The mixtures were dispersed by ultrasonics at intervals of 30 sec (to avoid heating) with a maximum period of 2 minutes. The dispersion was left stirred overnight using magnetic stirrers in a constant

temperature cabinet at 25 ± 1°C. The samples were allowed to stand for one hour at the same temperature and the supernatant liquid quickly filtered at room temperature (about 22°C) through a 0·22 μm millipore filter. The first few ml were discarded to eliminate errors from possible adsorption on the filter paper. The equilibrium concentration of CTABr solutions was determined by two phase titration against standard sodium lauryl sulfate (5). That of NaDBS was determined by titration against standard CTABr (5). For measurement of adsorption of PVA, 50 ml of PVA in the concentration range 0-300 ppm were added to 1 g of ethirimol powder and the same adsorption procedure applied. The method of analysis of PVA is based on measurement of colour intensity of the complex formed with a reagent of iodine + potassium iodide + boric acid, at 670 nm (6, 7). However, the presence of trace quantities of dissolved ethirimol in the equilibrium solution (ethirimol has a solubility of ∼160 ppm in water that is only very slightly increased in the presence of up to 300 ppm PVA), a turbidity in solution after addition of the I_2 + KI + boric acid reagent obscured measurement of the colour intensity. Extraction of ethirimol using polar and non-polar organic solvents could not be applied as it was found from blank experiments that some PVA was also extracted into the organic phase. Finally, we found that the turbidity could be removed by adjusting the pH of the equilibrium solution to 9-10 before adding the iodine reagent. For that purpose the pH of the PVA solutions used for the calibration curve were also adjusted to pH 9-10. PVA adsorption was also measured in the presence of NaDBS (2 x 10^{-3} mole/l) and the same procedure of analysis of PVA was adopted. As mentioned before (1), NaDBS did not interfere with the colorimetric method of analysis of PVA.

5. <u>Electrophoresis Experiments</u>. The electrophoretic mobility was measured using a commercial microelectrophoresis apparatus (Rank Bros., Bottisham, Cambridge, England). A cylindrical cell was used and the mobility of 20 particles was determined at each stationary level, reversing the polarity for successive timings. The average electrophoretic mobility was calculated from the average reciprocal time for all 40 measurements. Mobilities were measured as a function of CTABr or NaDBS concentration (choosing concentration ranges below and above the cmc) and in the same solutions containing varying amounts of PVA (100-400 ppm). Also the mobility of ethirimol particles was measured as a function of PVA concentration (0-300 ppm without addition of surfactants). In all cases, the ionic strength was controlled to 10^{-2} with NaCl. From the electrophoretic mobility the zeta potential was calculated using the classical Smolukowski equation (this is justified since $\kappa a \gg 100$).

6. <u>Flocculation Kinetics</u>. Flocculation kinetics was followed by manual counting technique using a Haemocytometer cell

(Hawskley) having an etched graticule of 25×10^{-4} mm^2 and a depth of 0·1 mm. A Leitz Ortholux microscope was used under bright field illumination and a magnification of 500 x. The eye piece of the microscope was connected to a television camera and the field of view shown on a television screen. In this way counting was less tedious and the number of particles in 40 squares of the graticule which were chosen at random was counted.

For these experiments the milled stock dispersion was diluted in the appropriate solution to give a particle concentration of $\sim 10^8$ particles/ml. The dispersion was sonified and kept in a constant temperature bath at $25 \pm 0.05^\circ C$ for 15 minutes to attain the temperature of the bath. It was then quickly sonified and the initial count taken immediately after redispersion by ultrasonics, sampling 1 cm below the surface of the dispersion. Counting was then repeated at various intervals of time (8-12 minutes) over a period of 1-2 hours, an aggregate being counted as one particle. Prior to sampling the dispersion was gently inverted 3-4 times to counteract sedimentation. The experiments were carried out at the same concentration range of CTABr, NaDBS and PVA used for electrophoresis. Also the ionic strength was controlled to 10^{-2} with NaCl.

In order to measure the critical flocculation concentration and the fast flocculation rate, the kinetics of flocculation of the aqueous ethirimol dispersion was followed as a function of NaCl concentration in the range $10^{-3} - 10^{-1}$ mole/l.

Results and Discussion

1. **Adsorption Isotherms.** Figure 1 shows the adsorption isotherm of CTABr on ethirimol, a two step isotherm with saturation adsorption of $1\cdot8-2\cdot0 \times 10^{-6}$ mole/g and $3\cdot5 \times 10^{-6}$ mole/g ethirimol respectively. Using the BET surface area (0·29 m^2/g), the area for CTA$^+$ ion was calculated to be 0·27 nm^2 and 0·14 nm^2 at the first and second plateau respectively. The stepwise isotherm could be due to bilayer adsorption of CTA$^+$, the first layer with the charge pointing towards the surface and the second layer forming by hydrophobic bonding between the hydrocarbon chains of the surfactant ions. However, the results of electrophoresis (see below) has shown the isoelectric point to occur at much lower concentration (1×10^{-5} mole/100 ml) than that at which the step occurs (8×10^{-5} mole/100 ml). We would therefore favour the following picture for adsorption of CTA$^+$ ions on ethirimol till the first plateau. Initially the surface has a negative surface charge (as indicated by a negative zeta potential of $-$ 22 mV) which could be due to some specifically adsorbed anions. In the initial adsorption stages these charges are neutralised with CTA$^+$ ions, with the positive charge pointing towards the surface, leaving uncovered hydrophobic regions in between. The subsequent CTA$^+$ ions will now adsorb on the hydrophobic sites, with the positive charges now pointing towards

the solution. The adsorption process continues until a vertically oriented monolayer of CTA^+ ions is reached which is complete at the first plateau. However, the area/CTA^+ ion at this plateau (0.27 nm^2) is lower than expected from models of a close packed trimethylammonium ions namely 0.35 nm^2. Robb and Alexander (8) found on polyacrylonitrile latices an area of 0.35 nm^2/CTA^+ ion, whereas Conner and Ottewill (9) found an area of 0.47 nm^2 on polystyrene latex in 10^{-3} M KBr and 0.35 nm^2 in 5 x 10^{-2} mole/l KBr. In view of the possible error of the BET area our value of 0.27 nm^2 seems reasonable. As the surface of ethirimol becomes fully saturated with CTA^+ ions, there is still the possibility of association of CTA^+ ions on the surface which could result in a rapid increase in adsorption. This explains the step in the isotherm. Therefore, association or 'hemimicelle' formation as depicted by Fuersteanau et al. (10) could account for this rapid increase.

Figure 2 shows the adsorption isotherm of NaDBS on ethirimol. The isotherm is more or less of a Langmuir type with a saturation adsorption of ∼ 3.5 x 10^{-6} mole NaDBS/g ethirimol which is similar to that obtained with CTABr at the second plateau. However, this saturation adsorption is reached at a much lower equilibrium concentration of NaDBS and there is a tendency of further increase in adsorption beyond this value as indicated by the dotted line on the isotherm. Using the BET area, the area/DBS^- ion at the initial plateau is ∼ 0.14 nm^2 which is smaller than that to be expected for a vertically oriented close packed monolayer of dodecyl benzene sulfonate ions. However if one uses the area calculated from CTABr adsorption (see below) an area/DBS^- ion of ∼ 0.18 nm^2 that is close to that for a close packed monolayer of DBS^- ions (0.20 nm^2) is obtained. Adsorption of an anionic surfactant on the negative ethirimol surface should have occurred by the hydrophobic part of the molecule pointing towards the hydrophobic ethirimol surface. Association or hemimicelle formation could account for the increased adsorption beyond the amount required to form a close packed monolayer. A similar type of isotherm has been found by Kudryavtseva and Larionov (11) for adsorption of sodium lauryl sulfate on hydrophobic gold hydrosol. These authors found an adsorption amount corresponding to 3 monolayers of sodium dodecyl sulphate assuming an area of 0.20 nm^2 per dodecyl sulphate ion.

Figure 3 shows the adsorption isotherm of PVA on ethirimol (curve A) and the effect of addition of NaDBS on PVA adsorption (curve B). In the absence of NaDBS, the initial part of the isotherm is what to be expected for the adsorption of a nonionic macromolecule, being of the high affinity type. However, as the concentration of PVA is increased, the adsorption continues to increase and a sudden increase in PVA adsorption is observed above an equilibrium concentration of 140 ppm. The shape of the isotherm is markedly different from that recently obtained on polystyrene latex (12). A similar increase in PVA adsorption on

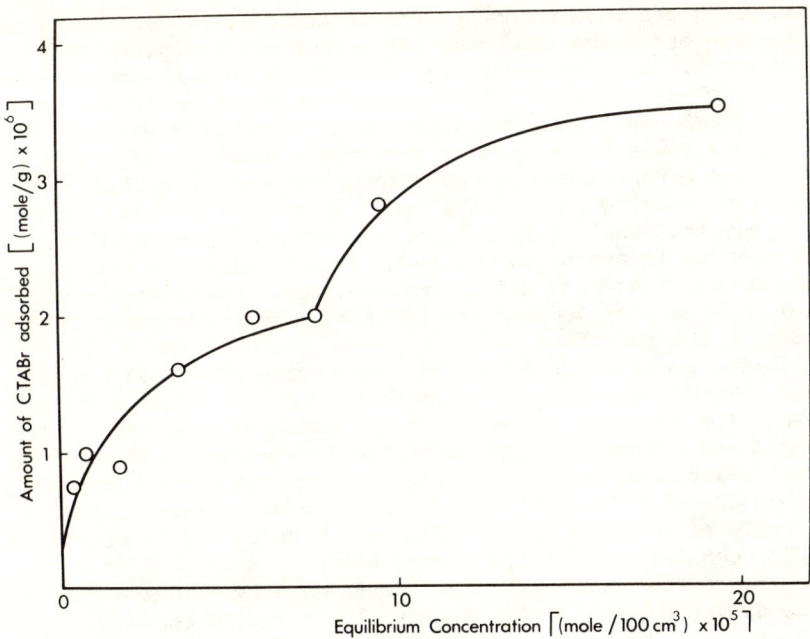

Figure 1. Adsorption isotherm of CTABr on ethirimol

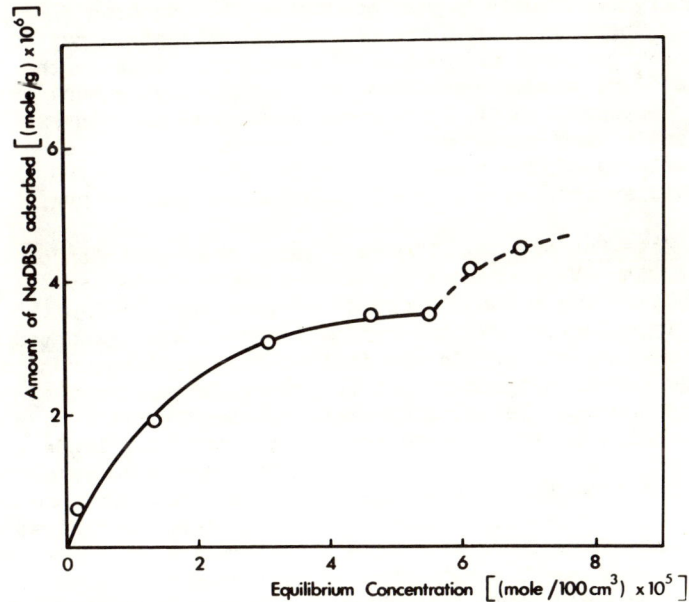

Figure 2. Adsorption isotherm of NaDBS on ethirimol

carbon black has been found by Johnson and Lewis (13). Using the BET surface area, the amount of adsorption at the initial plateau is ~ 8 mg/m^2, which is considerably higher than that recently obtained on polystyrene (12) for a similar PVA sample, namely 2.8 mg/m^2. However, the BET area could be underestimated in view of the low surface area of the sample which could result in significant error. An area could be calculated from CTABr adsorption assuming an area/CTA$^+$ of 0.35 nm^2 for a close packed monolayer; this would give 0.38 m^2/g. Using this value PVA adsorption would amount to ~ 6 mg/m^2, about twice the value obtained on polystyrene. It seems therefore from these results as well as those of Johnson and Lewis (13) that PVA adsorption depends on the substrate.

In the presence of NaDBS, the adsorption of PVA at the initial plateau has dropped to half its value in absence of NaDBS. However, the shape of the isotherm is maintained and the two curves A and B cross each other at the intermediate PVA concentration. The final value of adsorption is lower in presence of NaDBS than in its absence. The drop in adsorption of PVA is what to be expected in view of the more hydrophilic nature of the PVA-DBS$^-$ complex. Recently, Fleer, Koopal and Lyklema (14) demonstrated that the adsorption of PVA on silver iodide decreases as the degree of hydrolysis of the polymer (which leads to an increase in polymer-solvent interaction) increases. The decrease in adsorption as the chain solvent interaction increases is readily explained in terms of Hoeve (15) and Silberberg (16) models. According to these theories, relatively larger loops and consequently relatively higher adsorption is expected from poorer solvents. The lower adsorption in a good solvent is due to the high osmotic pressure created within the loops (due to the repulsive force between the segments in the loop region) which hinders the formation of thick loops and large amounts of adsorption in good solvents. One of course cannot rule out the weaker affinity of the more hydrophilic PVA-NaDBS to the hydrophobic surface of ethirimol relative to that of PVA.

2. *Zeta Potentials*. Figure 4 shows the variation of zeta potential with PVA concentration in the range 0-300 ppm (in 10^{-2} mole/l NaCl). The gradual decrease of zeta potential with increase in extent of PVA adsorption is what to be expected for a nonionic macromolecule. In the initial stages of adsorption the macromolecule will attain a flatter conformation, which as adsorption proceeds leads to thicker adsorbed layers. As a result of this a shift in shear plane towards the solution leads to a lowering of zeta potential and eventually a limiting value (-7 mV) is reached at the limiting adsorption value. However, one cannot rule out the possibility of displacement of some specifically adsorbed ions from the Stern plane on adsorption of PVA segments.

Figure 5 shows the variation of zeta potential with CTABr concentration in presence and absence of PVA, where Figure 6

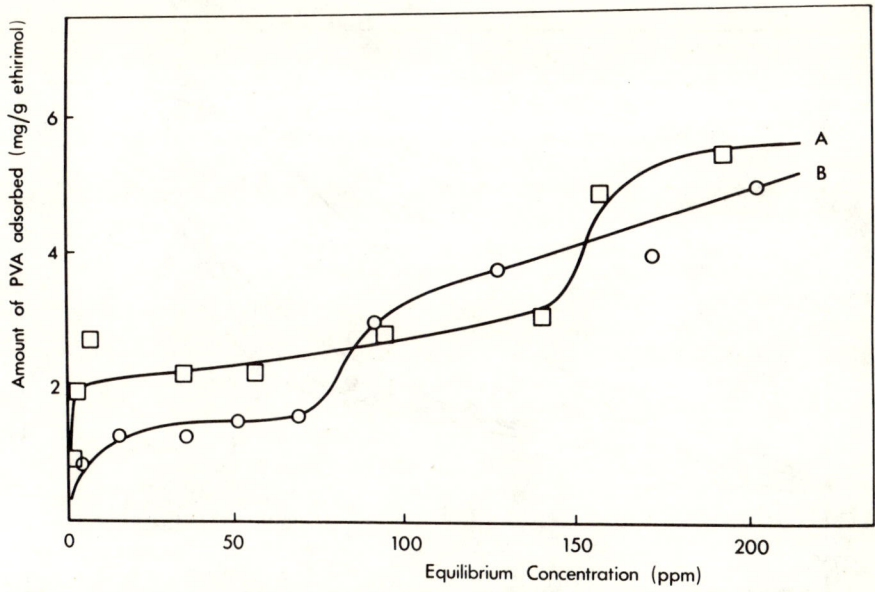

Figure 3. Adsorption isotherm of PVA on ethirimol in the presence (curve B) and absence of NaDBS (curve A)

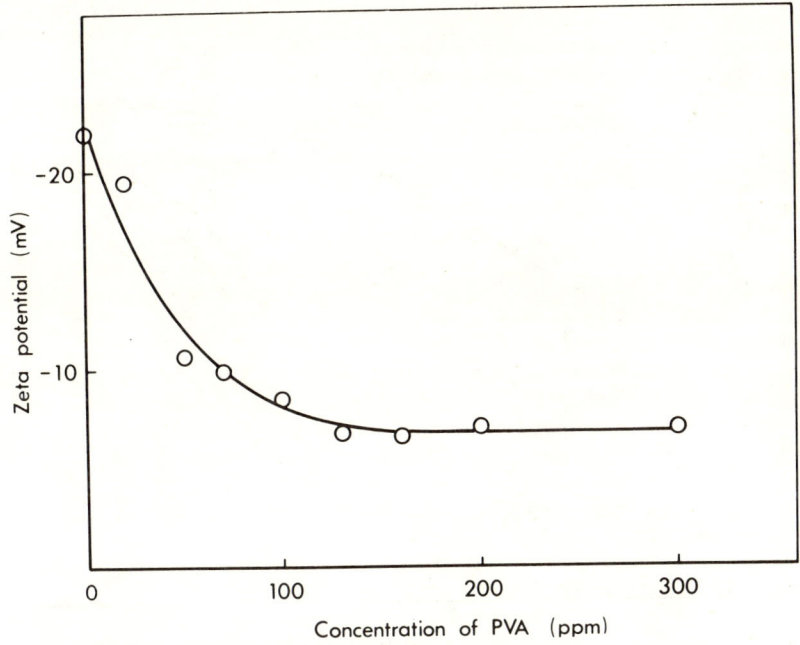

Figure 4. Variation of zeta potential with PVA concentration

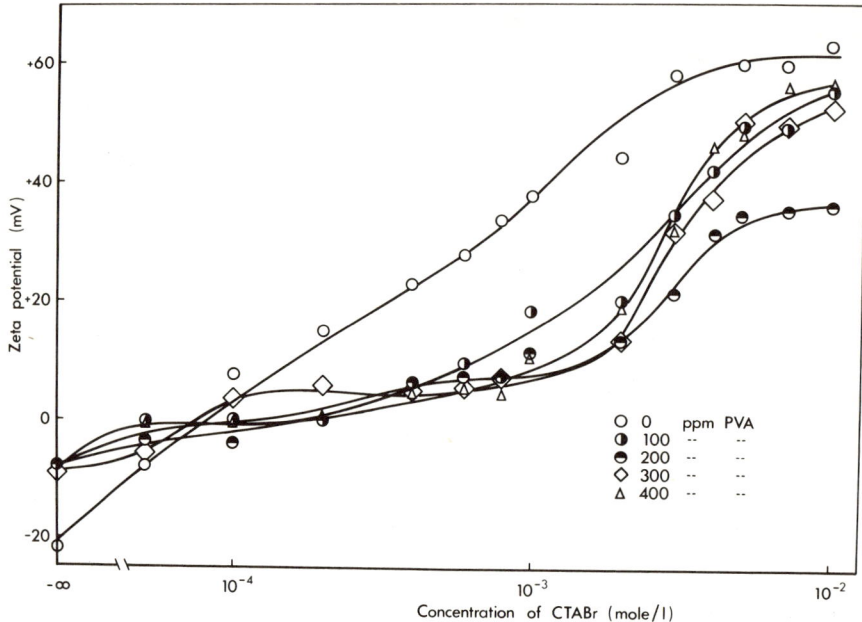

Figure 5. Zeta potential—log C_{CTABr} at various PVA additions

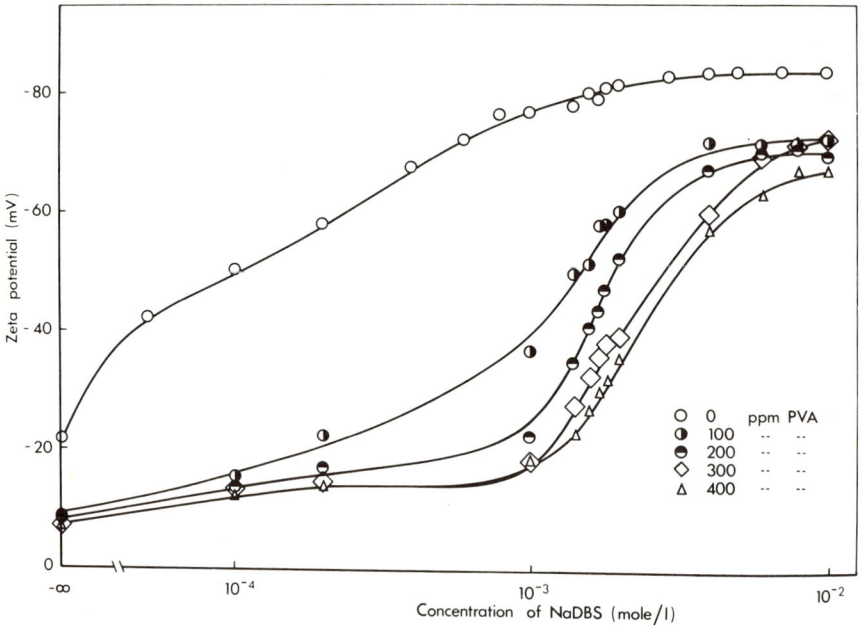

Figure 6. Zeta potential—log C_{NaDBS} at various PVA additions

shows the results for NaDBS. In absence of PVA, the initial negative zeta potential decreases in magnitude with increase in CTABr concentration and at a concentration of $\sim 1 \times 10^{-4}$ mole/l reversal of charge occurs and the positive zeta potential increase gradually and then more rapidly at and beyond the cmc of CTABr and eventually it reaches a plateau above 5×10^{-3} mole/l. With NaDBS, the initial negative zeta potential continues to rise with increase in concentration and a limiting value (-85 mV) is reached at 10^{-2} mole/l.

Addition of PVA, at a given surfactant concentration, always resulted in lowering of zeta potential to a value below that obtained in absence of PVA. With CTABr, the reduction in zeta potential was higher at 200 ppm PVA than at 100 ppm. However, the zeta potentials started to increase again at 300 and 400 ppm PVA, above the value obtained in 200 ppm. With NaDBS, there was a continual decrease of zeta potential with increase in PVA concentration until 400 ppm PVA. The decrease in zeta potential on addition of PVA to a given surfactant solution could be due to several reasons. Partial displacement of CTA^+ or DBS^- ions from the surface could be the casue of this drop. However, there is enough evidence to show that this has not occurred. Firstly, in the case of CTABr-PVA system, the drop in zeta potential did not continue by gradually increasing the PVA concentration. If displacement of CTA^+ ions would have occurred, one would expect more displacement at higher PVA concentrations. Secondly, our recent adsorption measurements on silica (1) has shown that adsorption of CTA^+ or DBS^- ions is enhanced by addition of PVA. The drop in zeta potential could therefore be best explained in terms of interaction between PVA and CTA^+ or DBS^- ions at the ethirimol/solution interface with the result of shift in shear plane. As a result of this interaction an 'association' polyelectrolyte between PVA chains and CTA^+ or DBS^- is formed; the resultant complex now adsorbing at the interface. This picture is supported by the fact that in the case of PVA-CTA^+ ions, the particles always maintained their positive charge above the i.e.p. of ethirimol. Due to the thicker adsorbed layer formed by the polymer surfactant complex compared to that in absence of polymer, the shear plane is shifted towards the solution and a lower zeta potential is obtained. The thicker the layer, the more the extent of the drop, as illustrated for NaDBS-PVA system. However, with CTABr-PVA system, there is a tendency of increase in zeta after reaching a minimum at 200 ppm PVA. Owing to the complexity of polymer surfactant interaction with the result of having various conformations and/or mixed micelle formation, a change in adsorbed layer thickness or overall charge of the complex could explain this apparent rise.

3. <u>Stability Measurements</u>. The quantitative assessment of stability against flocculation of a dispersion is usually complex, particularly when extended over a long period of time. However,

the direct particle counting method usually carried in the initial period is the simplest method to interpret if one considers flocculation as a bimolecular process. In this case the rate of decrease of particle number with time in the initial period follows the Equation

$$-\frac{dN}{dt} = K_2 N^2 \quad (1)$$

where N is the number of particles per unit volume of dispersion, after time t, and K_2 is the second order rate constant. On integration,

$$\frac{1}{N} = \frac{1}{No} + K_2 t \quad (2)$$

A plot of 1/N versus t gave lines which were initially straight and then started to curve at longer periods of time and if the counting was carried long enough, the curve reached a limiting value with an apparent zero slope. This is the behaviour to be expected for secondary minimum flocculation. From the initial slopes, values of K_2 were calculated for every dispersion studied.

It is the normal practice in colloid stability studies to express stability in terms of a stability ratio W, that is equal to the ratio between the rate constant $K_{2,o}$ obtained in absence of any barrier (fast flocculation rate) to the experimental value K_2. $K_{2,o}$ can be calculated using von Smolukowski theory ([17](#)) of rapid flocculation which considers flocculation to be a diffusion controlled process with $K_{2,o} = 4\pi DR$, where D is the diffusion coefficient and R the collision radius (2 x particle radius a). Replacing D by the Einstein coefficient ($D = kT/6\pi a\eta$, where η is the viscosity, k the Boltzmann constant and T the absolute temperature) gives $K_{2,o} = 6.15 \times 10^{-12}$ cm^3 sec^{-1} at 25°C. When the long range van der Waals attraction is taken into consideration ([18](#)), $K_{2,o}$ increases by about 3% using the measured A_{12} value of ethirimol of 4.6×10^{-21} J. However, taking the hydrodynamic correction ([19](#)) into consideration reduces $K_{2,o}$ to about 3×10^{-12} cm^3 sec^{-1}.

It is well known that polydispersity and asymmetry of the particles increase $K_{2,o}$ significantly ([20](#),[21](#)). For that reason we could not use the theoretical value of $K_{2,o}$ for calculation of the W value. Instead, we measured the fast flocculation rate for our polydisperse system, by measuring the rate of flocculation as a function of NaCl concentration. The fastest rate was obtained at NaCl concentration equal to or greater than 10^{-2} mole/l. A plot of log W versus log C_{NaCl} is shown in Figure 7. The critical flocculation concentration is 0.01 mole/l NaCl and the fastest rate was found to be 1.05×10^{-10} cm^3 sec^{-1}, a value that is about 20 times higher than the theoretical value if one neglects Spielman effect ([19](#)) and about 40 times higher than the theoretical value if the Spielman hydrodynamic correction is taken into consideration. This is not surprising in view of the polydispersity and asymmetry of the particles in our system.

Figure 8 shows the variation of W versus PVA concentration. As is clear W increases gradually with increase in PVA concentration and then reaches a limiting value of $\sim$ 31 above 150 ppm PVA. This is what to be expected from a build-up of adsorbed layers. The concentration at which W reaches its limiting value is nearly the same at which the zeta potential has reached its limiting value and also roughly the same at which the limiting value of adsorption of PVA has occurred before the sudden increase in adsorption (see Figure 3). If one assumes that the lowering of zeta potential on polymer adsorption to be solely due to shift in shear plane i.e. assuming Ψ_d to be the same before and after adsorption, one can calculate the thickness of the adsorbed layer from the magnitude of lowering of zeta potential using the equation

$$\tanh \frac{e\zeta}{4kT} = \tanh \frac{e\Psi_d}{4kT} e^{-\kappa(\Delta - \delta)}$$

Ψ_d is taken to be equal to the zeta potential before PVA adsorption and ζ is the measured zeta potential after PVA adsorption, Δ is the thickness of the adsorbed layer and δ is the thickness of the Stern layer taken to be equal to 0.4 nm, the other terms have their usual meaning. Proceeding in this manner, a value of Δ of 0.32 nm was calculated at the plateau of the W - C_{PVA} curve. This value seems to be an underestimate of the thickness of the PVA layer, in view of our recent (12) values of thickness of adsorbed layer of PVA with similar molecular weight on polystyrene. On polystyrene particles with a radius of 35.9 nm, a Δ value of $\sim$ 22.6 nm was obtained. Considering the effect of radius of curvature of the particles and assuming a similar thickness of adsorbed layer on ethirimol as on polystyrene a value of 40 nm is a reasonable guess for Δ. Using this value and the adsorption isotherm results the concentration of PVA in the absorbed layer would be $\sim$ 0.2 g/ml if one uses the BET surface area and $\sim$ 0.15 g/ml if one uses the area obtained from CTA^+ adsorption. Using these values and the polymer-solvent interaction parameter (χ) determined for PVA, namely 0.465, it is possible to calculate the Fischer osmotic term for repulsion (22). In view of the low value of the zeta potential (- 7 mV or possibly lower) V_R could be neglected. V_A was also calculated, taking retardation (23) into consideration and assuming A_{22} for adsorbed layer to be the same as water, namely 4×10^{-20} J. The details of the calculations have been described before (23). Proceedings in this way and using equations for flat plates, we obtained a secondary minimum depth of 8×10^7 kT/cm^2 in presence of adsorbed PVA and 8.2×10^8 kT/cm^2 in its absence at particle separation distances of 80 nm and 31.5 nm respectively.

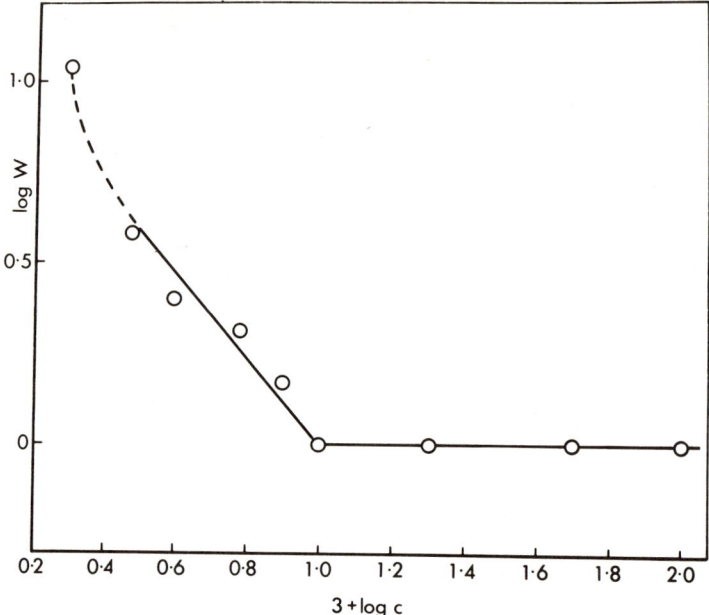

Figure 7. Log W (stability ratio)—log C_{NaCl} for ethirimol

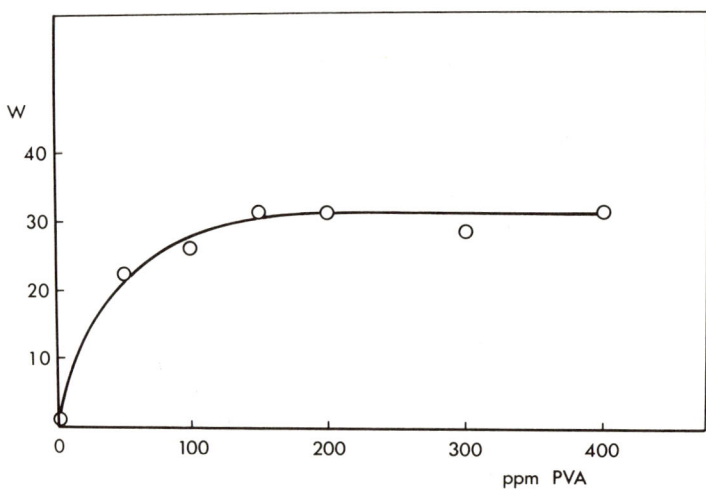

Figure 8. Variation of W with PVA concentration

Assuming an area of contact of particles on flocculation of ~ 1 μm^2 these values correspond to 0.8 and 8.2 kT for two particles interacting in a secondary minimum.

Figure 9 shows the variation of W with C_{CTABr} in presence and absence of PVA. In absence of PVA, fast flocculation (W = 1) is maintained until a concentration of $\sim 4 \times 10^{-4}$ mole/l CTABr, after which it increases gradually and a maximum is reached above the cmc after which W drops again. In the initial part of the stability curve, neutralisation of charge has occurred and a zero zeta potential is obtained at $\sim 1 \times 10^{-4}$ mole/l. This explains why the fast flocculation process was maintained up to this concentration. With further increase in CTABr concentration up to 4×10^{-4} the particles acquired a small positive charge which was not sufficient to impart stability to the particles. The following increase in W is what to be expected from the increase in the positive zeta potential of the particles. The energy of interaction between two particles, assumed to be represented by flat plates, covered with a surfactant layer of thickness 2 nm has been calculated at various zeta potential values. In these calculations, the Vold effect (24) and retardation were taken into consideration (23), assuming A_{22} for the adsorbed layer to be roughly equal to that of liquid paraffin namely $\sim 6 \times 10^{-20}$ J (25). The Fischer osmotic term was neglected as the adsorbed layer thickness is rather small. Proceeding in this way, the secondary minimum depth was estimated at various values of zeta potentials. The results are summarised in Table 3, which also gives the secondary minimum depth assuming an area of contact of 1 μm^2 between the particles. As is clear a secondary minimum depth below that obtained in absence of surfactant (8.2 kT) is obtained as the zeta potential increases significantly above 20 mV.

Table III. Secondary Minimum Depth for Interaction Two Ethirimol Particles Assumed to be Represented by Flat plates.

Zeta potential (mV)	Secondary minimum depth		Separation at which the minimum occurs (nm)
	kT/cm^2	$kT/\mu m^2$	
20	8.1×10^8	8.1	34.5
40	5.2×10^8	5.2	40.5
60	4.2×10^8	4.2	44.0
80	3.8×10^8	3.8	45.5

The drop in W after reaching a maximum value is difficult to explain. Some years ago, Ottewill and Rastogi (23) observed a drop in stability of AgI after reaching a maximum, with increase in cationic surfactant concentration. This drop occurred at a concentration below the cmc of the surfactant. The authors explained the drop as due to a mutual coagulation process between larger positive micelles induced by the sol particles and the negative sol particles. In the present case, however, both the ethirimol particles and micelles are positively charged and therefore coagulation by this mechanism does not take place.

In the presence of PVA, W is higher at CTABr concentration lower than the cmc. However, above the cmc of the surfactant, W is lower in presence of PVA than in its absence except for the 100 ppm PVA curve. As is clear from Figure 9, the pattern of W–log C_{CTABr} curves is rather complex. At the low CTABr concentration, addition of PVA is expected to increase the stability of the particle by forming a polymer-surfactant adsorbed layer, which in absence of PVA was only caused by a low positive charge from CTA^+ ions. However, at higher CTABr concentrations the presence of a thick adsorbed layer of PVA–CTA^+ complex did not compensate for the reduction in charge due to shift in shear plane. This is difficult to explain. From the above calculations, the secondary minimum depth was higher in presence of a surfactant layer even at the maximum value of zeta potential, as compared to that obtained with a nonionic PVA layer 40 nm thick. Of course one would expect a thinner layer for the PVA–CTA^+ layer, but it is unlikely that this layer will be thin enough to deepen the secondary minimum beyond that obtained in presence of CTABr alone. One possible reason for the decrease in stability could be due to the presence of the polymer-surfactant mixed micelles which increase the number of collisions between the various particles in solution and hence increase the rate of flocculation.

Figure 10 shows the variation of W versus C_{NaDBS} in the presence and absence of PVA. In the absence of PVA, W is much higher at low NaDBS as compared to the W values at comparable CTABr concentration. This is to be expected since addition of NaDBS always resulted in an increase in the negative charge on the particles. Also, a maximum, although shallower than that observed with CTABr is observed in the W-log C curve. The higher stability of the ethirimol dispersion obtained with NaDBS as compared to that obtained with CTABr is clear from the secondary minimum depth results (Table 3). Owing to the higher measured zeta potentials obtained with NaDBS, a shallower minimum is obtained, leading to a decrease in the rate of flocculation. Addition of PVA always resulted in an increase in the W values above that in the absence of PVA except in few cases. Moreover, the maximum in the W-log C curve became more pronounced on addition of PVA. An enlargement of the graphs in the region of the maximum is shown in the insert of Figure 10. Most striking. is the sudden increase in stability of ethirimol particles at low NaDBS concentration in the presence of 400 ppm PVA. Actually the W values in this range (10^{-4} - 10^{-3} mole/l) are greater than those obtained in absence of NaDBS. This is inspite of the fact

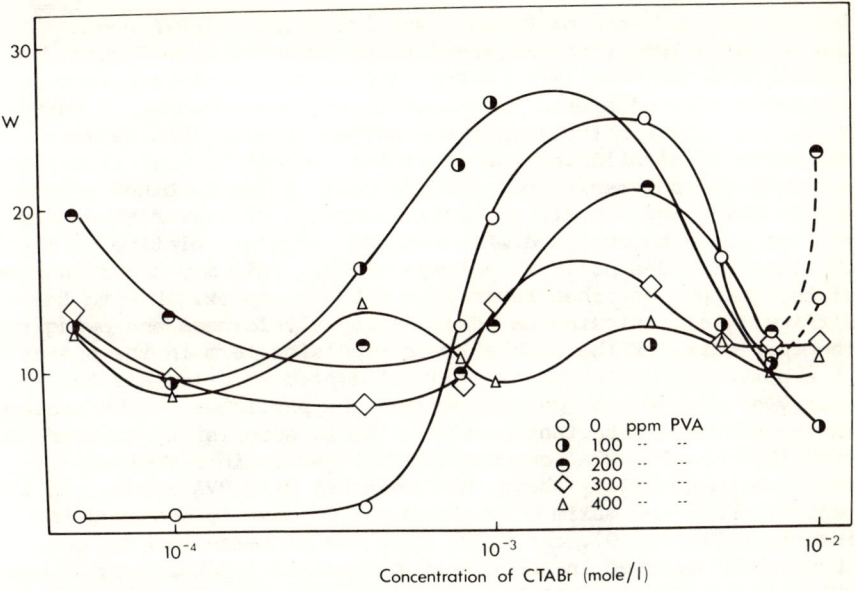

Figure 9. W—log C_{CTABr} at various PVA additions

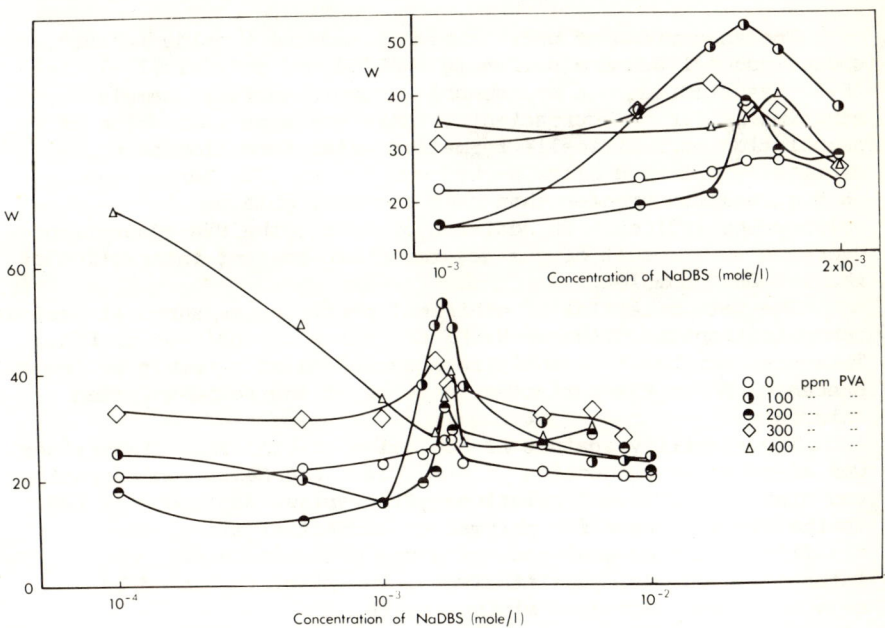

Figure 10. W—log C_{NaDBS} at various PVA additions

that in the presence of NaDBS, less adsorption of PVA and consequently less thick layer is to be expected (see Figure 3). The increase in stability must be due to the combined effect of electrostatic and steric force due to the presence of a polymer-surfactant polyelectrolyte like adsorbed layer. This is more effective in stabilising the particles against flocculation than the nonionic polymer alone. As a result of the combined effect of double layer and osmotic repulsive terms, a shallower secondary minimum could be obtained with PVA-DBS$^-$ complex relative to that obtained with the nonionic polymer alone. This may occur inspite of the thinner adsorbed layer of PVA-DBS$^-$ complex. Due to the difficulty of assigning an adsorbed layer thickness and owing to the complexity of the double layer repulsion term in the presence of a polyelectrolyte, we have not attempted to calculate the energy of interaction curves between the particles in the presence of the polymer-surfactant complex. It is surprising, however, to find that the PVA-CTA$^+$ complex did not result in a similar increase in stability above that observed with PVA alone. In the region, where the maximum in the W-log C curve is observed (see insert of Figure 10), one finds that the increase in W values above that observed in absence of polymer is greatest at 100 ppm PVA. Above this concentration, W begins to decrease again. Possibly the greater decrease in electrostatic force at higher PVA concentrations is the result of this decrease in W values.

Summary

The adsorption of cetyltrimethylammonium bromide (CTABr), sodium dodecyl benzene sulfonate (NaDBS) and poly(vinyl alcohol) (PVA) was studied on a hydrophobic organic surface, namely ethirimol. For the surfactants, there was some indication of association or hemimicelle formation after formation of a vertically oriented close packed monolayer. PVA adsorption (6-8 mg/m^2) was greater than that recently obtained on polystyrene latices. On addition of NaDBS, the PVA adsorption decrease as a result of the greater chain-solvent interaction of the PVA-DBS$^-$ complex.

The zeta potential of ethirimol particles measured at various concentrations of CTABr or NaDBS decreases on addition of PVA. This was ascribed to a shift in shear plane as a result of the presence of the PVA-surfactant complex at the solid-solution interface.

The stability against flocculation of ethirimol dispersions was measured as a function of PVA, CTABr and NaDBS concentration and also in surfactant solutions with various additions of PVA. In the simpler cases for polymer or surfactant alone, the stability could be analysed using the potential energy of interaction curves. Qualitatively, the stability could be related to the secondary minimum depth.

At low surfactant concentrations (below the cmc), addition of

PVA always increased the stability above that obtained with the surfactant alone. With the PVA-DBS⁻ system, the stability ratio W exceeded in some cases the value obtained with PVA alone, inspite of the expected thinner adsorbed layer of the PVA-DBS⁻ complex. At high surfactant concentration (above the cmc) W-log C curves were complex leading to an increase followed by decrease in W values as the PVA concentration increases.

Acknowledgements

The author is indebted to Mr. D. Heath for carrying out most of the experiments described in this paper. He is also grateful to Mr. M.J. Garvey for his help in the computation of the potential energy of interaction curves. Permission by the Management for publication of this work is greatly appreciated.

Literature Cited

1. Tadros, Th. F., J. Colloid Interface Sci., (1974), **46**, 528.

2. Ethirimol is 5-n-butyl-2-ethylamino-4-hydroxy-6-methyl pyrimidine. For further details see: "Pesticide Manual" British Crop Protection Council 1972, H. Martin, Ed., p. 243.

3. Wahlstrom, E. E., "Optical Crystallography" 4th ed. Wiley, N.Y., (1969).

4. Gregory, J., Adv. Colloid Interface Sci. (1969), **2**, 396.

5. Barr, T., Olivier, J. and Stubbings, W. V., J. Soc. Chem. Ind. (1948), **67**, 45.

6. Zwick, M. M., J. Appl. Polym. Sci. (1965), **9**, 2393; J. Polym. Sci. A-1, (1966), **4**, 1642.

7. Horacek, J. Chem. Prumysl., (1962), **12** 385; C.A. 58, 10305 f. Chene, M., Martin-Barret, O. and Clery, M., (1966), 273; Monte-Vobi, A. J., (1969), J. Ass. Offic. Anal. Chem. 52 891.

8. Robb, D. J. M. and Alexander, A. E., Society of Chemical Industry Monograph No. 25, 292 (1967).

9. Conner, P., and Ottewill, R. H., J. Colloid Interface Sci. (1971) **87**, 642.

10. Gaudin, A. M. and Feursteanau, D. W., Trans. AIME (1955) **202**, 958.

11. Kudryavtseva, N. M. and Larionov, O. G., Russian J. Phys. Chem. (1971) 45, 657.

12. Garvey, M. J., Tadros, Th. F. and Vincent, B., J. Colloid Interface Sci., (1974) in press.

13. Johnson, G. A. and Lewis, K. E., British Polym. J. (1969) 1, 266.

14. Fleer, G. J., Koopal, L. K. and Lyklema, J., Kolloid Z.Z. polym. (1972) 250, 689.

15. Hoeve, C. A. J., J. Polym. Sci. (C), (1971) 34 1.

16. Silberberg, A., J. Chem. Phys. (1968) 43, 2835.

17. Von Smoluchowski Z-physik. Chem. (1917) 92, 129.

18. McGown, D. N. L. and Parfitt, G. D., J. Phys. Chem. (1967) 71, 449.

19. Spielman, L. A., J. Colloid Interface Sci. (1970) 33, 562; Honig, E. P., Roebersen, G. J. and Wiersema, P. H., J. Colloid Interface Sci. (1971) 36, 97.

20. Müller, H., Kolloid Z, 38, 1 (1926); Kolloid Beih (1928) 26, 257.

21. Booth, Faraday Soc. Discussions (1954) 18, 104.

22. Fischer, E. W., Kolloid Z. (1958) 160, 120.

23. Garvey, M. J. and Tadros, Th. F., in Proceedings VI International Congress Surface Activity Vol. II (2) p.715, published 1973.

24. Vold, M. J., J. Colloid Sci. (1961) 16, 1.

25. Visser, J., Adv. Colloid Interface Sci. (1972) 3, 331.

26. Ottewill, R. H. and Rastogi, M. C., Trans. Faraday Soc. (1960) 56, 866.

Rheological Studies of Polymer Chain Interaction

ROBERT J. HUNTER, PAUL C. NEVILLE, and BRUCE A. FIRTH
Department of Physical Chemistry, University of Sydney, Sydney, N.S.W. 2006, Australia

Introduction

In this paper we examine some aspects of the flow behaviour of latex particles coated with a hydrophilic polymer. The core particle is a poly-(methyl methacrylate) sphere and the adsorbed tri-block copolymer is poly-(ethylene oxide-b-methyl methacrylate) in which the end hydrophobic moieties act as anchors and permit the ethylene oxide units to form a loop which stretches into the surrounding solution. These systems are thermodynamically stable (1) below a certain temperature, called the critical flocculation temperature (c.f.t.), because the close approach of the core particles is prevented by a rapid rise in Gibbs Free Energy as the adsorbed polymer chains on the colliding particles begin to interact (2-4). The total energy of interaction, V_T, consists of the van der Waals energy of attraction between the core particles, V_A, and the polymer interaction energy $V_R = \Delta G_R = \Delta H_R - T\Delta S_R$ where ΔH_R and ΔS_R are the enthalpy and entropy changes which occur when the adsorbed polymer chains interpenetrate. Both are positive in this system so that at the c.f.t, ΔG_R has presumably been reduced to a value comparable with V_A.

The rheological behaviour of a colloidal suspension is directly affected by the interaction energy between the particles. Systems which are unstable or only marginally stable, exhibit pseudoplastic or plastic behaviour (5,6) and the total energy dissipation, E_T, consists of a purely viscous part, E_V, and a part which may be attributed to the overcoming of interparticle interactions, E_I. Likewise the shear stress at high shear rates can be divided into a viscous part, τ_V, and an additional term, τ_B, called the Bingham yield value since:

$$E_T = E_V + E_I = \tau D = \tau_V D + \tau_B D \tag{1}$$

Figure 1 shows the behaviour which is anticipated. To characterise this curve we need to be able to interpret (a) the yield value, τ_B, (b) the slope of the linear part of the graph (usually called the plastic viscosity, η_{pl}), (c) the shear rate at which the relation becomes linear and (d) the detailed shape of the non-linear part of the τ-D curve.

In some cases the systems showed at low shear rates, a more complicated, Ostwald type (7) flow behaviour (Figure 2), the origin of which is discussed below.

Theory

(i) <u>Ostwald Flow Behaviour</u>. The gradual reduction in viscosity with increase in shear rate is usually attributed to a breakdown in "structure" in the system and various attempts have been made to quantify this idea. We propose that Ostwald behaviour occurs when the shear field has to destroy not only the shear-induced doublets, but also some of those created by Brownian motion. Doublets which separate by Brownian motion (irrespective of how they are formed) make no contribution to the interaction energy.

$$\text{Thus } E = n_D(D) \, E_S \tag{2}$$

where $n_D(D)$ is the number of doublets destroyed by shear per unit volume per second at shear rate D, and E_S is the energy which the shear field must supply in order to separate the particles in the doublet.

The total energy dissipation per unit volume per second is then

$$E_T = D^2 \eta_S (1 + 2.5\phi) + n_D(D) \, E_S \tag{3}$$

where η_S is the solvent viscosity and ϕ is the volume fraction of particles. Here we have assumed that E_V is given by the simple Einstein relation.

Smoluchowski (8) showed that the total number of doublets produced by both Brownian motion and shear was

$$\frac{3\phi^2}{\pi^2 a^3} \left(\frac{kT}{4\eta_S a^3} + D \right) \tag{4}$$

where a is the particle radius. In order to determine

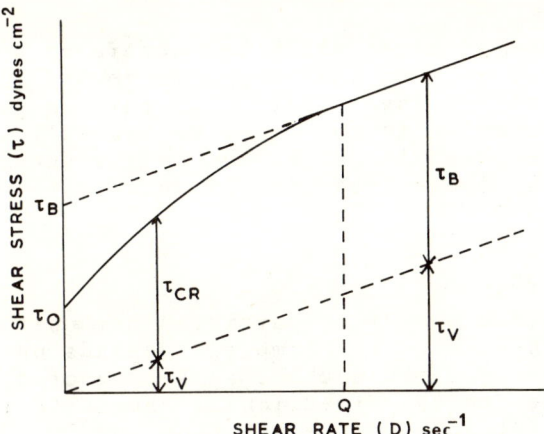

Figure 1. The basic shear diagram for a plastic material showing the contributions to the total stress, as suggested by Michaels and Bolger (21)

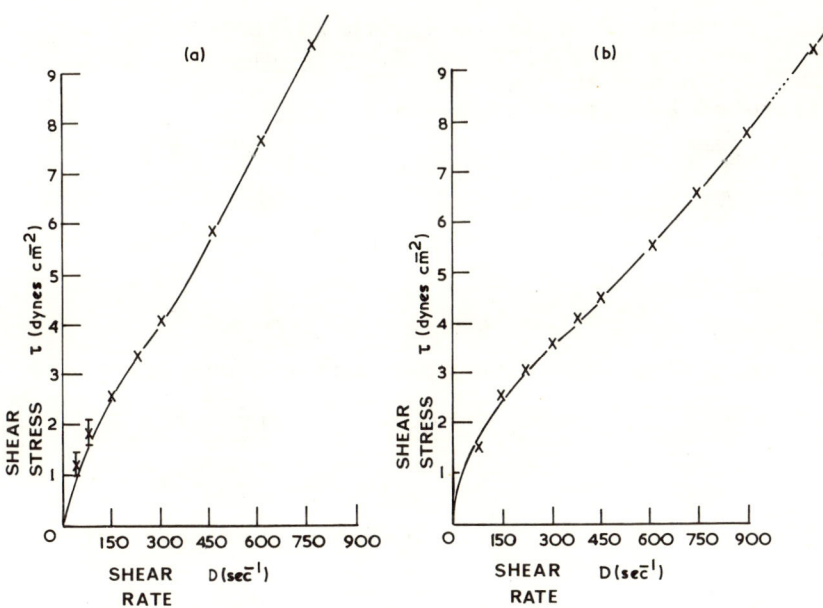

Figure 2. Ostwald flow curves for thermodynamically stable suspensions of the same radius (116 nm), showing the effect of temperature on the curved region. The temperatures were (a) 35°C and (b) 43°C respectively and the theta temperature 47°C for $MgSO_4$ (0.39M).

the fraction of these which are destroyed by the shear field, we must know the average lifetime of a doublet in the shear field. If this is greater than the Brownian motion lifetime, we will assume that that doublet is not broken by the shear field. Smoluchowski showed that the Brownian motion lifetime was

$$t_B = 3\eta_S/4kTN \tag{5}$$

where N is the number of particles per unit volume.

The shear-flow lifetime t_S, depends on the angle, Φ, between the doublet axis projected onto the shear plane and the direction of shear. Goldsmith and Mason (9) derived the relation

$$t_S = (5/D)\tan^{-1}(\tfrac{1}{2}\tan\Phi) \tag{6}$$

Thus there exists a critical angle Φ_0 below which the shear lifetime is shorter than t_B and this will be given (from Equations 5 and 6) by:

$$\Phi_0 = \tan^{-1}(2\tan(3\eta_S D/20\ kT\ N)) \tag{7}$$

The fraction of collisions with angle less than Φ_0 was shown by Goldsmith and Mason (10) to be given by $\sin^2\Phi_0$. Hence the fraction, B, of collisions destroyed by the shear field is:

$$B = \sin^2(\tan^{-1}(2\tan(3\eta_S D/20\ kT\ N))) \tag{8}$$

The value of $n_D(D)$ is then

$$n_D(D) = \frac{3\phi^2}{\pi^2 a^3}\left(\frac{kT}{4\ \eta_S a^3} + D\right) B \tag{9}$$

and combining this with Equations (1) and (3) gives

$$\tau = D\eta_S(1 + 2.5\ \phi) + \frac{3\phi^2}{\pi^2 a^3}\left(1 + \frac{kT}{4\ \eta_S a^3 D}\right) B\ E_S \tag{10}$$

At high shear rates this equation degenerates to a straight line:

$$\tau = D\eta_S(1 + 2.5\ \phi) + \frac{3\phi^2}{\pi^2 a^3} E_S \tag{11}$$

since $kT/4\eta_S a^3 D \to 0$ and $B \to 1$.

Extrapolation of this straight line to $D = 0$ gives the Bingham yield value:

$$\tau_B = \frac{3\phi^2}{\pi^2 a^3} E_S \qquad (12)$$

Equation 12 has been derived previously (11) Substitution of Equation (12) in Equation (11) gives

$$\tau = D\eta_S (1 + 2.5 \phi) + \tau_B \left[1 + \frac{kT}{4 \eta_S a^3 D} \right] \qquad (13)$$

which is the equation we would expect to describe Ostwald flow. τ_B is determined by extrapolation of the experimental data and ϕ is calculated as a hydrodynamic volume fraction, ϕ_H, from the measured differential viscosity at high shear rate, using the Einstein relation:

$$\eta_{PL} = \eta_S (1 + 2.5 \phi_H) \qquad (14)$$

The interpretation of ϕ_H in terms of the expected volume fraction will now be dealt with.

(ii) <u>The Hydrodynamic Volume Fraction</u>. The effective volume of the particle is augmented by the presence of the adsorbed polymer and the hydrodynamic volume fraction, ϕ_H, is related to the volume fraction of core particles, ϕ_p, by the simple relation

$$\phi_H/\phi_p = [(a_o + \delta^1)/a_o]^3$$

where δ^1 is the extension of the polymer loops into the solution; δ^1 depends on the balance between the heat of dilution (κ) and the entropy of dilution (ψ_1) factors for the polymer - solvent interaction. For the solution to behave ideally these parameters must be equal and at this point the change in free energy for interpenetration and compression of the polymer is equal to zero, and hence, so too is the repulsion energy V_R. These are called the <u>theta conditions</u> for the polymer in that solvent and it then assumes its unperturbed root-mean-square end-to-end distance $<r_o^2>^{1/2}$ in free solution. The dimensions under other conditions may be represented thus

$$<r^2>^{1/2} = \alpha <r_o^2>^{1/2} \qquad (16)$$

where α, the intramolecular expansion factor, is a

measure of the solvency of the dispersion medium for the stabilizing macromolecules. The PEO polymer is under theta conditions in 0.39 M $MgSO_4$ at a temperature of 320 K and in 0.45M K_2SO_4 at 308 K and these are the dispersion media used in the present studies.

Adsorption of the polymer onto an impenetrable surface restricts the segments in such a way as to increase the mean extension. We may write

$$\delta^1 = \delta \, \alpha < r_o^2 >^{1/2} = \delta < r^2 >^{1/2} \qquad (17)$$

where δ is a dimensionless parameter which elementary statistical notions would suggest should be equal to about 2.

(iii) <u>The Bingham Yield Value, τ_B</u>. In order to use Equation (12) we now substitute $a = (a_o + \delta^1)$ where a_o is the radius of the core particle and $\phi = \phi_H$. The energy of separation of a pair of particles is equal to the difference between the van der Waals and the repulsion energy:

$$E_S = \frac{A \, a_o}{12 H_o} - V_R \qquad (18)$$

and the analysis of Evans and Napper (1) provides an expression for V_R which when combined with Equations (12) and (18) gives:

$$\tau_B = \frac{3\phi_H^2}{\pi^2 a^3} \left[\frac{A a_o}{12 H_o} - \frac{(2\pi)^{5/2}}{27} <r_o^2>^{3/2} \frac{\omega^2 a_o N_A^2 (\alpha^5 - \alpha^3) kTS_L}{M^{1/2}} \right] \qquad (19)$$

where ω = concentration of stabiliser in g cm^{-2} of surface (assuming 100% coverage);

N_A = Avogadro's number; M = molecular weight of PEO (here 96000).

S_L = a dimensionless segment density function.

H_o, the distance of closest approach can be expressed as a multiple of the unperturbed r.m.s. length:

$$H_o = \delta_o \, \alpha < r_o^2 >^{1/2} = \delta_o < r^2 >^{1/2}$$

since S_L is calculated for various values of the distance parameter, δ, and δ_o is the minimum value of δ corresponding to closest approach.

(iv) The van der Waals constant, A. At the theta-temperature, θ, the value of α is unity and so the repulsion energy is zero. The yield strength at this temperature can, therefore, be used to estimate a value for the van der Waals constant:

$$A = \frac{4\pi^2 a^3}{\phi_H^2 a_o} H_o \tau_B(\theta) \tag{21}$$

provided that H_o can be estimated. Alternatively, the value of A/H_o may be obtained from the value of τ_B at the theta temperature ($\tau_B(\theta)$) and this used for calculations at other temperatures. We could then write

$$\tau_B = \tau_B(\theta) - \frac{3\phi_H^2}{\pi^2 a^3} \frac{(2\pi)^{5/2}}{27} \frac{<r_o^2>^{3/2} \omega^2 a_o N_A^2 (a^5-a^3) kT S_L}{M^{\frac{1}{2}}}$$

$$= \tau_B(\theta) - B(T) f(\alpha) \tag{22}$$

where $B(T)$ is a slowly varying function of temperature and

$$f(\alpha) = \left[1 + \left(\frac{\delta <r_o^2>^{\frac{1}{2}}}{a_o}\right) \alpha\right]^3 \left[\alpha^5 - \alpha^3\right]$$

varies rapidly with temperature, chiefly because of the second term. For $T > \theta$, $\alpha < 1$ and so $f(\alpha)$ is negative. The yield strength should, therefore, always be greater than its value at the theta temperature, but since $f(\alpha)$ goes through a maximum (in absolute value), so too does τ_B. It occurs at about the same point as the minimum in $\alpha^5 - \alpha^3$ (i.e. at $\alpha \doteq (3/5)^{\frac{1}{2}}$). It should be pointed out, however, that the theory from which the value of V_R is calculated is unlikely to be very reliable above the θ-temperature. Polymer scientists show little interest in the region where $\alpha < 1$ since in free solution this corresponds to such poor solvency that the polymer is thrown out as a separate phase. Here it merely settles down onto the particle surface.

The physical reason for the maximum in τ_B is not difficult to find. If τ_B is written in the form

$$\tau_B = g(\phi^2) E_S \tag{23}$$

we see that although E_S increases with increase in temperature, the collision factor $g(\phi^2)$ begins to decrease above the θ-temperature as the effective particle radius decreases.

(v) The Apparent Viscosity. Discussions of non-Newtonian behaviour of suspensions are usually conducted in terms of the equation

$$\eta = \eta_S(1 + k_1\phi + k_2\phi^2) \qquad (24)$$

where η is the apparent viscosity ($\eta = \tau/D$) and $k_1 = 2.5$ for non-interacting spheres at infinite dilution. Increased values of k_1 for charged spheres are attributed to the primary electroviscous effect (12) which we may here treat as negligible because of the high ionic strength.

k_2 is evaluated by considering (i) the particle selfcrowding effect and (ii) the secondary electroviscous effect due to the interaction between particles. Particle selfcrowding becomes important for $\phi_p > 0.20$ approximately (13) and will not be considered further here. A model for the secondary electroviscous effect has been proposed by Chan, Blachford and Goring (14, 15). It attributes the enhanced viscosity to an increase in the size of the rotating doublet due to the repulsion between the particles. This term would be contained within the parameter E_S which we evaluate from the experimental τ_B. Equation (10) and (24) combine to yield

$$k_2 = \frac{3B}{\pi^2 a^3 D \eta_S} \left(1 + \frac{kT}{4\eta_S a^3 D}\right) E_S \qquad (25)$$

Experimental. The PMMA latex particles were prepared by a modified Kotera procedure (16) and monodisperse spherical particles with sizes in the range 0.09 - 0.2μm were obtained. The triblock stabilising polymer was made by the method of Napper(3) using polyethylene oxide of viscosity average molecular weight 96,000 (polyox WSRN10, Union Carbide). The copolymer was adsorbed on to the PMMA particles by adding approximately 100 times the amount of PEO required for 100% coverage to a cold dispersion of PMMA. After a 24-hour equilibration the excess was removed by centrifugation.

Critical flocculation temperatures of the suspensions were determined by turbidimetric measurements at 655 nm and were found to be very close to the theta temperatures of the free polymer in the corresponding solution.

The expansion factor, α, required for Equation (19) was calculated from the approximate Flory (<u>17</u>) equation:

$$\alpha^5 - \alpha^3 = 2 C_M \psi_1 (1 - \theta/T) M^{1/2} \qquad (26)$$

for temperatues below the θ temperature; C_M is a constant. Above the theta temperature α was determined from the relationship (<u>18</u>)

$$\alpha = ([\eta]_T/[\eta]_\theta)^{1/3}$$

where $[\eta]$ is the limiting viscosity number.

Results

(i) <u>Ostwald Flow Behaviour</u>. Figure 2 shows the behaviour exhibited by a stabilised suspension in 0.39 M $MgSO_4$ for which the θ-temperature is 320K. At lower temperatures these samples showed a small yield value and an Ostwald Type flow curve. The full lines in Figure 2, are the theoretical curves from Equation (13) and they obviously give a fair representation of the experimental data, considering the possible errors involved. Figure 2 shows the same suspension at two different temperatures below the theta temperature, whilst Figure 3 shows a comparison of two different suspensions with different values of N and a_o at the same temperature.

(ii) <u>Hydrodynamic Volume Fraction</u>. The slope of these curves at high shear rate gives an experimental value for η_{PL} which when substituted in Equation (14) gives an experimental value for ϕ_H. This value was used in all other equations in place of the particle volume fraction ϕ_p. When substituted in Equation (15) it corresponded to a δ^1 value of 55 nm which gives a value of = 2.1 from Equation (17). This is in very satisfactory agreement with the expected value of about 2.

(iii) <u>The Bingham Yield Value, τ_B</u>. According to Equation (19) τ_B should be proportional to ϕ_H^2, and Figure 4 shows that this is approximately true even for the $\tau_B - \phi_p^2$ relation. Figures 4, 5 and 6 show how important the effect of temperature is on τ_B in the neighbourhood of the theta temperature. The results for MgSO4 were less reliable because

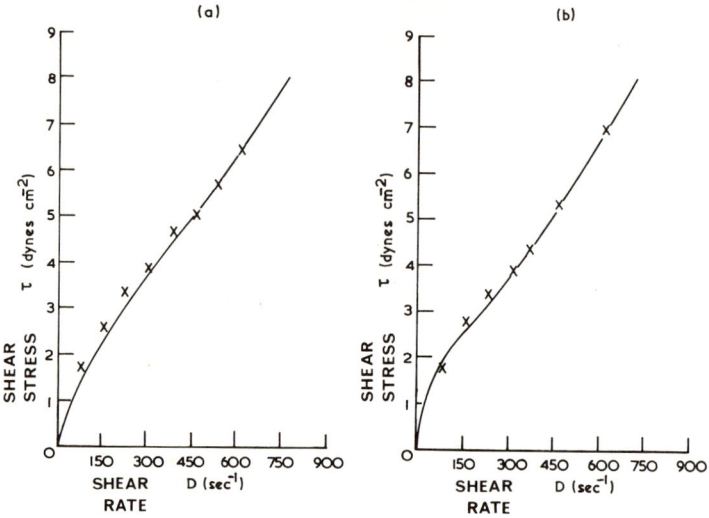

Figure 3. Comparison of Ostwald flow curves for particles of different radii (116 nm (a) and 164 nm (b), respectively) at the same temperature (35°C). Note that the shear rate at which the linear region begins decreases with increase in particle size.

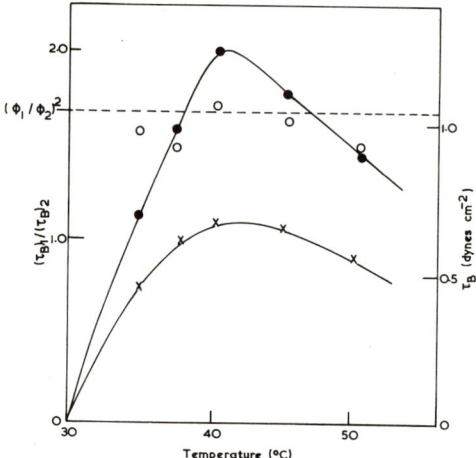

Figure 4. The effect of volume fraction on yield value for two systems of the same particle radius (167 nm); $\phi_1 = 0.072$, $\phi_2 = 0.055$. The dotted line shows the square of the ratio of the gravimetric volume fractions which should be equivalent to the τ_B ratios at that particular temperature. The points (○) give the experimental values for this ratio.

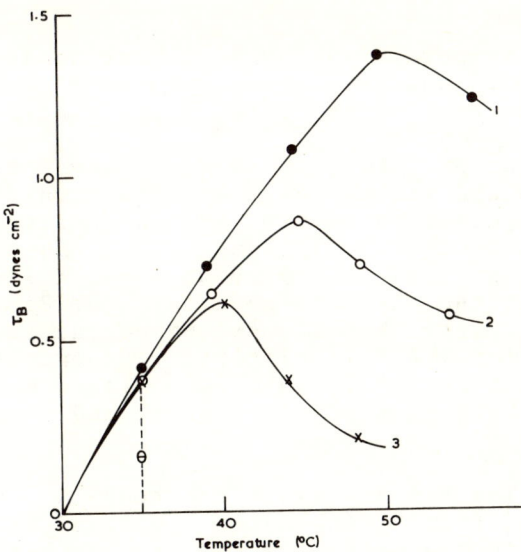

Figure 5. τ_B—temperature curves for a series of particles in K_2SO_4 (0.45M). The effect of particle radius on the position of the maximum in the curve is described in the text. The particle radii were $(a_o)_1 = 164$ nm; $(a_o)_2 = 116$ nm; $(a_o)_3 = 103$ nm.

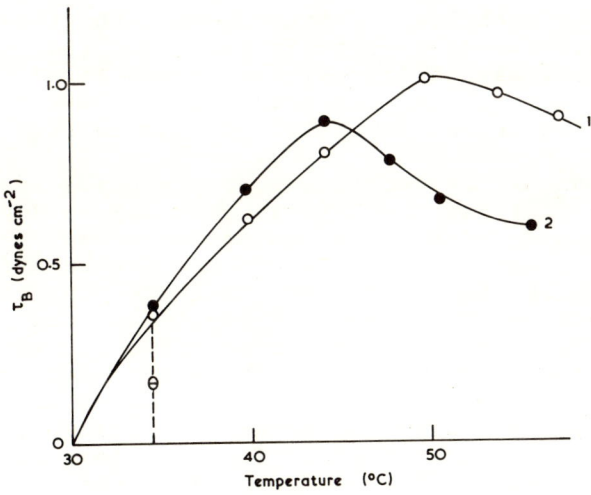

Figure 6. Comparison of theoretical (1) and experimental (2) τ_B–temperature data for a particular size (116 nm) particle using a Hamaker constant of 1×10^{-20} J

the higher temperatures involved made measurements more difficult but they showed the same qualitive behaviour pattern as the K_2SO_4 systems illustrated in Figures 4 - 6.

The rise in the value of τ_B at about the theta temperature and its reaching a maximum value at slightly higher temperatures are in qualitative accord with Equations (19) and (22). Note that for most of the systems the τ_B value never falls below its value at the theta temperature (cf. Equation 22). Note particularly that the extent of the rise and fall in τ_B for $T > \theta$ is critically dependent on the core particle radius (Figure 5). Figure 6 gives a comparison between the experimental and theoretical curves.

(iv) The van der Waals constant, A. Values of A calculated from $\tau_B(\theta)$ using Equation (21) depend critically on the choice of the distance of closest approach. The smallest possible value is set by the condition that the intervening polymer chains are compressed to their normal density in the solid. This would occur at H_o = 4 nm which is obviously too small for it would require the expulsion of all solvent from around the polymer chains. For a value of 5 nm the van der Waals constant comes out to be 1×10^{-20} J which is very reasonable. The much larger values of H_o which are calculated for stable systems by equating the hydrodynamic compressive force with the electrostatic repulsion (14), would require correspondingly higher values for A. That model, however, does not attribute the extra energy dissipation to van der Waals attraction (see below).

Discussion. The separation energy which appears in Equation (10) for the Ostwald flow curves does not arise from van der Waals attraction energy between core particles. These systems are stable and so the repulsion should in any case dominate over attraction. Also, Figures 2 and 3 and Table I show that the separation energy is independent of core radius but dependent on temperature which is the opposite of the behaviour expected of a van der Waals energy.

We believe that the separation energy is due to the necessity for the shear field to tear particles apart at a rate which is faster than the relaxation time of the interacting polymer chains. The relaxation time for the PEO chains can be expected to be (19, 20) about 5×10^{-4} sec. The life-time of a

Table 1. Energy Required for Separation of Doublets by Shear Field.

Fig. No.	Latex Preparation Method	Temperature °C	a_o (nm)	$a = a_o + \delta^1$ (nm)	ϕ_p	τ_B Nm^{-2}	E_S/kT
2	SP*	25	116	170	0.15	$0.03^{+}_{-}.01$	$4^{+}_{-}1$
3	SP*	35	116	160	0.14	0.05	6
2	SP*	43	116	150	0.12	0.06	8
3	EP*	35	160	190	0.20	0.05	6

* SP: suspension polymerisation   *EP: Emulsion polymerisation.

shear doublet, t_S, is given by Equation (6) and even for $D = 10^3$ sec^{-1} it is about 7×10^{-3} sec; it will be even longer for lower shear rates, so there is ample time for the chains to find their time averaged configurations as the particles approach and rotate around one another in a doublet.

Suppose that the particles are being separated by a simple laminar shear field with a velocity gradient of 300 sec^{-1}. If the particle centres are displaced by one diameter in the direction of the velocity gradient, then the velocity of one particle with respect to the other is $3 \times 10^{-5} \times 300 = 10^{-2}$ cm sec^{-1}. If the distance of polymer interpenetration is 5 nm then the time for pulling the doublet apart is $5 \times 10^{-7}/10^{-2} = 5 \times 10^{-5}$ sec or an order of magnitude faster than the polymer chains can relax. Hence the shear field has to "tear" the particles apart. On the other hand, when separation occurs by Brownian motion, the process can occur slowly enough to allow the polymer chains to relax ($t_B > 10^{-2}$ sec.). In our analysis only the energy required for shear separation is important and Table I shows that it has a value of a few times kT.

As a further piece of circumstantial evidence for the hypothesis that chain:chain interactions are responsible for the Bingham yield value, we show in Figure 7 a plot of E_S against $(1 - \theta/T)$. According to Flory's analysis (17), the free energy change which occurs on mixing polymer chains is

$$\delta(\Delta G/kT) = \text{constant } \psi_1 (1 - \theta/T) \delta V$$

and although the data is subject to a wide margin of error it does appear to correlate with Flory's equation.

The suggestion that the fall off in τ_B at temperatures above θ is due to a reduction in the effection collision frequency (and hence $f(\phi^2)$ in Equation (23)) is supported by the data in Figure 5. For small particles, increases in temper. have a proportionately larger effect than fo large particles because δ^1/a_0 is larger and hence $f(\alpha)$ in Equation (22) undergoes larger changes with temperature.

The interpretation of enhanced viscosity in terms of the usual description (14,15) of the secondary electroviscous effect (Equation (25)) presents some difficulties. Although the electrostatic repulsion forces are essentially zero in

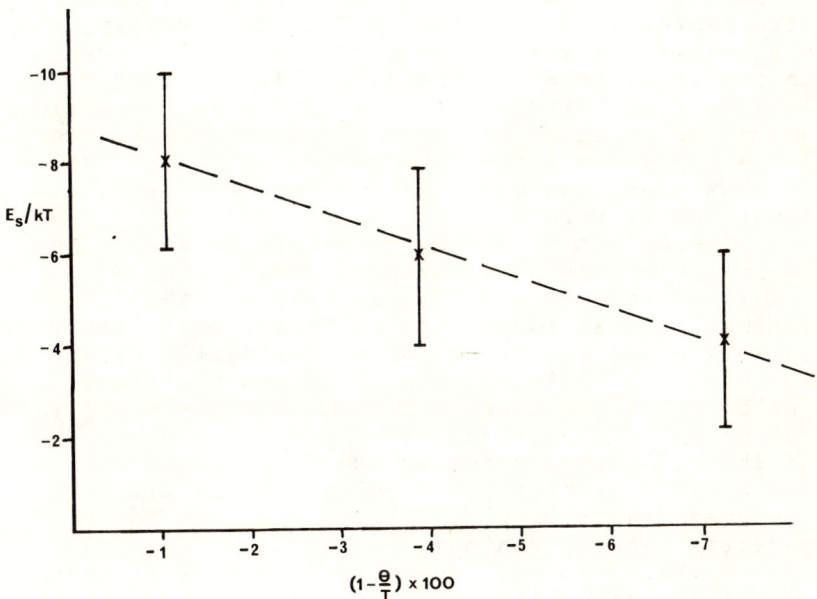

Figure 7. Energy of separation E_s vs. $(1 - \theta/T)$. The straight line suggests that the energy of separation is due to polymer chain interactions.

these systems there are still repulsive forces between the particles, as is evident from the stability of the sol. According to Chan et al (14) this would cause an increase in the hydrodynamic energy dissipation because the rotating doublet is increased in size and the effect should increase with increase in V_R. Yet we find that raising the temperature (which reduces V_R) causes an increase in the value of E_S in Equation (24) (see Figure 7). This is especially true of the unstable suspensions above the theta temperature where the interaction is wholly attractive. The collision doublet would in that case presumably be smaller than it is at lower temperatures.

One effect of the hydrodynamic forces is to limit the approach of two particles during a collision because the outflow time for the intervening fluid is long compared to the collision time. This is especially true for larger particles and above $a_0 = 5$ µm it becomes the dominant effect. The behaviour of smaller particles ($a_0 < 1$ µm) is not known with such certainty but it is clear that if the hydrodynamic forces are to be large enough to separate a collision doublet, the particles involved cannot be permitted to approach to the potential energy minimum.

Perhaps the difference in behaviour of these sterically stabilised systems can be traced to the fact that the repulsive force does not appear until polymer chain interpenetration is significant. The detailed calculation of the energy dissipation as the doublet rotates and separates would pose very considerable difficulties in this case, since the flow behaviour of the intervening fluid is undoubtedly influenced by the local polymer segments.

Some aspects of this work are published elsewhere (22, 23).

Summary

Latex particles stabilized by adsorbed hydrophilic polymers exhibit both pseudoplastic and Ostwald type flow behaviour in the neighbourhood of the critical flocculation temperature (c.f.t.), which corresponds closely to the theta-temperature of the adsorbed polymer. Ostwald flow occurs below the theta temperature in stable systems with a finite, though small, yield value. It can be explained quantitatively by calculating the excess energy which the shear field must supply to destroy all doublets not broken down by Brownian motion.

At and above the c.f.t. all systems exhibit a yield value, τ_B, and the dependence of τ_B on temperature can be qualitatively explained by considering both the interaction energy and the probability of collision between particles. The energy of interaction is independent of core particle radius and is of the order of a few times kT.

The plastic viscosity of the system obeys the Einstein relation with an effective volume fraction which suggests that the adsorbed polymer chains are stretched to about twice their normal r.m.s. length.

Acknowledgements.

This work was supported by grants from the Australian Research Grants Committee and by an Australian Government Post Graduate Studentship for B.A.F.

Literature Cited.

1. Evans, R., Napper, D.H. Kolloid-Z.Z.Polym. (1973), 251, 329, 409.
2. Napper, D.H. J. Colloid Interface Sci. (1969), 21, 168.
3. Napper, D.H. J. Colloid Interface Sci. (1970), 32, 106.
4. Napper, D.H., Netschey, A. J. Colloid Interface Sci. (1971), 37, 528.
5. Hunter, R.J., Nicol, S.K. J. Colloid Interface Sci., (1968), 28, 250
6. Friend, J.P., Hunter, R.J. J. Colloid Interface Sci. (1971), 37, 548.
7. Ostwald, W. Kolloid-Z.Z.Polym. (1925), 36, 99.
8. Smoluchowski, M. von Z. phys. Chem. (Leipzig) (1917), 92, 129.
9. Goldsmith, H.L., Mason, S.G., in "Rheology Theory and Applications", F.R. Eirich, Ed.Vol.4, p. 167, Academic Press, New York, 1967.
10. ibid, p. 173.
11. Gillespie, T., Wiley, R., J. Phys. Chem.(1962) 66, 1077.

12. Conway, B.E., Dobry-Duclaux, A., in "Rheology Theory and Applications", F.R. Eirich Ed., Vol. 3 Chap.3, Academic Press, New York, 1960.

13. Krieger, I.M., Advan. Colloid Interface Sci. (1972), 3, 111.

14. Chan, F.S., Blachford, J., Goring, D.A.I. J. Colloid Sci. (1966), 22, 378.

15. Stone-Masui, J., Watillon, A., J. Colloid Interface Sci. (1968), 28, 187.

16. Kotera, A., Furusawa, K., Takeda, Y. Kolloid-Z.Z. Polym. (1970), 239, 677.

17. Flory, P.J. "Principles of Polymer Chemistry", Chapter 14, Cornell University press, New York, 1953.

18. ibid. p. 606

19. Thurston, G.B., Schrag, J.L. J. Polym. Sci. (1968), Part A-2, 6, 1331.

20. Thurston, G.B., Morrison, J.D. Polymer (1969) 10, 421.

21. Michaels, A.S., Bolger, J.C. (1962) Ind. Eng. Chem. Fundam. 1, 153.

22. Neville, P.C., Hunter, R.J. J. Colloid Interface Sci. (1974) (in press).

23. Firth, B.A., Neville, P.C., Hunter, R.J. J. Colloid Interface Sci. (1974) (in press).

14

Nonaqueous Dispersion: Carbon Black in Heptane Solution of Manganese and Zinc Salts of 2-Ethylhexanoic Acid

B. R. VIJAYENDRAN

Copier Products Division, Pitney Bowes, Inc., Norwalk, Conn. 06852

Introduction

Nonaqueous dispersions have drawn considerable amount of attention recently both from the academic and industrial chemists. Several papers (1-10) have been published in this area covering various aspects such as coagulation, origin of charge, stability, theoretical studies, etc., of the interesting as well as industrially useful nonaqueous dispersions. Patent literature is abundant with various compositions of nonaqueous dispersions useful in organic coatings and paints. One area where nonaqueous dispersions have found some new applications has been in electrostatic copying as well as in electrodeposition of paints. The electrical characteristics of the dispersed particles are of paramount importance in such applications. We shall restrict to this aspect of nonaqueous dispersions in this study.
In the electrostatic copying process using liquid development (11), the latent electrostatic images are toned from a suspension (commonly called toner) consisting of charged colored particles (carrying electrical charge opposite to that of the latent image) in a low dielectric and insulating solvent. The charged toner particles migrate to the oppositely charged image areas on the photoconductor through electrophoretic mobility. In some cases, there is also some dielectrophoresis due to the inhomogenous electric field around the image. Electrostatic color copying has been ahcieved by a similar process using the photoelectrophoresis property of some materials (12). As one can see, the charge on the toner particle is very important for electrostatic copying process. Besides, the charge on the particle also imparts the necessary

stability to the toner dispersions. Table I lists some of the typical properties of toners based on carbon black.

Table I (13, 35). Typical Properties and Process Parameters for Electrophoretic Development of Latent Images

Particle size	0.05 - 2 micrometers
Average toner charge	10^{-10} - 10^{-13} micro coul
Zeta potential	20 - 75 mV
Dielectric constant of medium	2 - 3
Volume resistivity of medium	10^{10} ohms-cms
Electrophoretic mobility	5-40 $\frac{cm^2}{sec.volt}$ × 10^{-5}
Apparent surface potential	150 - 250 volts

The actual mechanism responsible for charging the suspended particles in low dielectric constant and highly insulating liquids is not well understood. In general, charges on particles in liquids are produced by one of three phenomena (14): Contact electrification, preferential adsorption of ions and proton transfer. Contact potential occurs when electrons move between two solids, two liquids, or a solid and liquid as a result of a difference in the work functions. Electrons move from the material with the lower work function to the material with the higher work function, and the surfaces acquire positive and negative charges, respectively. Such a charging mechanism is responsible for the fire hazard in the petroleum industry (15).

There are several experimental studies to show that selective adsorption of ions generates charge on the particle. Just to mention a few: Koelmans and Overbeek (16) reported electrophoretic mobility values of several substances in xylene and found Fe_2O_3 and $BaSO_4$ in xylene became positively charged by the adsorption of oleic acid, fatty acids, Cu Oleate and Aerosol OT; SiO_2 and $CaCO_3$ (17, 18) became positive by the adsorption of Aerosol OT; Carbon black in benzene became positive by the adsorption of calcium salt of di-isopropyl salicylic acid (19). The nature and magnitude of the charge is influenced strongly by the presence of trace amounts of water in the system. This is very pronounced for particles having polar groups such as TiO_2, Al_2O_3, and Fe_2O_3. Papers relating to this aspect (20, 21) of nonaqueous suspensions have been published. It is interesting to note that the normal zeta potential of 50 - 125 mV observed in nonaqueous systems is generated by something like 10 - 30 ions adsorbed per particle. Because of such low concentrations needed for the charging of particles, very small quantities of impurities will affect the results drastically. This is probably one reason why rather contradictory experimental results on similar systems have been reported in the literature.

Fowkes and associates (22) have proposed a proton transfer mechanism in an attempt to explain the nature of surface charges. The mechanism appears to explain the electrophoretic behavior of particles suspended in hydrocarbon liquids in the presence of polymers capable of undergoing a proton transfer with the particle surface. This view is confirmed in a recent study (23).

The aim of this paper is to present the data obtained in the study of a carbon black (Sterling R) in heptane in the study of manganese and zinc octoates (salts of 2-ethylhexanoic acid). Data on the electrical conductivities of the heptane solutions of the two octoates, adsorption isotherm of the two octoates on Sterling R as well as zeta potential of Sterling R in the heptane solutions containing the salts are reported. It is accepted in this work, that ion pairs are in equilibrium with free ions despite the fact that the dielectric constant of the medium is low. Zinc octoate was chosen for this study, since conductivity studies of metal soaps in nonaqueous media (24, 25, 26) have shown that zinc soaps behave differently because of the weaker interactions of the zinc-carboxylate bond. Since this property of zinc soaps in nonaqueous liquids plays a role in determining its solution properties, viz., equivalent conductivity, it appears that it is reasonable to assume here that it should also affect the electrophoretic property of the suspended particles. Manganese octoate was chosen as the representative of metal soaps having fairly strong polar bond. Hopefully, the differences in the electrophoretic behavior of carbon black in the two metal octoates, if any, would shed some light on the mechanism of charging of carbon black in the present system.

Experimental

Materials. Manganese 2-ethylhexanoate (referred to as manganese octoate - $Mn(oct)_2$) and zinc 2-ethylhexanoate (zinc octoate - $Zn(oct)_2$) were obtained from Research Organic/Inorganic Chemical Corp., Sun Valley, California. These were used without any further purification. The metal content of the two octoates determined using atomic absorption agreed well with the theoretical metal content assuming $Mn(oct)_2$ and $Zn(oct)_2$ to be their compositions. The results were as follows (theoretical values in parenthesis): $Mn(oct)_2$: Mn 15.9% (16.11%); $Zn(oct)_2$: An 18.2% (18.5%). The salts were kept over P_2O_5 prior to being used. Reagent grade n-heptane was refluxed over P_2O_5 and fractionally distilled. Sterling R of the Cabot Co. was used as the sample of carbon black. The original sample of carbon black was extracted with acetone and heptane. The washed sample was dried at 100 - 110°C for 2 - 3 hours before the preparation of the suspension. The BET surface area of Sterling R has been found to be 21 m^2/g and the surface showed presence of some volatile >CO groups.

Procedure. Solutions of the metal octoates in heptane were prepared by adding known amounts of the salt to a convenient volume of heptane. Electrical conductivities of the metal octoate solutions in heptane were measured using a 1620A General Radio capacitance assembly and Balsbaugh precision three terminal cell at 25°C. Extreme care was taken to eliminate any possible moisture contamination during the conductivity as well as other measurements.
Adsorption isotherms of the two octoates from heptane on Sterling R at 25°C were determined. Carbon black suspensions in heptane containing various known amounts of manganese and zinc octoates were prepared by adding a small quantity of the carbon black to 100 ml of heptane solution and then subjecting to a brief ultrasonic radiation. Then the suspensions were shaken for 24 hours to attain equilibrium. The concentration of the metal octoate in the solution after the attainment of adsorption equilibrium was determined. In the case of manganese octoate, a simple absorption vs. concentration plot at 600 nm was used to determine the residual concentration of manganese octoate. In the case of zinc octoate, atomic absorption was used to determine the zinc content.
Microelectrophoresis on the Sterling R suspensions were carried out to obtain zeta potential.

The flat cell used was constructed of quartz and has been described elsewhere (27). Suspensions of Sterling R in heptane containing various concentrations of manganese and zinc octoates for the electrophoresis experiments were prepared by diluting the suspensions made for adsorption studies with the appropriate metal octoate solution such that the concentration is about 10^6 particles/cm^3. Immediately before filling the cell the dispersions were subjected to ultrasonic radiation for 5 minutes. Carbon black particles dispersed by the metal octoates were aggregates which were nearly spherical and the mean diameter of which were 1 - 3 μm. Experiments were done at 25°C.

The criteria for true electrophoresis were observed and the distribution of velocities, from wall to wall was parabolic in all cases (28). Only after a steady state had been reached within the cell (≃30 - 45 minutes) measurements of the electrophoretic mobility were made. Between 6 - 10 particles were tracked over a distance of 48 μm and the field polarity was often reversed to ensure absence of polarization. Fields as high as 150v cm^{-1} were used in the present work without any evidence of polarization. This is in agreement with the findings of van der Minne and Hermanie (28) and Parreira (27) who used high fields with carbon black in benzene.

The electrophoretic mobility μ is related to the zeta potential ζ by the Equation,

$$\zeta = K \cdot f(\kappa a)^{-1} \cdot \mu \cdot \pi \cdot \eta / \varepsilon \qquad (1)$$

where $f(\kappa a)$ is the Henry factor, η is the viscosity of the medium, ε is the dielectric constant of the medium, and K is the correction factor due to the relaxation effect of the double layer.

The appropriate value of $f(\kappa a)$ depends on the shape of the particle and also the relative magnitudes of the "thickness of the double layer" κ^{-1} and the particle radius a. In nonaqueous systems the double layer is quite extended (κ^{-1} is of the order of several micrometers) and as a result κa is found to be small. Under these conditions, Henry (29) has shown that $f(\kappa a) = 1/6$ for spheres and is 1/4 or 1/8 for cylinders with their axes parallel or perpendicular to the field, respectively. The correction factor due to relaxation (K) is close to unity when Ka is small and the zeta potential is of the order of 100 mV (30).

Hence Equation (1) reduces to Henry's equation.

$$\zeta = 6\pi\eta \cdot \mu/\epsilon \qquad (2)$$

for spheres. Most authors have used $f(\kappa a) = 1/6$, which implies a spherical shape for the particles, even though van der Minne and Hermanie ([28](#)) assumed a value of 1/4 without an apparent reason. Recently, Lewis and Parfitt ([31](#)) felt that a value of 1/6 could not be justified since there was electron microscope evidence to show that linear aggregation of particles resembling rigid cylinders with their axes parallel to the field were formed, before the electrophoretic measurements were made. Hence they used a value of $f(\kappa a) = 1/4$. As mentioned above $f(\kappa a)$ can vary from 1/4 to 1/8, therefore, the true value of $f(\kappa a)$ must lie between those limits. Thus, the value of 1/6 used in this, and other works, appears to represent best value for $f(\kappa a)$ despite an assumed spherical shape for the carbon black particles.

Results

Figure 1 shows the variation in conductance with concentration of $Mn(oct)_2$ and $Zn(oct)_2$ salts in heptane. Figure 2 shows the effect of concentration on the equivalent conductance. The most striking features of the results are, first, the relatively high conductivities of the metal octoates in such a non polar solvent as heptane and, secondly, the unusual form of the equivalent conductance concentration relation in the case of $Zn(oct)_2$. We shall discuss this later. Conductance of the magnitude observed here has been reported by others in the following systems: Aerosol OT in heptane ([20](#)); metal soaps in toluene ([25](#), [26](#)) and others ([24](#)); this shows that the conduction is due to ions and/or other charged species such as micelles or triple ion formation of the metal octoates in heptane.

Figure 3 shows the adsorption isotherm of the two metal octoates on Sterling R. The manganese salt appears to be adsorbed more so than the zinc salt.

Figure 4 gives the zeta potential of Sterling R in the two octoate solutions. Sterling R is charged positively in all the solutions. In pure heptane no electrophoretic mobility is observed. This is in agreement with the observation of others ([31](#), [32](#), [33](#)). The zeta potential is found to decrease, after an initial sharp increase, with increasing concentration of the octoates. The maximum in the zeta potential at low concentration has been observed by

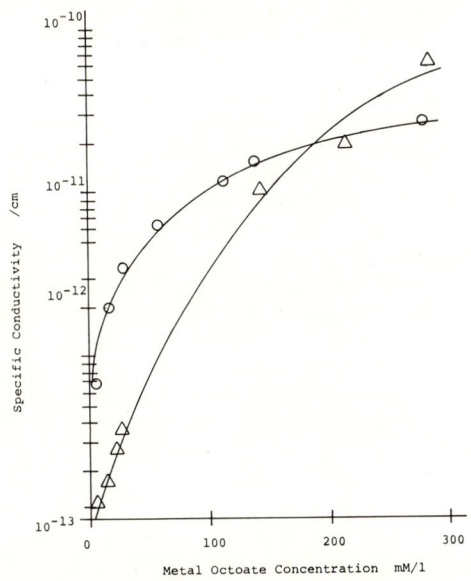

Figure 1. Effect of metal concentration of metal octoate on conductivity. ○: $Mn(oct)_2$; △: $Zn(oct)_2$.

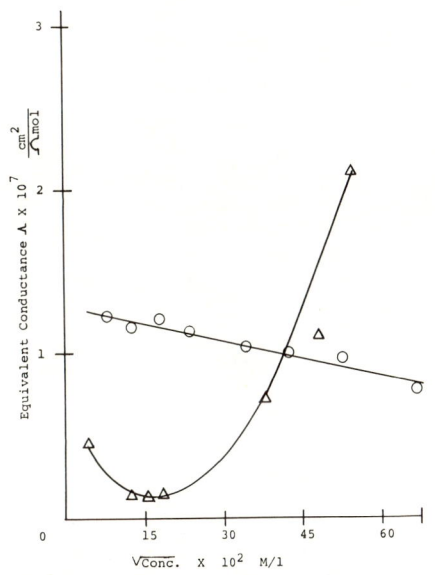

Figure 2. Effect of metal concentration on equivalent conductance. ○: $Mn(oct)_2$; △: $Zn(oct)_2$.

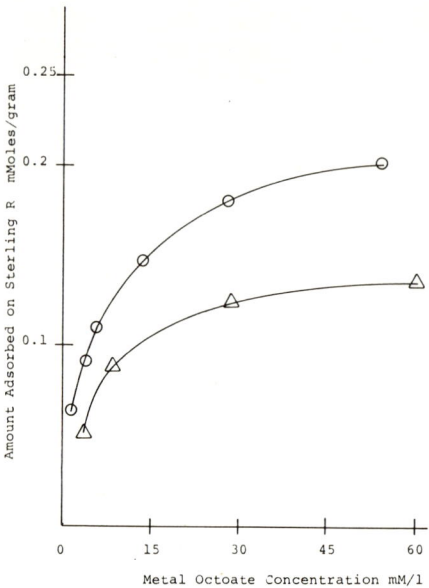

Figure 3. Adsorption of metal octoates on Sterling R in heptane. ○: $Mn(oct)_2$; △: $Zn(oct)_2$.

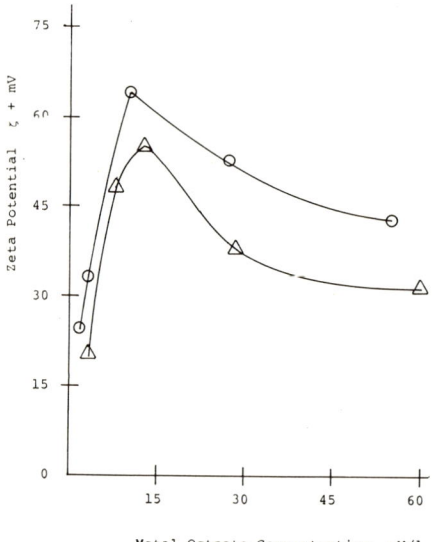

Figure 4. Effect of metal octoate concentration on the zeta potential of Sterling R in heptane. ○: $Mn(oct)_2$; △: $Zn(oct)_2$.

others working with carbon black in nonaqueous media (19, 31, 32, 33). The maximum has not been fully understood, even though a few hypothesis have been advanced. Bell and Levine (34) ascribed the maximum to a discreetness of charge effect. The presence of water has been invoked. It is also possible that it may be caused by the differences in conductances between the bulk and the surface phases, particularly at low salt concentrations, where the maximum generally occurs. An interesting speculation (32) that may be worth considering here is whether the critical micelle concentration (cmc) of the salts is related to the maximum. In the case of $Mn(oct)_2$ the maximum is seen to occur at a lower concentration as compared to that in $Zn(oct)_2$. The zeta potential is seen to be lower in $Zn(oct)_2$ than in $Mn(oct)_2$.

Discussion

The relatively high conductivities of the metal octoates in heptane clearly indicate there is some degree of dissociation of these salts in heptane. Such occurence of ionic conductance in metal soap-hydrocarbon systems have been reported exhaustively (24, 25, 26). Conductivity behavior and micelle formation of metal soaps in nonaqueous liquids results from the interplay of a wide variety of intermolecular forces. The conductivity of soaps depend on the metal-carboxylate bond (whether it is polar, ionic or non-polar) and the tendency to form micelles. Increasing polarity of the metal-carboxylate bond has been found to increase conductivity (25, 26), but soaps having ionic type metal-carboxylate bond are often found to be insoluble in nonpolar liquids. The micelle formation in nonaqueous liquids (the micelles are of inverse structure, namely, the polar groups are in the core and the non polar hydrocarbon chains facing the solvent) depends on the acid (soft or hard) and base characteristics of the soap as well as the solvent. These factors governing micelles formation in nonaqueous liquids have been treated (24).

Nelson and Pink (25, 26) measured conductivities and aggregation numbers of divalent soaps in toluene and found that the conductivity increased in the order of increasing polarity of the metal-carboxylate bond in the approximate order of zinc, copper, nickel, manganese, magnesium and calcium. They also found that the aggregation numbers to be less in zinc soaps as compared to nickel or magnesium soaps. The higher degree of association in nickel or magnesium soaps as opposed to zinc has been explained by stronger acid-base interactions in the former case leading to larger micelles (24).

In our case, we observe similar conductance behavior in $Mn(oct)_2$ and $Zn(oct)_2$ solutions of heptane. The manganese octoate appears to behave like a salt with fairly polar Mn-carboxylate interactions leading to higher conductance at lower concentration and probably larger micelles. In the case of zinc octoate, the conductance is low (compared to that in $Mn(oct)_2$ up to a certain concentration but increases rapidly (more so than $Mn(oct)_2$) at higher concentration. This is more apparent in Figure 2 where it is seen that the equivalent conductance increases dramatically with concentration in the case of $Zn(oct)_2$ only. This is probably due to the much lower tendency of zinc soaps to dissociate and form micelles in nonaqueous media (micelle sizes are smaller and micelles are not observed at lower concentrations). In the case of $Mn(oct)_2$ the dissociation as well as micelle formation appears to be more favorable. The much lower tendency to form micelles in the case of zinc is due to the fact that it is a soft type lewis acid (zinc is a borderline case and is a less stronger lewis acid than manganese) and cannot bind strongly to the carboxylate groups.

The adsorption isotherm of the two octoates on Sterling R reflects the differences in solution properties of the two salts in heptane. $Mn(oct)_2$ being more dissociated than $Zn(oct)_2$, it is adsorbed more strongly than the latter.

It is reasonable to assume from the conductivity data that the metal octoates dissociates in heptane as follows:

$$M(oct)_2 = M(oct)^+ + oct^{-1}$$

Since the particles are positively charged as observed from microelectrophoresis, it is also reasonable to assume that the adsorption is brought about by the interaction of the positively charged ions of the salts and the negatively charged > CO groups present on Sterling R. This again would explain why $Mn(oct)_2$ is adsorbed more strongly than $Zn(oct)_2$. Of course, there will be adsorption of other species such as micelles on the carbon black which also has some hydrophobic sites.

Now let us look at the zeta potential in the two octoate solutions. Both exhibit a maximum in the zeta potential but the maximum is found to occur at a higher concentration in $Zn(oct)_2$ as compared to $Mn(oct)_2$. As mentioned earlier, if the maximum is related to the cmc of the salts then this would be consistent with the observation that micelles are formed more easily in $Mn(oct)_2$ than in $Zn(oct)_2$. However, the maximum in this case is found to be around 10^{-2} moles/l whereas the cmc in nonaqueous media are around 10^{-4} moles/l. This discrepancy may be due to the fact the octoates having shorter chain length probably have a higher cmc. Further work is needed to understand the maximum in zeta potential at low concentration, which has been observed in several nonaqueous systems (19, 31, 32, 33).

It is also seen that the zeta potential is lower in $Zn(oct)_2$ as compared to that in $Mn(oct)_2$. This is consistent with the adsorption isotherm as well as conductivity behavior of the salts, namely, $Mn(oct)_2$ is more dissociated and adsorbed strongly on Sterling R surface thus creating a higher charge.

Parreira (32) in his studies on the electrophoresis of carbon black in benzene in the presence of alkaline earth sulfosuccinates salts found carbon black to be positively charged and the positive charge to increase in the order of Ba<Ca<Mg. He explained his results by arguing that the ionic adsorption of the salts onto the surface is caused by the ion induced interaction between the surface >CO groups and the metal. Further, he showed that the interaction energy for the three alkaline earth metals is in the order of $E_{Mg} > E_{Ca} > E_{Ba}$. The energy of interaction is dependent on the ionic radius of the metal element, viz., smaller ions have a higher interaction enery in keeping with the theory that the small ion is the preferentially adsorbed species (16). The same type of argument can be made to explain the adsorption as well as the zeta potential results obtained in this study. Greater adsorption and higher zeta potential are observed with $Mn(oct)_2$ as compared to $Zn(oct)_2$ because Mn^{++} being smaller than Zn^{++}, the former is preferentially adsorbed and the energy of the ion induced dipole interaction with >CO groups of the carbon black will be such that $E_{Mn} > E_{Zn}$. Other similar studies (35) of carbon black in nonaqueous media in the presence of various metal soaps seem to agree with this explanation.

Of course this is a very simplistic picture of the various interactions taking place at the interface of Sterling R and heptane in the presence of metal octoates. One should also probably consider the negative ion adsorption as the concentration increases.

Careful examination of Figure 4 suggests that zeta potential decreases with increase in concentration and this cannot be completely due to double layer compression effects since such effects are negligible in low permitivity liquids. This is probably why Lewis and Parfitt (31) and others (2) have found carbon blacks containing >CO groups to negatively charged by Aerosol OT. One cannot also ignore other hydrophobic interactions at the interface. These factors are difficult to separate and study individually, because of the complex nature of the systems.

Acknowledgements

The author wishes to thank Mr. G. Brana for some of the experimental work and Mr. H. Hazelton for preparing the graphs. Thanks are due to Pitney Bowes for permission to publish this work.

Summary

Adsorption isotherm, micro electrophoresis of a Sterling R carbon black dispersed in heptane, in the presence of various concentrations of manganese and zinc salts of 2-ethylhexanoic acid (octoate) are reported. Conductivity of the two octoates in heptane are also reported. Differences observed in the conductivity, adsorption and zeta potential characteristics between the two octoates are ascribed to the differences in the metal-carboxylate interactions as well as ionic size. The zeta potential is found to be positive and decreasing with increase in concentration with an initial sharp increase. The generation of charge on the carbon black has been found to be due to the preferential ionic adsorption from solution caused by an ion induced surface dipole mechanism.

Literature Cited

1 Lyklema, J., Adv. Colloid Interface Sci., (1968), 2, 65.
2 Meadus, F. W., Puddington, I.E., Sirianni, A. F., and Sparks, B. D., J. Colloid Interface Sci., (1969), 30, 46.
3 Koelmans, H., Phillips Research Reports, (1955), 10, 161.
4 Parfitt, G. D., and Willis, E., J. Colloid Interface Sci., (1966), 22, 100.

5 Lawrence, S. G., and Parfitt, G. D., J. Colloid Interface Sci., (1971), 35, 675.
6 Parfitt, G. D. and Wodd, J. A., J. Chem. Soc. Faraday Trans. I., (1973), 69, 1908.
7 Bagchi, P. and Vold, R. D., J. Colloid Interface Sci. (1970), 33, 405.
8 Beuche, F., J. Colloid Interface Sci. (1972) 41, 374.
9 Baylis, R. L., Trans. Inst. Metal Finishing (1972), 50, 80.
10 Dowbenko, R., and Hart, D. P., Ind. Eng. Chem. Prod. Res. Dev., (1973), 12, 14.
11 Metcalfe, K. A., J. Sci. Instr., (1955), 32, 74.
12 Tulagin, V., J. Opt. Soc. Amer., (1969), 59, 328.
13 Envall, A. D. Rev. Pure Appl. Chem., (1970), 20, 118.
14 Reif, R. B., Ind. Research., (1971), Dec., 48.
15 Klinkerberg, A., and van der Minne, J. L. "Electrostatics in the Petroleum Industry", Elsevier, Amsterdam, 1958.
16 Koelmans, H. and Overbeek, J. Th. G., Disc. Faraday Soc., (1954), 18, 52.
17 Darmerell, V. R., and Urabanic, A., J. Phys. Chem., (1944), 48, 125.
18 Darmerell, V. R., and Mattson, R., J. Phys. Chem., (1944), 48, 134.
19 van der Minne, J. L. and Hermanie, P. H. J., J. Colloid Sci., (1953), 8, 38.
20 Kitahara, A., Karasawa, S. and Yamada, H., J. Colloid Interface Sci., (1967), 25, 490.
21 McGown, D. N. L., Parfitt, G. D., and Willis, E., J. Colloid Sci., (1965), 20, 650.
22 Fowkes, F. M., Anderson, F. W., and Moore, R. J., Preprints, 150th American Chemical Society Meeting, Atlantic City, Sept. 1965.
23 Tamaribuchi, K., and Smith, M. L., J. Colloid Interface Sci., (1966), 22, 404.
24 Shinoda, K., Ed., "Solvent Properties of Surfactant Solutions", Chap. 3, Marcel Dekker, Inc., N.Y., 1967.
25 Nelson, S. M., and Pink, R. C., J. Chem. Soc., (1952), 1744.
26 Nelson, S. M. and Pink, R. C., J. Chem. Soc., (1954), 4412.
27 Parreira, H. C., J. Colloid Interface Sci., (1969), 29, 432.
28 van der Minne, J. L., Hermanie, P. H. J., J. Colloid Sci., (1952), 7, 600.
29 Henry, D. C., Proc. Roy. Soc. London, (1931), A 133, 106.

30. Wiersema, P. H., Loeb, A. L., and Overbeek, J. Th., G., J. Colloid Interface Sci., (1966), 22, 78.
31. Lewis, K. E., and Parfitt, G. D., Trans. Faraday Soc. (1966), 62, 1652.
32. Parreira, H. C., J. Electroanal. Chem., (1970), 25, 69.
33. Parreira, H. C., J. Colloid Interface Sci., (1973), 43, 382.
34. Bell, G. M. and Levine, S., Disc. Faraday Soc., (1966), 42, 97.
35. Vijayendran, B. R., (1971), unpublished work.

…

Micelle Formation in Nonaqueous Media

AYAO KITAHARA and KIJIRO KON-NO

Science University of Tokyo, Kagurazaka, Shinjiku-ku, Tokyo, Japan

Introduction

Experimental facts about micelle formation of oilsoluble surfactants in nonaqueous media have been recently compiled(1-3). Thus micelle formation in nonpolar solvents such as hydrocarbons ia markedly affected by the kind of polar parts of surfactants (4). For example, ionic surfactants make larger micelle than nonionic ones, anionic sulfates making larger micelle than cationic ammonium salts(4). On the other hand, the effect of hydrocarbon parts of surfactants on micelle formation is not so much remarkable than the effect in aqueous media. However, the considerable effect of hydrocarbon parts on micelle formation in nonpolar systems has been recognized. Thus the aggregation number of micelle decreased with increase of the carbon number of hydrocarbon part of quarternary ammonium halides or sodium di-(n-alkyl) sulfosuccinates(4). This trend was also shown by metal soaps in toluene(5).

In order to consider a theory of micelle formation in nonpolar media, the interaction was divided into two parts, that is, the interaction among polar parts of surfactants and that of hydrocarbon parts themselves and hydrocarbon parts and solvent molecules. The former can be estimated by ionic force for ionic surfactants or hydrgen bonding for nonionic surfactants(3,6,7). As for the effect of hydrocarbon parts, Singleterry and Fowkes suggested the steric hinderance(1,6). However, no quantitative explanation about the latter has yet been proposed.

In this paper, the idea of the entropic and enthalpic parameters for the interaction of hydrocarbon parts with solvent molcules at micelle formation will be introduced. The idea has been proposed by Flory for dissolution of polymers(8). This idea can be essentially utilized for low molecular weight substance. The idea has been used for elucidation of the protective effect on colloidal dispersion, that is, by Fischer(9), Meier(10), Napper(11) and Bagchi and Vold(12) for polymer dispersants and by Ottewill and Walker(13) for surfactant dispersants.

Theory

The model as shown in Figure I was assumed to make a theory for micelle formation of a surfactant in nonpolar media. It was assumed that the monomer solution of an oil-soluble surfactant is ideal and micelle formed becomes nonideal in the area which hydrocarbon parts exist.

Excess chemical potential of the solvent resulting from non-ideality can be expressed as follows (8,13):

$$\mu_1 - \mu_1^0 = -\bar{R}T(\psi_1 - \kappa_1)v_2^2 = -\bar{R}T(\psi_1 - \kappa_1)c^2/D^2 \quad (1)$$

where ψ_1 and κ_1 are entropic and enthalpic paramters of mixing proposed by Flory and v_2 being volume fraction of hydrocarbon part of the surfactant in the hydrocarbon area which can be reduced to c/D (c is the concentration of the hydrocarbon part in the area, kg l^{-1} and D the density of hydrocarbon). Equation (1) was alternatively deduced by Fischer with use of second virial coefficient B as Equation (1'):

$$\mu_1 - \mu_1^0 = -\bar{R}TB\bar{V}c^2 \quad (1')$$

where $\bar{V}$ is the partial molar volume of the solvent.

On the other hand, the excess osmotic pressure π_{ex} can be related to excess chemical potential by the following equation thermodynamically:

$$\mu_1 - \mu_1^0 = -\pi_{ex}\bar{V} \quad (2)$$

Equation (3) can be obtained from Equation (1) and (2):

$$\pi_{ex} = \bar{R}T(\psi_1 - \kappa_1)c^2/\bar{V}D^2 \quad (3)$$

Hence the free energy change of hydrocarbon parts, ΔG_h, resulting from micelle formation can be expressed by Equation (4):

$$\Delta G_h = \int \pi_{ex} dV = \{\bar{R}T(\psi_1 - \kappa_1)c^2/\bar{V}D^2\}\int dV \quad (4)$$

where $\int dV$ shows the volume of the area which hydrocarbon parts exist.

We concider ionic surfactants in this paper. Interaction among polar parts can be expressed by electrostatic force in ionic parts. It can be written in micelle as follows:

$$\Delta G_i = -\lambda e^2/2 \cdot 4\pi\varepsilon_{ef}^0 d \quad (5)$$

where λ is the aggregating number of micelle, e the elementary charge, ε_{ef} the effective permittivity and d interionic distance in the interior of micelle. The numeral 2 is the factor for pair counting.

Total free energy change following micelle formation ΔG is the sum of Equations (4) and (5). That is,

$$\Delta G = \Delta G_h + \Delta G_i = \frac{\bar{R}T(\psi_1-\kappa_1)c^2}{\bar{V}D^2}\int dV - \frac{\lambda e^2}{8\pi\varepsilon^0_{ef}d} \quad (6)$$

Furthermore, c can be rewritten as follows:

$$c = N\rho_0\lambda/\int dV \quad (7)$$

where N and ρ are the number and the mass of CH_2 group in the hydrocarbon part of the surfactant. Then Equation (6) is reduced to Equation (8) with use of Equation (7).

$$\Delta G = \frac{RT(\psi_1-\kappa_1)N^2\rho_0^2\lambda^2}{\bar{V}D^2\int dV} - \frac{\lambda e^2}{8\pi\varepsilon^0_{ef}d} \quad (8)$$

Numerical Calculations

A few assumptions were made in order to calculate numerically:
1) We assumed that micelle is a concentric sphere as shown in Figure 2 in which the polar ionic part makes an inner sphere having radius R and the hydrocarbon area makes an outer sphere having thickness r. For a surfactant having longer hydrocarbon, thickness of the hydrocarbon area is thicker than the ionic part, so that r≫R was assumed as an extreme case. Then we can obtain the following equation:

$$\int dV = 4\pi r^3/3 \quad (9)$$

Contrary, for a surfactant of shorter hydrocarbon, r≪R can be assumed as another extreme case. Then Equation (10) can be obtained:

$$\int dV = 4\pi R^2 r \quad (10)$$

2) Next, we assumed two limiting cases for r. A polymer model was taken for longer hydrocarbon, that is,

$$r = (2Nl)^{1/2} \quad (11)$$

where l is the carbon-carbon distance. On the other hand, we took a rod-shaped model for shorter hydrocarbon, that is,

$$r = Nl' \quad (12)$$

where l' is the length proportional to l. For a real surfactant, r would have the value between Equations (11) and (12), that is, Equation (13) would be true:

228　　　　　　　　　　　　COLLOIDAL DISPERSIONS AND MICELLAR BEHAVIOR

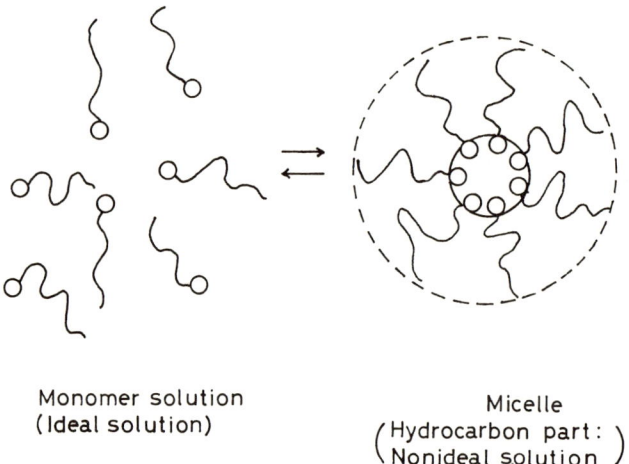

Monomer solution　　　　　　　Micelle
(Ideal solution)　　　　　(Hydrocarbon part:)
　　　　　　　　　　　　　 (Nonideal solution)

Figure 1. Model of micelle formation in nonaqueous media

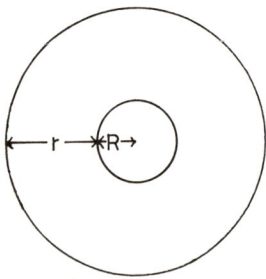

Figure 2. Model of micelle depicted in Figure 1

$$r = kN^n l \qquad 1/2 < n < 1 \qquad (13)$$

where k is a constant.

For a <u>longer hydrocarbon surfactant,</u> ΔG is rephrased as Equation (14) from Equations (9) and (11).

$$\Delta G = \Delta G_h + \Delta G_i = K\lambda^2 \sqrt{N} - \lambda e^2/8\pi\varepsilon_{ef}^0 d \qquad (14)$$

where
$$K = 3\bar{R}T(\psi_1 - \kappa_1)\rho_0^2/4\sqrt{2\pi}\bar{V}D^2 l^3 \qquad (15)$$

The equilibrium aggregation number λ_0 can be obtained from $\partial(\Delta G)/\partial\lambda = 0$, that is,

$$\lambda_0 = e^2/16\pi K\sqrt{N}\varepsilon_{ef}^0 d \qquad (16)$$

We took a benzene system tentatively. Here numerical values are as follows:

$\psi_1 - \kappa_1 = 0.1^*$, $\rho_0 = 0.014/6.02 \times 10^{23}$ kg, $\bar{V} = 97.5 \times 10^{-6}$ m^3 mol^{-1}, $D = 700$ kg m^{-3}, $l = 1.54 \times 10^{-10}$ m, $T = 300$ K.

Then $K = 1.3 \times 10^{-22}$ was obtained. In order to compare the theory with the experiment on the benzene solution of dialkyl dimethyl ammonium bromide, the following value of d was assumed:

$$d = r_{Br^-} + r_{(CH_3)_2NH} = 5.3 \times 10^{-10} \text{ m.}$$

Hence Equation (16) is reduced to the following equation with use of above values:

$$\lambda_0 = 840/\varepsilon_{ef}\sqrt{N} \qquad (17)$$

where ε_{ef} is effective dielectric constant.

An experimental point of the Reference 4 ($\lambda_0 = 6.2$ at $N=20$) was utilized to determine the value of ε_{ef} with use of Equation (17). The value of 30 obtained as ε_{ef} is reasonable from view of polar property in micelle. Then Equation (17) is reduced to Equation (18):

$$\lambda_0 = 28/\sqrt{N} \qquad (18)$$

The theoretical curve following Equation (18) was drawn as a full line in Figure 3 in which open circles indicate experimental results in the Reference 4.

For <u>shorter hydrocarbon surfactants,</u> we took another limiting case to which Equations (10) and (12) are applied.

---

* We referred to the value of Napper (<u>11</u>) as $\psi_1 - \kappa_1$.

Here Equation (8) can be expressed as follows:

$$\Delta G = K'\lambda^2 N - \lambda e^2/8\pi\varepsilon^0_{ef}d \qquad (19)$$

where
$$K' = \bar{R}T(\psi_1-\kappa_1)\rho_0^2/4\pi\bar{V}D^2R^2 1' \qquad (20)$$

Similar to the way of obtaining Equation (16), the following equilibrium aggregation number λ_0 can be obtained:

$$\lambda_0 = e^2/16\pi K'N\varepsilon^0_{ef}d \qquad (21)$$

Now no direct estimation of K' was done, because values of R and 1' are difficult to be estimated. With use of the ε_{ef} value obtained above and the λ_0 value at lower N (for example, λ_0=8.8 at N=16), the value of K' was calculated from Equation (21). Then $K'=2.6\times10^{-23}$. Hence Equation (21) is reduced to the following equation:

$$\lambda_0 = 140/N \qquad (22)$$

This theoretical curve was depicted as a dotted line in Figure 3. The real theoretical curve should be between Equations (17) and (21), because both equations express two limiting cases.

Equation (14) or (19) indicates that the longer hydrocarbon surfactant has lower equilibrium aggregation number. We can graphically elucidate the theoretical result by Figure 4. This corresponds with the experimental results that increase of carbon number of surfactant hydrocarbon decreases the aggregation number (4,5). Since ΔG_i is considerably negative for ionic interaction, minima appear in total ΔG and the equilibrium aggregation numbers λ_0 are present as shown in Figure 4. However, if polar interaction is small, that is, ΔG is low, total ΔG is always positive and no aggregates are presint. This can be seen in benzene solution of nonionic surfactants (4,14).

Literature Cited

1. Singleterry, C.R., J. Am. Oil Chem. Soc., (1955), 32, 446.

2. Fowkes, F.M., in "Solvint Properties of Surfactants" K. Shinoda, Ed. P.65, Marcel Dekker, New York, 1967.

3. Kitahara, A., in "Cationic Surfactants" Jungermann, Ed. p.287 Marcel Dekker, New York, 1970.

4. Kon-no, K. and Kitahara, A., J. Colloid Interface Sci., (1971), 35, 636.

5. Nelson, S.M. and Pink, R.C., J. Chem. Soc., (1952), 1744.

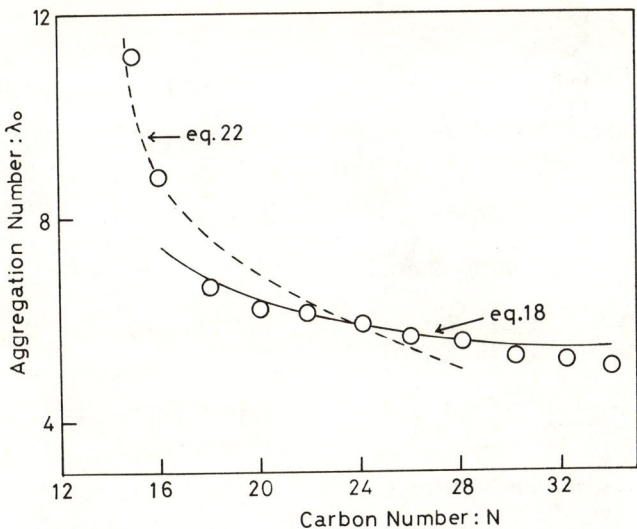

Figure 3. Relation of aggregation number and carbon number of $RR'N(CH_3)_2Br$ in benzene. ○: experimental; ———, - - - - -: theoretical.

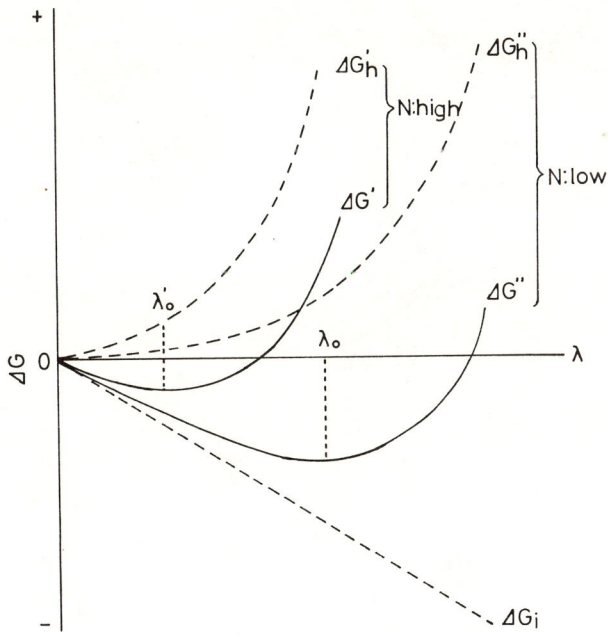

Figure 4. Qualitative explanation of equilibrium aggregation number (λ_0)

6. Fowkes, F.M., J. Phys. Chem., (1962), $\underline{66}$, 1843.

7. Debye, P. and Coll, H., J. Colloid Sci., (1962), $\underline{17}$, 220.

8. Flory,P.J., "Principles of Polymer Chemistry" Ch.12, Cornell Univ. Press, Ithaca New York, 1953.

9. Fischer,E.W., Kolloid-Z., (1958), $\underline{160}$, 120.

10. Meier,D.J., J. phys. Chem., (1967), $\underline{71}$, 1861.

11. Napper,D.H., Trans. Faraday Soc., (1968), $\underline{64}$, 1701; J. Colloid Interface Sci., (1970), $\underline{32}$, 106.

12. Bagchi,P. and Vold,R.D., J. Colloid Interface Sci., (1972), $\underline{38}$, 652; (1970), $\underline{33}$, 405.

13. Ottewill,R.H. and Walker,T., Kolloid-Z. Z. Polym. (1968), $\underline{227}$, 108.

14. Sirianni,A.F. and Coleman,R.D., Can. J. Chem., (1964), $\underline{42}$, 682.

Thermodynamics of Micelle Formation

K. S. BIRDI

Fysisk-Kemisk Institut, Technical University of Denmark, Lyngby, Denmark 2800

Introduction

The understanding of the thermodynamics of micelle formation is of much theoretical and practical importance. Even though a great many studies on the thermodynamics of micelle formation have been reported in the literature (1-14), the data available are not completely consistent. The purpose of this study is to report on the current theories and to comment on the data reported in the literature on the thermodynamics of micelle formation. The variation of critical micelle concentration (CMC) and the aggregation number (N) of a nonionic surfactant (Triton-X-100) with temperature were measured, since these data are essential in order to discuss the current theories reported on the micelle formation in the literature.

Experimental

Triton X-100 (OPE_{10}) was used as supplied by Rohm and Haas Co. The sample is reported to be polydisperse with respect to the ethyleneoxide adducts. Critical micelle concentration (CMC) of OPE_{10} in water and 0.025 M-KBr was determined at three different temperatures, viz., 25, 35 and 45°C by the U.V. difference spectrophotometric method, as reported in literature (10). The values of CMC in water and in 0.025 M-KBr were identical, i.e. within the experimental accuracy. Furthermore, these results agreed with the data reported by other investigators in water at these temperatures (10).
The micellar molecular weights (number average, M_n) of OPE_{10} were determined in 0.025 M-KBr aqueous solutions by using membrane osmometry, as reported by us elsewhere in detail (15,16). The micelle molecule weights, and subsequently the aggregation numbers, N, were determined at various temperatures, e.g., 5, 10, 15, 25, 30, 35, 40 and 45°C.

Results and Discussion

It is of interest to describe the micellar system of a nonionic surfactant, before discussing the data on the variation of aggregation number of micelles, N, and the CMC with temperature. The process of micellization involves the reversible aggregation of N molecules of the amphiphile to form a micelle (2,17,18,19) as given below:

$$N\text{-}m \rightleftharpoons M \tag{1}$$

The equilibrium constant of this reaction is given by:

$$K_a = A_M / (A_m)^N \tag{2}$$

where A_m and A_M are the activities of monomer and micelles respectively. At low surfactant concentration, however, it is reasonable to replace the activities of monomer and micelles by their respective concentrations (19):

$$K_c = C_M / (C_m)^N \tag{3}$$

where C_m and C_M are the concentrations of monomer and micelle respectively.

Assuming ideality, the chemical potential of micelles is given by:

$$\mu_M = \mu_M^o + RT \ln C_M \tag{4}$$

and for monomer:

$$\mu_m = \mu_m^o + RT \ln C_m \tag{5}$$

At equilibrium we have:

$$N \mu_m = \mu_M \tag{6}$$

From these relations the standard free energy of micellization, ΔG^o, per monomer is found to be given by (19):

$$\Delta G^o = RT \ln C_m - RT / N \ln C_M \tag{7}$$

$$= RT \ln C_m + RT / N \ln N - RT / N \ln C_M' \tag{8}$$

where $C_M' = N \cdot C_M$. It is generally assumed that if N is large, then at CMC, the Equation (8) reduces to (2,19):

$$\Delta G^o \cong RT \ln C_m = RT \ln CMC \tag{9}$$

Many investigators have further assumed that if the variation of N with temperature is negligible (1,2,6 -10, 19),the enthalpy of micelle formation,ΔH^o,can be obtained by using a Clausius-Clapeyron type of relationship:

$$\Delta H^o = - T^2 \left(\partial \Delta G^o/T / \partial T\right)_p \qquad (10)$$

$$= - RT^2 \left(\partial \ln CMC / \partial T\right)_p \qquad (11)$$

A great number of investigations on the determination of ΔH^o by applying Equation (11) have been reported in the literature for both the ionic and nonionic micellar systems (1,2,6 - 10). These ΔH^o values reported,however,do not all give a strictly consistent description of the micellar formation. For instance, the heat of micellization of N-dimethyldodecylamine oxide as determined from Equation (11) is reported to be 1900 cal/mole (1), while the calorimetric value is reported to be 2600 cal/mole (4). Another typical example is that of the n-dodecylpyridinium bromide in 2M urea solutions ; the ΔH^o values by Equation (11) and by calorimetric method are reported to be -798 cal/mole and -3496 cal/mole,respectively (13). This indicates that the method of using the relation in Equation (11) to obtain $\Delta H^\bullet$ is imprecise,probably due to certain assumptions used in the derivation of this relation,as discussed further below. Further,in many investigations reported on the determination of ΔH^o by using Equation (11),the data obtained does not give a strictly consistent description of the micelle formation (7,8,10-13).This analysis therefore clearly indicates that the usage of the relationship given in Equation (11),is not strictly valid for obtaining the heat of micellization, as also pointed out by other investigators (19).

It was therefore considered of interest to reconsider the assumptions made in the derivations of Equations (9) and (11). Since μ_M^o is a function of T,p and N (19), the variation of CMC and N with temperature of a nonionic surfactant, OPE_{10},was determined. The values of CMC were found to agree with the values reported by other investigators (10). The aggregation number,N,was determined by using membrane osmometry as described under experimental section.

The relationship in Equation (8) at the CMC,can then be rewritten as :

$$\Delta G^o = RT \ln CMC + RT / N \ln N \qquad (12)$$

It is thus seen that the relation given in the above Equation (12) differs from that given in Equation (9),in that while former is a function of both CMC and N, latter is only a function of CMC.

The results of CMC and N versus temperature of OPE_{10} are given in Figure 1. It is seen that the CMC decreases with

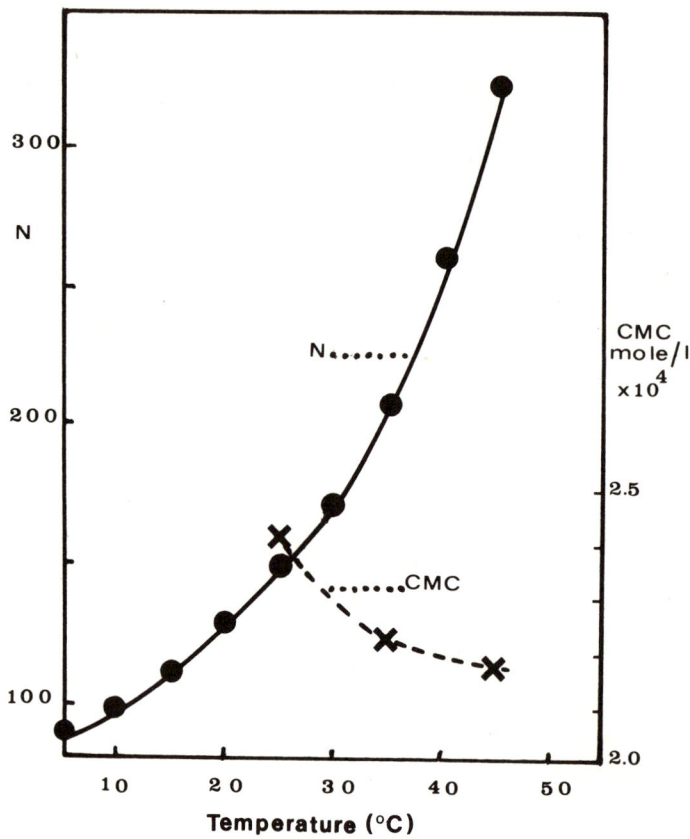

Figure 1. Aggregation number (N) and CMC vs. temperature of OPE_{10} in 0.025M KBr

temperature, and finally around 40-45°C no change in CMC with temperature is observed,as also reported by other investigators (10).The aggregation number,N, on the other hand increases with temperature from 5 - 45°C. This is typical for the non-ionic micelles,as reported in literature, and discussed elsewhere (16).The results in Figure 1 thus clearly show that the heat of micellization is not zero around 40 - 45°C, as has been reported in literature (10)by using the Equation (11).The second term in the Equation (12) relates to the transfer of material to and from already exsisting micelles. Since the Equations (12) and (9) differ by the term:

$$RT / N \ln N \qquad (13)$$

we find that the difference in ΔH^o determined after making this correction for ΔG^o, is 130-170 cal/mole,in the range of 5-45°C (from the data given in Figure 1). The aggregation number ,N, of $C_{10}E_6$, $C_{12}E_6$, $C_{14}E_6$ and $C_{16}E_6$ have been reported (5)to change with temperature as expressed below:

$$(\partial \ln N / \partial T)_p = 0.109 \qquad (14)$$

This shows that the N increases with increasing temperature, and that the alkyl chain has no effect on this change. It is also reported that the rate of change of N with temperature decreases as the number of ethyleneoxide adducts increases (5,19).Thus in Figure 1, the change of N for OPE_{10} is about four times lesser than that reported for the alkyl chain with six ethyleneoxide adducts,as mentioned above (5).In other words, the ΔH^o values will differ by about 600 cal/mole due to the term in Equation (13) in the case of these surfactants with six ethyleneoxide adducts.
To summarize,we have shown that the aggregation number ,N,of nonionic micelles changes appreciably with temperature.Preliminary results of another nonionic surfactant,nonylphenol with 10 ethyleneoxide units,also indicates that N changes with temperature quite appreciably,as determined by membrane osmometry. Since the relation given in Equation (9) for ΔG^o does not seem to give consitent results for the heat of micellization,it is clear that at this stage the thermodynamics of micelle formation is far from well understood. In the case of nonionic micelles,the determination of N as a function of temperature gives a more consistent description. It shows that this process has positive enthalpy over a large temperature range (from 5 - 45°C ,for both OPE_{10} and nonylphenol with 10 ethyleneoxide adducts).

Acknowledgements

It is a pleasure to thank Prof. Jørgen Koefoed for many helpful suggestions. The excellent technical assistance of Mrs. H. Birch is acknowledged.

Literature Cited

1. Herman,K.W.,J.Phys. Chem.(1962)66, 295
2. Schinoda,K. and Hutchinson,E.,J.Phys.Chem.(1962)66 ,577
3. Balambra,R.R.,Clunie,J.S.,Corkill,J.M. and Goodman,J.F., Trans. Faraday Soc. (1962) 58, 1661
4. Benjamin,L.,J.Phys.Chem. (1964)68, 3575
5. Balambra,R.R.,Clunie,J.S.,Corkill,J.M. and Goodman,J.F., Trans. Faraday Soc. (1964)60 ,979
6. Adderson,J.E. and Taylor,H.,J.Colloid Sci.(1964)19, 495
7. Corkill,J.M.,Goodman,J.F. and Harrold,S.P.,Trans. Faraday Soc.(1964)60 ,202
8. Corkill,J.M.,Goodman,J.F.,Robson,P. and Tate,J.R.,Trans. Faraday Soc.(1966)62, 987
9. Corkill,J.M.,Goodman,J.F.,Harrold,S.P. and Tate,J.R., Trans. Faraday Soc. (1967) 63, 240
10. Ray,A. and Nemethy,G.,J.Phys. Chem.(1971)75, 809
11. Elias,Hans-Georg,J.Macromol.Sci.(1973)A-7,(3)601
12. Piercy,J.,Jones,N.M. and Ibbotson,G.,J.Colloid Interface Sci.(1971)37, 165
13. Jones,M.N.,Agg,G. and Pilcher,G.,J.Chem. Thermodynamics, (1971)3, 801
14. Poland,D.C. and Scheraga,H.A.,J.Phys.Chem.(1965)69,2431 and 4425.
15. Birdi,K.S. and Kyllingsbæk,H.,J.Pharm. Pharmacol.(1971) 23, 900
16. Birdi,K.S.,Kolloid-Z. Z. Polym. (1972) 250, 731
17. Murray,R.C. and Hartley,G.S.,Trans.Faraday Soc.(1935)31, 183
18. Vold,M.J.,J.Colloid Sci.(1955)51, 561
19. Hall,D.G. and Pethica,B.A.,in "Nonionic Surfactants", M.J.Schick,Ed.,pp.516-557,Marcel Dekker,Inc.,New York, 1967.

17

Anomalies of Partially Fluorinated Surfactant Micelles[*]

PASUPATI MUKERJEE
School of Pharmacy, University of Wisconsin, Madison, Wis. 53706

KAROL J. MYSELS
8327 La Jolla Scenic Drive, La Jolla, Calif. 92037

Introduction

The behavior of amphipathic compounds such as ordinary surfactants having a hydrophobic hydrocarbon moiety and a hydrophilic polar one or of the perfluoro surfactants in which all the hydrogens of the hydrophobic moiety are replaced by fluorines is now reasonably well understood. Partially fluorinated surfactants, in which the hydrophobic moiety contains both hydrogens and fluorines, present, however, anomalies to which we wish to call attention. Our main point is that these anomalies can be readily understood once it is realized that the partially fluorocarbon parts, though both hydrophobic, are not mutually philic but phobic and tend to exclude each other. The lipophilic hydrocarbon part is fluorophobic and the fluorocarbon part is lipophobic.

The anomalies of the partially fluorinated surfactants involve both macroscopic phenomena such as their critical micellisation concentrations (c.m.c.) and microscopic ones as revealed by fluorine-NMR results. Both are in accord however with the bulk behavior of fluoro- and hydrocarbons and with the mixed micelle formation of ordinary and perfluorosurfactants.

The Mutual Phobicity of Hydrocarbons and Fluorocarbons

Though often neglected, this has been known and well documented since the early days of fluorocarbon development. In 1950 Simons and Dunlap (1) reported that n-pentane and n-perfluoropentane were completely miscible only above $265.5°K$

---

[*]The main conclusion of this paper was presented at the Fall 1970 ACS meeting in Chicago in connection with a paper by Mukerjee, P. and Cardinal, J. R. (unpublished work). The principal arguments were submitted in a paper received July 19, 1971, but not accepted by the Journal of Physical Chemistry. This material has been reorganized and amplified for presentation at the Volds symposium.

whereas regular solution theory would predict 62°K for their consolute temperature. The corresponding carbon compounds form two phases up to 40°C (2). Hildebrand, Prausnitz, and Scott devote several pages of their book (3) to this tendency of perfluorocarbons and hydrocarbons to unmix.

Even though completely miscible at room temperature, the C_4 compounds form extremely non-ideal solutions, the activity coefficient of the perfluorobutane reaching the value of 10 at infinite dilution (4). In other words, the tendency of an isolated perfluorobutane molecule to escape from n-butane is ten times greater than when it is surrounded by its own kind. For n-butane the activity coefficient reaches 5 (4).

Thus clearly hydrocarbons and fluorocarbons, though they will mix if the temperature is high enough, show a considerable reluctance to do so and a considerable mutual phobicity.

Mixed Micelle Formation Between Ordinary and Perfluoro Surfactants

Klevens and Raison (5) had published about 20 years ago what seems to be the only study thus far of the micellization of such a mixed system, namely sodium dodecyl sulfate (SDS) and perfluorooctanoic acid (FOA). Figure 1 shows their results for surface tension measurements on the pure systems (though the SDS itself contains minor impurities as shown by the minimum) and at four intermediate mole fractions, plotted as a function of total surfactant concentration. They noted the peculiar character of their results but an interpretation had to wait till a better understanding of solubilization and mixed micelle formation of ordinary surfactants had been gained. Details of such an interpretation will be given elsewhere (6) and only the feature pertinent to the present discussion will be brought out here.

It may be noted that ordinary surfactants generally form micelles which are mutually completely miscible. Thus commercial non-ionic surfactants which are generally mixtures of many homologues behave very much as their average, and the phase diagram of sodium decyl and dodecyl sulfates (7) (Figure 2) shows only minor deviations from ideality.

If the results of Klevens and Raison for systems containing 1.0, 0.75 and 0.20 mole fraction FOA are plotted against the concentration of the FOA as shown in Figure 3, several points become apparent. All three systems reach substantially the same low and constant surface tension after a sharp change of slope which indicates the formation of FOA micelles after a definite c.m.c. for them. For the 1.0 and 0.75 mole fraction systems, the values of the c.m.c. are indistinguishable at about 8.5 mM showing that the solubility of the SDS in these micelles is sufficiently small to be undetectable. In the 0.2 mole fraction system these micelles do not form until the FOA concentration reaches 30 mM, i.e. 21.5 mM higher than in the other two. This system also shows a change of slope and perhaps a small minimum around 2 mM FOA where

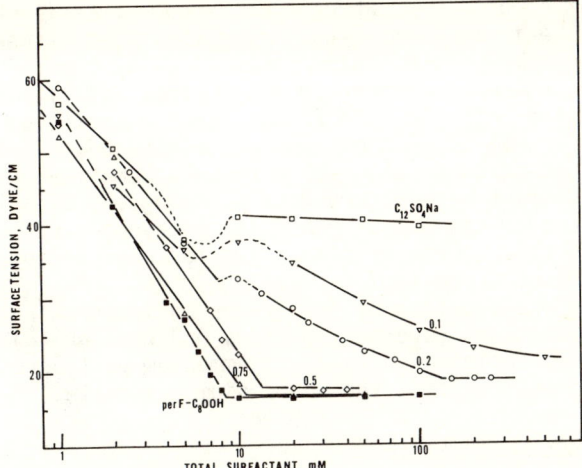

Figure 1. Surface tensions of sodium dodecyl sulfate, perfluoro octanoic acid and their mixtures as a function of the total surfactant concentration. This is Figure 4 of reference (5) replotted from the data given in that reference.

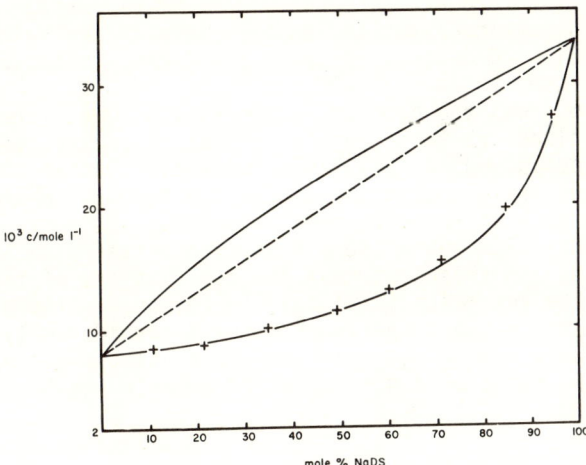

Journal of Colloid Science

Figure 2. The equilibrium compositions of micelles and monomers in the sodium decyl and dodecyl sulfate system. Solid lines represent the behavior of ideal solutions of both monomers and micelles. $\times$ = experimental c.m.c.'s. ---- = the composition of mixed micelles as deduced from conductivity data (7).

the concentration of SDS is about 8 mM. Hence this is where SDS micelles must begin to form. Thus it appears that between these concentrations some of the FOA is solubilized by the SDS micelles but the rest continues to increase its monomer activity till its c.m.c. is reached. The 21.5 mM FOA difference indicates upon closer analysis that the SDS micelle dissolves about 15 moles% of FOA. The resulting phase diagram is shown schematically in Figure 4. The contrast with Figure 2 is striking and clearly can only be ascribed to the mutual phobicity of hydro- and perfluorocarbons. Thus this characteristic of macroscopic behavior is also found in the micellar realm.

Following presentation of this paper an interesting confirmation of this point was reported by Tiddy and Wheeler (8) who found that the solubilization of octanol by ammonium perfluorooctanoate was about 4 times less than that by sodium octanoate.

C.m.c.'s of Perfluorinated Surfactants

Perfluoro alkanes are more hydrophobic than the hydrocarbon ones. Though comparative solubility data are sparse, the effect is clear. CF_4 is almost seven times (9) less soluble in water on a mole basis than CH_4 and water is also almost seven times (10) less soluble in perfluoroheptane than ordinary heptane on a weight basis and this becomes a factor over 25 on a mole basis. This greater hydrophobicity leads to a more pronounced amphipathic character of the perfluoro surfactants as shown by their well known greater surface activity both in terms of surface tension lowering at same concentration, and by the ultimate value attained which goes below 20 dyne/cm for many perfluoro surfactants. This is also reflected in their lower c.m.c.'s.

Figure 5 shows the available comparisons (11). The best documented values are those for the potassium octanoates. Here the perfluorocompound has a 13-fold lower c.m.c. than the ordinary one. At lower chainlengths the difference seems to decrease as shown by the only three-fold lower values for the very poorly documented potassium hexanoate and for the better established butyric acids. In this last case the c.m.c.'s are of the order of 1 M so that the intermicellar liquid is no longer simply aqueous. Since in the following comparisons we will consider only compounds having 8 or more carbon atoms in the chain, we can conservatively take 13 as the ratio of c.m.c.'s in hydrocarbon and perfluorocarbon systems.

C.m.c.'s of Partially Fluorinated Surfactants

If partially fluorinated chains were to behave ideally, one would expect the c.m.c.'s of these compounds to lie between those of the corresponding perfluoro and ordinary surfactants varying approximately linearly with the fraction of hydrogens replaced by fluorines as indicated by the solid line of Figure 6. As shown

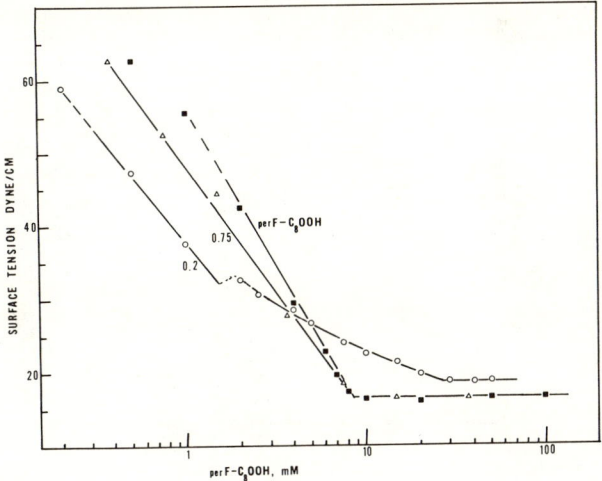

Figure 3. Data for selected systems of Figure 1 as a function of the concentration of perfluorooctanoic acid present

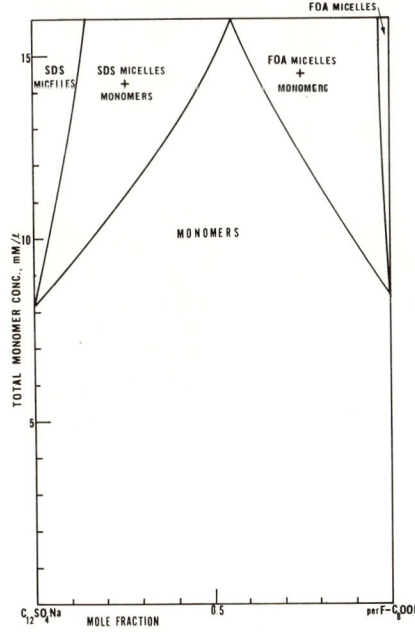

Figure 4. Schematic of the equilibrium compositions of micelles and monomers in the sodium dodecyl sulfate and perfluoro octanoic acid system. Note the presence of two types of micelles having limited mutual solubility. Based on Ref. (6).

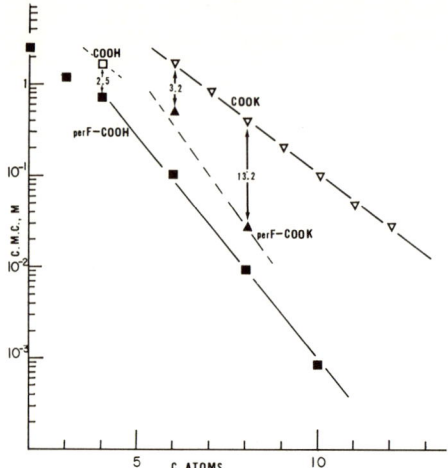

Figure 5. Critical micelle concentrations of perfluoro and corresponding ordinary surfactants as a function of carbon atoms. Data from Ref. (11).

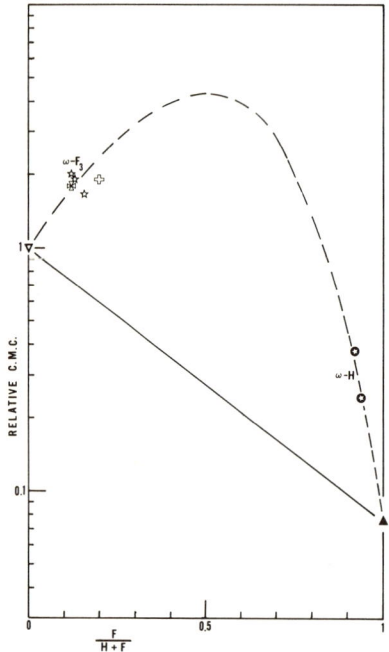

Figure 6. Relative critical micelle concentrations of ordinary, perfluoro, and partially fluorinated surfactants as function of the F/H ratio in the molecule. Based on Figures 5, 7, and 8. Solid line indicates ideal behavior.

also on that figure, the real situation is quite different.

The two groups of intermediate points of Figure 6 correspond to the ω-hydro perfluoro and the ω-trifluoro compounds. The comparison of c.m.c.'s on which these points are based are shown in Figures 7 and 8. It should be noted that the replacement of a single fluorine by a hydrogen increases the c.m.c. by a factor of about 3 or 5, i.e. brings it within a factor of 3 or 4 of the ordinary surfactant. Replacement of three ω-hydrogens by fluorines in the ω-trifluoro surfactants actually raises the c.m.c. instead of reducing it as would be expected from the simple mixture rule.

These anomalous c.m.c.'s of the partially fluorinated surfactants are readily understandable, however, in terms of the mutual phobicity of the hydrocarbon and fluorocarbon portions of their chains which counteracts the aggregating tendency due to the hydrophobic character of the hydrocarbon portion and the still more hydrophobic character of the fluorocarbon part. Additional factors reinforcing the mutual phobicity at the water interface include the dipole moment of the partially fluorinated chains (12) and the possible tendency of the ω-hydrogen to form hydrogen bonds with water (13).

Fluorine NMR Results

The NMR line of an atom shifts generally with the solvent and thus can be used to investigate the surroundings of an atom within the micelle. Muller (14) developed the concept of the "degree of hydrocarbon-like character," Z, of the environment of such an atom. $Z = (\delta_m - \delta_h)/(\delta_w - \delta_h)$ where δ is the line shift and the subscripts indicate m, the micelle, h hydrocarbon, and w, water, respectively. This can be readily generalized to a "degree of organic character" by replacing δ_h by δ_o for the shift in the appropriate organic solvent as Muller has effectively done later (15).

Perfluoro Surfactants. There are data for two perfluorosurfactants, the C_4 carboxylic acid reported by Bailey and Cady (16) and the C_8 sodium carboxylate studied by Muller and Simsohn (15). The former authors did not interpret their results in terms of environment effects but their data permit the calculation of Z factors using the anhydrous perfluorobutyric acid itself as the basis for the δ_0 value and obtaining the δ_m values by extrapolation of the linear portion of their graph. Table I shows the Z values so obtained as well as those given by Muller and Simsohn.

It is apparent from Table I that Z is higher, i.e. the environment is more organic for the ω-carbon than for the β-carbon and this is higher than for the α-carbon. That the α-carbon of a micellized surfactant should be exposed to water for a significant fraction of the time is to be expected from the fact that it is attached to the hydrophilic group which, along with its hydration

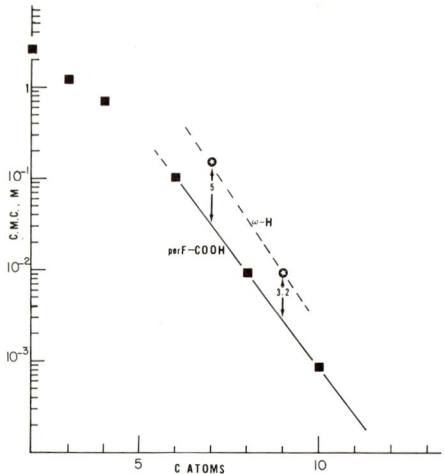

Figure 7. The c.m.c.'s of ω-H-fluoro and corresponding perfluoro surfactants as a function of the number of carbon atoms. Data from Refs. (11, 32).

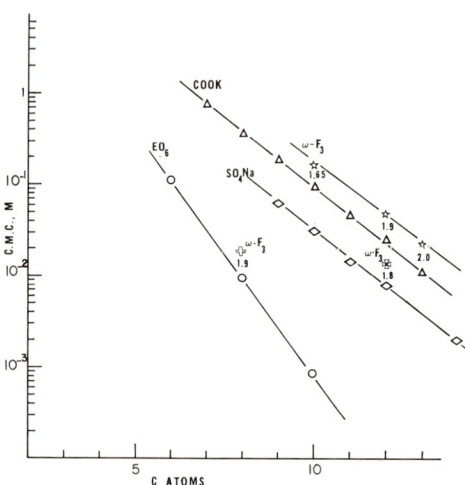

Figure 8. The c.m.c.'s of ω-trifluoro and corresponding ordinary surfactants as a function of the number of carbon atoms. Data from Refs. (11, 14, 22, 23).

Table I. The Degree of Organic-like Character as Determined by Fluorine-NMR Line Shifts for Perfluoro and Some ω-trifluoro Surfactants.

$$Z = (\delta_m - \delta_w)/(\delta_0 - \delta_w)$$

	α-F	β-F	ω-F
Perfluoro C_4OOH (from data of ref. 16)	0.71	0.77	0.82
Perfluoro C_8OONa (15)	0.37	0.66	0.84
ω-F_3(C_{10},C_{12},C_{13})OONa (14)	--	--	0.53

sphere, resists strongly immersion in the hydrophobic part of the micelle. Thus the α-carbon is held at the water-core interface and any thermal fluctuation will expose is even more. Direct evidence for the high degree of wetting of this carbon atom comes from the work of Kurz (17) who found that the spontaneous hydrolysis of the dodecyl sulfate ester anion by water, known to involve the cleavage of the carbon-oxygen bond, was hardly affected by micellization. That the β-carbon is somewhat exposed to water because of its attachment to the α-one is not surprising.

That even the ω-carbon seems to be hydrated to a definite though slight degree cannot be ascribed to experimental uncertainties alone. There could be a contribution due to the fact that any point within the micelle is always close to the water--the micelle is all a surface monolayer--whereas the reference δ_0 is taken for bulk systems, but more important certainly is the fact that the polar heads do not cover completely the surface of the micelle but leave some of its hydrophobic surface exposed. This is inherent in the fact that micelles are the result of two opposite tendencies: that of the hydrophilic groups to diffuse away from each other, and that of the hydrophobic ones to aggregate and reduce their contact with water. The balance has to involve some yielding on both accounts and therefore some organic-water interface. Furthermore, for the longer chain-lengths and especially for the bulkier perfluoroalkanes, the surface of a micelle limited in its spherical or cylindrical radius by the length of the chain, has a larger surface than can be covered by the polar groups. Figure 9, taken from the work of Stigter (18) shows a scale drawing of the probable surface aspect of the micelle of sodium dodecyl sulfate with the crosshatched areas indicating the core surface accessible to the center of the counterions and hence certainly also to water molecules. Supporting evidence comes from the study of volume changes accompanying micellization in homologuous series (19) which suggests that from 2 to 4 carbons are hydrated and also from proton NMR chemical shifts (20, 21) which point to as much as 4

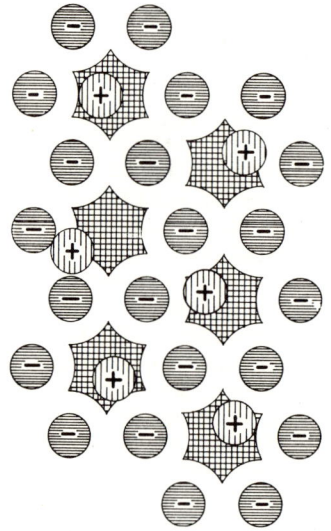

Journal of Physical Chemistry

Figure 9. Schematic surface view of the micelle of sodium dodecyl sulfate showing to scale the relative size of the ions, polar groups, and (crosshatched) the exposed hydrocarbon surface. Reproduced from Ref. (18).

hydrated carbon atoms. Both lines of argument involve however somewhat questionable assumptions and the fact that butyric acid does form micelles shows that the latter value must be an overestimate in some cases. In the perfluoro derivatives the existence of micelles of the C_3 acid shows that the number of hydrated carbons is still smaller.

The ω-Trifluoro Surfactants. Muller and co-workers have studied the ω-trifluorinated C_{10} to C_{12} carboxylates (12), the C_{12} sodium sulfate (22) and a C_8 nonionic (23). In each case they found Z values close to 0.5, i.e. a semi-aqueous surrounding of the terminal CF_3 group. On the implicit assumption that this group fairly samples the hydrophobic core, they concluded that "There is considerable penetration of water into the interior of the micelles" (12).

This last conclusion not only was not anticipated on the basis of all the work on association colloids up to that time but clearly ran counter to it. The presence of water within the core of the micelle would seem to destroy what has been assumed heretofore to be the basic reason for micelle formation, namely the hydrophobicity of the tails, their tendency to escape from the aqueous environment or, more exactly, the tendency of the strongly bonded water to press them out into a small droplet. That the micelle is indeed a sufficiently permanent structure to apply this reasoning and not a highly ephemeral one for which macroscopic concepts are misleading (24) is shown by recent work of Oakes (25) who reported the resolution of the ESR spectra of solubilized and unsolubilized probes in a micellar solution. Hence the lifetime of the solubilized specie, and therefore of the micelle, had to be long on the ESR time scale of about 10^{-6} seconds. This is certainly also sufficient for diffusional and molecular rearrangements to reach equilibrium within the micelle.

More quantitatively, the solubility within the micellar core accounts presumably for the solubilization of non-polar molecules. Those that are only slightly solubilized and thus affect the least the structure of the micelle are most useful in exploring its nature. Hartley has already studied (26) the behavior of transazobenzene in cetylpyridinium chloride and found that the solubility in the micellar core was very close to that in n-decane. This led him to conclude in 1938 "... the paraffin chain salt... behaves as a <u>liquid</u> solvent upon solution in water, and that part of it which is responsible for the solvent action <u>is not diluted by the water</u>...." (the emphasis is Hartley's). The same conclusion can be drawn from the behavior (27) of the so-called "Orange OT" dye which is 1-o-tolylazo-2-naphthol (c.i. solvent orange 13). The data for both are shown in Table II.

The solubility of water in hydrocarbons is also known (29) and, as may also be seen in Table II, it is much lower than that of either azobenzene or of "Orange OT". Would it not be most surprising if its solubility in the core of the micelle were several

Table II. Solubilities in Hydrocarbons and in Micelles of Some Sparingly Solubilized Compounds and of Water.

	Saturation mole fraction at room temperature	
	in alkanes	in Micelles
tr-Azobenzene (26)	0.144 (C_{10}) 0.176 (C_{16})	0.148 (Cetylpyridinium sulfate)
"Orange OT" (1-o-tolylazo-2-naphthol)	0.0076 (C_{12}) (28)	0.008 (Sodium dodecyl sulfate) (27)
Water (29)	0.00061 (C_{12}) 0.00051 (C_7)	? ? ?

orders of magnitude greater? Indeed, a study of proton NMR shifts of the hydrophobic moiety of a nonionic surfactant has led Podo, Ray, and Nemethy (30) to conclude that there is no water or very little in the interior of the micelle. That the presence of fluorines does not affect this argument is evident from the fact that as already mentioned, water is even less soluble in perfluoroalkanes than in simple alkanes.

Thus, it is apparent that the apparent hydration of the ω-trifluoro methyl group in the partially fluorinated surfactants must be due to the fact that they are partially and not completely fluorinated, to their part hydrocarbon, part fluorocarbon nature. Indeed, there is no independent support for the implicit assumption that the trifluoromethyl group samples uniformly the interior of the otherwise hydrocarbon core of the micelle. The opposite assumption, that this group is concentrated on the core's surface, has already been proposed (31) and is an obvious consequence of the non-ideality of mixing of fluoro- and hydrocarbons, i.e. of their mutual phobicity and the additional factors pointed out above. This then explains both the difference between the surroundings of the ω-group in micelles of perfluoro and of ω-trifluoro surfactants and the highly aqueous environment in the latter case as shown by Muller, *et al's* experiments.

Conclusion

Partially fluorinated surfactants show at least two apparent anomalies: Their c.m.c. is higher than could be expected from the change produced by perfluorination, and the ω-trifluoro group has a much more aqueous environment than expected for the core average. Both facts are, however, in harmony with what is known about the mutual phobicity of hydrocarbon and perfluorogroups from the limited mutual solubility and high non-ideality of solutions of hydrocarbons and fluorocarbons and is in accord with the very limited mutual solubility of the micelles of corresponding surfactants. It is this resistance to mixing of these two kinds of groups that reduces their tendency to micellize when attached to the same chain and causes the ω-trifluoro group to sample predominantly the water-micellar core interface rather than the core average when this core is hydrocarbon-like.

Literature Cited

1. Simon, J. H. and Dunlap, R. D., J. Chem. Phys.(1950), 18, 335.
2. Hildebrand, J. H., Prausnitz, J. M., and Scott, R. L., "Regular and Related Solutions," p. 171, Van Nostrand Reinhold, New York, N. Y., 1970.
3. Ibid., p. 177.
4. Ibid., p. 18.
5. Klevens, H. B. and Raison, M., J. Chimie (1954), 51, 1.
6. Mukerjee, P. and Mysels, K. J., to be submitted.
7. Mysels, K. J. and Otter, R. J., J. Colloid Sci. (1961), 16, 474.
8. Tiddy, G. J. T. and Wheeler, B. A., J. Colloid Interface Sci., (1974), 47, 59.
9. Hildebrand, J. H., Prausnitz, J. M., and Scott, R. L., "Regular and Related Solutions," p. 204, Van Nostrand Reinhold, New York, N. Y. 1970.
10. Linke, W. F., "Solubilities," pp. 1135-1137, Van Nostrand Co., Inc., Princeton, N. J., 1958.
11. C.m.c. values unless otherwise noted are taken from Mukerjee, P. and Mysels, K. J., "Critical Micelle Concentrations of Aqueous Surfactant Systems," NSRDS-NBS 36, Superintendent of Documents, Washington, D. C., 1971.
12. Nelson, R. D. and Lide, O. R. jr., "Selected Values of the Electric Dipole Moments for Molecules in the Gas Phase" NSRDS-NBS-10, Superintendent of Documents, Washington, D. C., 1967.
13. Thorp, N. and Scott, R. L., J. Phys. Chem., (1956), 60, 1441.
14. Muller, N. and Birkhahn, R. H., J. Phys. Chem., (1967), 71, 957.
15. Muller, N. and Simpsohn, H., J. Phys. Chem., (1971), 75, 942.
16. Bailey, R. E. and Cady, G. H., J. Phys. Chem.,(1969), 73, 1612.
17. Kurz, J. L., J. Phys. Chem., (1962), 66, 2239.
18. Stigter, D., J. Phys. Chem., (1964), 68, 3603.
19. Corkill, J. M., Goodman, J. F., and Walker, T., Trans. Faraday Soc. (1967), 63, 768.
20. Clifford, J. and Pethica, B. A., Trans. Faraday Soc. (1964), 60, 1483.
21. Clifford, J., ibid (1965), 61, 1276.
22. Muller, N. and Johnson, T. W., J. Phys. Chem. (1969), 73, 2042.

23. Muller, N. and Platko, F. E., J. Phys. Chem. (1971), 75, 547.
24. Zisman, W. A., private communication, 1972.
25. Oakes, J., Nature (1971), 231, 38.
26. Hartley, G. S., J. Chem. Soc. (1938), 1968.
27. Mysels, K. J., in "Pesticidal Formulations Research," Advances in Chemistry Series No. 86, American Chemical Society, Washington, 1969.
28. McBain, M. E. L. and Hutchinson, E., "Solubilization and Related Phenomena," p. 111, Academic Press, New York, N. Y., 1955.
29. Schatzberg, P., J. Phys. Chem., (1963), 67, 776.
30. Podo, F., Ray, A., and Nemethy, G., J. Am. Chem. Soc. (1973), 95, 6164.
31. Cordes, E. H. and Gitler, C., in "Progress in Bioorganic Chemistry," E. T. Kaiser and F. J. Kezdy, Ed., Vol. 2, John Wiley, New York, N. Y., 1973.
32. Bernett, M. K. and Zisman, W. A., J. Phys. Chem., (1959), 63, 1911. This reference has been overlooked by reference 11 and its values are preferred.

18

Deuteron NMR Studies on Soap-Water Mesophases

HÅKAN WENNERSTRÖM, NILS-OLA PERSSON and BJÖRN LINDMAN
Division of Physical Chemistry 2, The Lund Institute of Technology,
Chemical Center, P.O.B. 740, S-220 07 LUND 7, Sweden

Introduction

Soap-water systems form in addition to isotropic solutions and crystalline phases several different types of mesophases which have properties intermediate between liquids and solids. The mesophases are often very viscous but the motion of the individual molecules is normally almost as rapid as in an ordinary liquid. The viscous appearence of the phases is caused by the formation of large aggregates and by the presence of long-range order. The various mesophase structures differ in the types of aggregates and in their packing characteristics. In a lamellar mesophase the soap molecules usually form double-layered lamellae which are separated by lamellae of variable thickness containing the water molecules and the counterions. In a normal hexagonal phase, on the other hand, the soap molecules form long rodshaped aggregates which are arranged in a hexagonal array. In addition to these two phases, which are best understood, a hexagonal phase of the reversed type and different types of cubic isotropic mesophases may form in a binary soap-water system. If a third component is added to a soap-water system the phase equilibria may become quite complex and in addition to those phases mentioned also other types of structures may occur. A thorough account of the phase equilibria in amphiphilic systems as well as of the present state of knowledge of the phase structures has recently been given by Ekwall (1).
     In mesophases the molecular arrangement is more ordered than in a solution but the ordering is less than in a crystalline phase. In order to understand the structure and the properties of the soap-water mesophases in more detail it is necessary to have information on the orientation effects on a molecular level. During recent years it has been demonstrated that some static magnetic resonance parameters may provide pertinent information in this respect. Previous studies have mostly been concerned with orientation effects in the soap lamellae whereas the aqueous part has attracted less interest, partly because orientation

effects here are small and consequently difficult to study. Recent studies have, however, demonstrated that static quadrupole effects in NMR spectra can give rather detailed information on orientation effects in the aqueous layers. The present study was aimed at providing an insight into water orientation effects, using deuteron NMR, for chiefly some lamellar liquid crystalline phases occurring in three-component systems of soap, alcohol and heavy water. Thereby we have studied how water orientation depends on sample composition, on temperature and on the counterion. The study also provides some information on the orientation effects in the amphiphilic lamellae by investigating the deuteron NMR signal of the -OD group of the alcohol.

Deuteron NMR studies on soap-water mesophases have been reported previously (for a review see Ref. 2) but since some misunderstanding of the theoretical basis of interpretation seems to exist in the literature we will start by briefly giving the theoretical background of deuteron NMR studies of anisotropic systems. Especially we want to point out some effects which may complicate the analysis of the NMR spectra.

Theory

The deuteron has the spin quantum number $I = 1$ and thus possesses an electrical quadrupole moment. Through the quadrupole moment the nucleus interacts with electrical field gradients in the surroundings. The main terms in the deuteron spin hamiltonian are the Zeeman term, H_Z, and the quadrupole coupling term, H_Q,

$$H = H_Z + H_Q = -\nu_o I_Z + \beta_Q \sum_q (-1)^q V_{-q} A_q \qquad (1)$$

Here the V_q's are the irreducible components of the electric field gradient tensor (of second rank) and the A_q's the standard components of a second rank spin tensor operator (3). β_Q is given by $eQ/2I(2I-1)h$. The hamiltonian is expressed in frequency units.

In an isotropic liquid the mean value of H_Q is zero and the quadrupole interaction contributes only to relaxation. In an anisotropic medium as, for example, a lamellar or hexagonal mesophase the mean value of H_Q is no longer zero and a quadrupole splitting appears in the NMR spectrum. To elucidate the effects of the anisotropic medium more clearly it is convenient to rewrite H_Q as (3)

$$H_Q = \beta_Q \sum_{qq'q''} (-1)^q V_{-q}^M A_{q''}^L D_{q'q}^{(2)}(\Omega_{DM}) D_{q''q'}^{(2)}(\Omega_{LD}) \qquad (2)$$

The indices M, L and D stand for the molecular, laboratory and director reference frames, respectively (see Figure 1). The $D^{(2)}$'s are second rank Wigner rotation matrix elements and Ω_{ij} denotes the eulerian angles that specify the transformation from coordinate system j to system i. The director is defined by the symmetry of the mesophase and is usually coincident with the optical axis. If it is assumed that a nucleus stays within a domain of a given orientation of the director over a time that is long compared to the inverse of the quadrupole interaction, the mean value of H_Q is

$$\overline{H_Q} = \beta_Q \sum_{qq''} (-1)^q V_{-q}^M A_{q''}^L \overline{D_{oq}^{(2)}(\Omega_{DM})} D_{q''o}^{(2)}(\Omega_{LD}) \qquad (3)$$

Here we have made use of the result that the mean value $\overline{D_{q'q}(\Omega_{DM})}$ is zero for $q' \neq 0$ if there is a threefold or higher symmetry around the director axis.

The quadrupole term is usually small compared to the Zeeman term and, to first order, it is only the secular part of $\overline{H_Q}$ that contributes to the time-independent hamiltonian, H_0.

$$H_0 = -\nu_o I_Z + \frac{\nu_Q S}{6}(3\cos^2\theta_{LD}-1)(3I_Z^2-2) \qquad (4)$$

The order parameter, S, is given by

$$S = \frac{1}{2}\{(3\cos^2\theta_{DM}-1) + \eta(\sin^2\theta_{DM}\cos 2\phi_{DM})\}$$

and ν_Q by

$$\nu_Q = \frac{3}{4}\frac{e^2 Qq}{h} \cdot (\frac{e^2 Qq}{h} \text{ is the quadrupole coupling constant.})$$

Here η is the asymmetry parameter of the electric field gradient tensor, eQ the nuclear quadrupole moment and eq is twice the largest component of the electric field gradient tensor. θ and φ are equal to the eulerian angles β and γ respectively.

The hamiltonian given in Equation (3) gives the three energy levels

$$E_1 = -\nu_o + \nu_Q S(3\cos^2\theta_{LD}-1)/6$$

$$E_o = -\nu_Q S(3\cos^2\theta_{LD}-1)/3$$

$$E_{-1} = \nu_o + \nu_Q S(3\cos^2\theta_{LD}-1)/6$$

and for an oriented sample with a uniform director orientation the deuteron NMR spectrum consists of two equally intense peaks separated by

$$\Delta(\theta) = |\nu_Q S(3\cos^2\theta_{LD}-1)| \qquad (5)$$

For a powder sample, where all values of $\cos\theta_{LD}$ are equally probable, the NMR spectrum consists of a broad absorption curve with two marked peaks separated by

$$\Delta = |\nu_Q S| \qquad (6)$$

We denote Δ (and Δ(θ)) the quadrupole splitting.

One point that may be made from these theoretical considerations is that the asymmetry parameter may affect the magnitude of the splitting but not the spectral shape (4). Therefore, conclusions in the literature about the magnitude of the asymmetry parameter from the spectral shape are not justified.

Chemical Exchange

The equations given in the previous section were derived under the assumption that the deuterons are fixed in the molecules. In systems containing heavy water and an alcohol, for example, the deuterons can exchange between the water and the alcohol molecules. Furthermore, deuteron exchange between water molecules interacting differently with the amphiphilic surfaces have to be taken into account. In order to include the possibility of deuteron exchange we have to modify the equations given above. The effect of chemical exchange on the deuteron NMR spectrum depends on the ratio between the exchange rate and the difference in quadrupole splitting between the different deuteron environments.

In the limit of slow exchange the NMR spectrum is a superposition of the NMR spectra of the different binding sites and for each site the Equations (5) and (6) are separately valid. In the opposite limit when the exchange is much faster than the splitting difference only a single quadrupole-split deuteron NMR spectrum is observed with a splitting given by

$$\Delta = \left| \sum_i p_i \Delta_i \right| \tag{7}$$

Here p_i is the fraction of deuterons in site i and Δ_i the splitting characterizing this site. By means of Equation (7) it may be possible to obtain significant information on the mode of molecular orientation. Thus if it is possible to observe the deuteron spectrum in both the exchange limits the relative signs of the order parameters of two different sites may be determined.

In the region of intermediate exchange rate a broadening of the NMR signal appears, which is analogous to the well-known chemical exchange effects in proton NMR. For a powder sample this broadening is less marked than in proton resonance since the signals are broad even in the absence of exchange. It is, however, possible to determine the exchange time from an exchange-broadened powder spectrum (5). Although the precision of such a determination is rather low the method is of interest since it allows studies of exchange phenomena which are much more rapid than those which are within reach of ordinary proton NMR methods. Also the precision may be improved by using macroscopically aligned samples, for which the peaks are more narrow and thus a more precise determination of the amount of exchange broadening is possible. Furthermore, in the case of macroscopically aligned samples the splitting difference, and thus also the range of exchange rates attainable, may be varied by changing the angle between the magnetic field and the constraint responsible for macroscopic alignment.

The effect of chemical exchange on quadrupole-split NMR spectra was deduced only recently and no systematic application of the method to chemical problems has yet been performed. For deuteron NMR spectra of soap-alcohol-heavy water mesophase samples the peaks are, however, often broadened due to deuteron exchange especially at temperatures slightly above room temperature. As an illustration we give in Figure 2 some spectra obtained for a lamellar mesophase sample composed of lithium n-octanoate, decanol and D_2O.

Double Quantum Transitions

The appearance of a central peak at the Larmor frequency in addition to the "powder pattern" in the deuteron spectrum of an unoriented sample has often been interpreted as due to phase inhomogeneities (6, 7), which cause part of the water molecules to reside in an essentially isotropic environment. This central peak can, however, be observed in spite of a careful sample preparation to avoid a two-phase system. This would seemingly indicate that there are deuterons moving isotropically even in the anisotropic mesophase structure (8). There is, however, an alternative explanation to the observation of a central peak. From the energies of the deuteron spin system given above it follows that $E_{-1} - E_1 = 2\nu_o$ independently of θ_{LD}. A double quantum transition at the frequency

$$\nu = \frac{E_{-1} - E_1}{2} = \nu_o$$

should then appear in the spectrum at sufficiently strong RF-fields. It has been shown for one system, sodium n-octanoate-n-octanoic acid-D_2O, that these double quantum transitions are clearly observable (9). As shown in Figure 3 the central peak is unobservable at low RF-field amplitudes while it is much more intense than the powder pattern at high RF-field strengths. As described in Ref. 9 it is easily investigated by determining the dependence of the signal intensity on the RF-field strength if a central peak is due to a double quantum transition or has another cause.

Macroscopic Alignment

The deuteron NMR spectrum from an anisotropic soap-heavy water mesophase usually shows the typical powder pattern arising from a random distribution of director orientations in the sample. It is though usually possible to orient at least lamellar mesophase samples by placing a thin layer of the sample between glass plates (10). From the angular dependence of the splitting it can

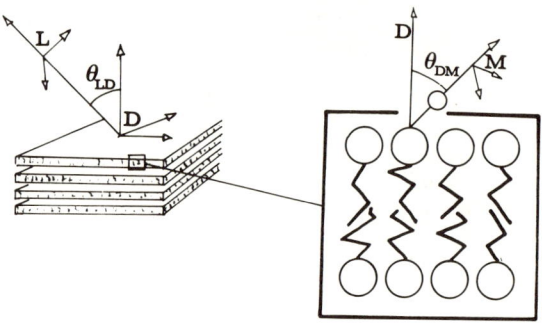

Figure 1. Schematic drawing of the geometry in the lamellar liquid crystalline phase. The different coordinate systems used in the text are outlined in the figure, laboratory frame (L), director frame (D), and molecular frame (M). θ_{LD} and θ_{DM} denote the angles between the Z axes of the laboratory and director systems and the director and molecular systems, respectively (3).

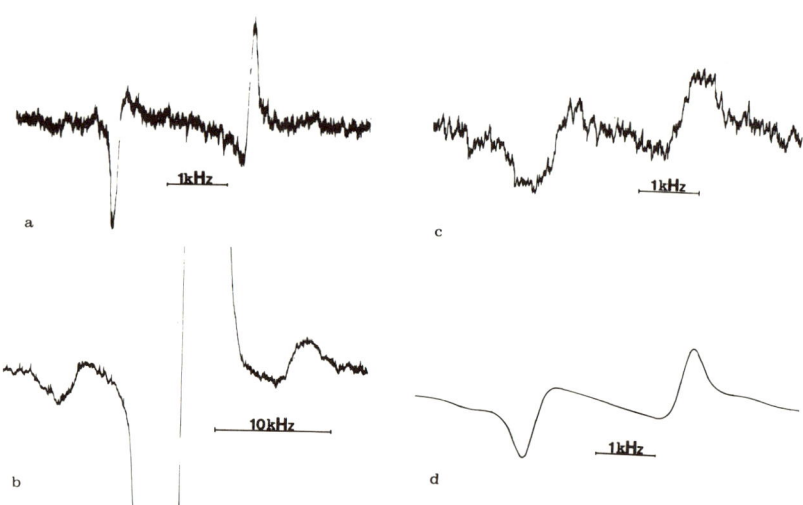

Figure 2. Deuteron NMR spectra illustrating the effect of deuteron exchange. The spectra were obtained for a lamellar mesophase sample with the molar ratio D_2O:lithium octanoate:decanol being 83.0:6.8:10.2. a) Spectrum for the water deuterons at 27°C. b) Spectrum for the decanol–OD group at 27°C. c) Spectrum at 40°C showing deuteron exchange effects. d) Computer-simulated spectrum, corresponding to spectrum c, obtained according to a procedure described in Ref. 5, with intermediate exchange rate (exchange time 23 μs) (5).

be deduced how the director is oriented relative to the glass plates. For lamellar mesophase samples composed of alkali soap, decanol or octanoic acid and heavy water and placed in layers between thin glass plates we have investigated how the splitting depends on the angle between the magnetic field and the normal to the glass plates. As an example we give in Figure 4 the angular dependence for a sample composed of sodium n-octanoate, n-octanoic acid and D_2O. A comparison with Equation (5) shows that the normal to the glass plates coincides with the director and therefore in this, and the other systems we have studied, the lamellae are aligned parallel to the glass plates.

Four-component systems of water, soap, alcohol and a simple salt do sometimes orient spontaneously in a magnetic field (11, 12). The structure of these phases has not yet been fully elucidated. One characteristic feature is that if the sample has been oriented in the magnetic field, with the sample tube in a fixed position, a rotation of the sample destroys the orientation. This effect disappears though if the procedure is repeated without removing the sample from the magnetic field (11). The most obvious, but not necessarily the only, explanation to these observations is that the director is oriented perpendicularly to the magnetic field (11). Initially this results in the directors being randomly distributed in the plane perpendicular to the magnetic field. The successive reorientations then lead to an orientation of the director along the rotation axis. If a sample that has been oriented in a fixed position in the magnetic field is rotated in the magnetic field and a NMR spectrum is taken before the new orientation has taken place the spectrum should consist of a broad absorption but not the usual powder pattern.

A calculation using the methods of Cohen and Reif (13) gives that the lineshape $g(\nu)$ is

$$g(\nu) = 2 P(\cos \theta) \frac{d}{d\nu} \cos \theta = \frac{2}{3\nu_Q S \cos \theta \sqrt{\sin^2 \alpha - \cos^2 \theta}} \quad (8)$$

for a sample that has been rotated an angle α after a single orientation in the magnetic field. Using the relation

$$\nu = \pm \frac{\nu_Q S}{2} (3 \cos^2 \theta - 1)$$

Equation (8) gives, using arbitrary intensity units and the abbreviation $x = \nu/\nu_Q$,

$$g_+(x) = (\frac{1+x}{3})^{-1/2} (\sin^2 \alpha - \frac{1+x}{3})^{-1/2} \quad ; -1 \leq x \leq 3\sin^2 \alpha - 1$$

$$g_-(x) = (\frac{1-x}{3})^{-1/2} (\sin^2 \alpha - \frac{1-x}{3})^{-1/2} \quad ; 1-3\sin^2 \alpha \leq x \leq 1$$

$$g(x) = g_+(x) + g_-(x) \quad (9)$$

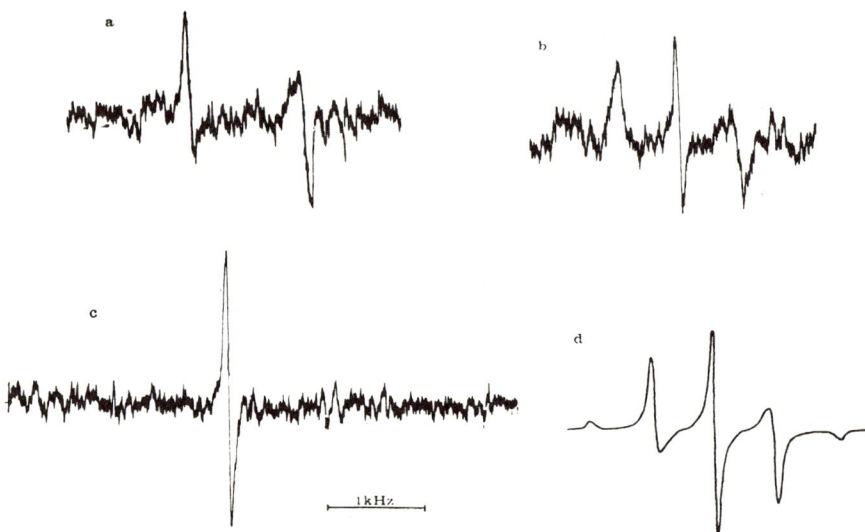

Figure 3. Deuteron NMR spectra illustrating the effect of double quantum transitions. Spectra were obtained for a lamellar mesophase sample with the molar ratio D_2O:sodium octanoate:octanoic acid being 77.2:12.6:10.3. The RF field amplitudes were a: 4.8 Hz, b: 39 Hz, and c: 159 Hz. d is a computer-simulated spectrum, with parameters corresponding to those of spectrum b, taking double quantum transitions into account. A detailed description of the calculations is given in Ref. 9 (9).

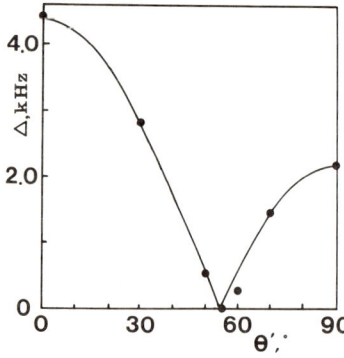

Figure 4. Deuteron quadrupole splittings for a lamellar mesophase sample aligned between glass plates. The sample composition was in mole fractions 0.770 D_2O, 0.147 sodium octanoate and 0.083 octanoic acid. The quadrupole splitting is given as a function of the angle, θ', between the normal to the glass plates and the magnetic field direction. Dots show observed quadrupole splittings whereas the solid curve gives the angular dependence expected for the case where $\theta' = \theta_{LD}$ (cf. text).

The spectrum, which is given in Figure 5 has $\alpha=90°$ and thus corresponds to a situation where the directors are randomly oriented in a plane containing the magnetic field direction and the rotation axis. (A normal powder pattern is also shown in Figure 5 for comparison.) An experimental confirmation of this predicted spectrum would give a strong support for the proposed model of orientation.

The spectral shape given in Equation (9) should also apply for the case where macroscopic alignment is achieved by glass plates but where the directors are oriented in parallel to the plates.

Amphiphile Orientation

For unoriented lamellar mesophase samples composed of alkali n-octanoate, decanol and heavy water we have recorded the deuteron NMR spectrum as a function of sample composition and temperature. For temperatures around 20°C the observed spectra consist of two quadrupole-split powder patterns, one for the decanol-OD group and one for the water deuterons. It is according to Equation (6) possible to calculate the absolute value of the order parameter from an assumed value of the quadrupole coupling constant (e^2qQ/h). Since the water molecules are distributed over several different sites in the system a knowledge of the effective S is of limited value without a further analysis (cf. below). The decanol molecules, on the other hand, most probably reside with the polar heads at the interface between the hydrocarbon and the water lamellae and a qualitative interpretation of S is directly possible.

For lamellar mesophase samples composed of alkali n-octanoate, decanol and heavy water we have determined the decanol-OD quadrupole splitting as a function of sample composition at 27°C. The experimental results are given in Figure 6 at two constant molar fractions of D_2O as a function of R_{ionic}, the molar ratio between ionic and total amphiphile. Since variations in the quadrupole coupling constant can be assumed to be small the data give information on the variation of the order parameter with sample composition. It is seen that the degree of decanol orientation is roughly independent of the counterion present but increases with decreasing water content and increasing R_{ionic}.

The observed decanol quadrupole splittings are within the range 20 - 30 kHz. The quadrupole coupling constant for the hydroxyl deuteron in methanol in the gas phase is $e^2qQ/h=303$ kHz (14), which is very close to the value for water. Hydrogen bonds which form in the liquid phase tend to reduce the quadrupole coupling constant and we estimate $e^2qQ/h=220$ kHz in the liquid. From the data we then obtain $|S|=0.12-0.17$. This is a factor of two smaller than the values obtained by Charvolin et al. (15) for the main part of the deuterons in the aliphatic chain of potassium n-dodecanoate in the lamellar phase. This indicates

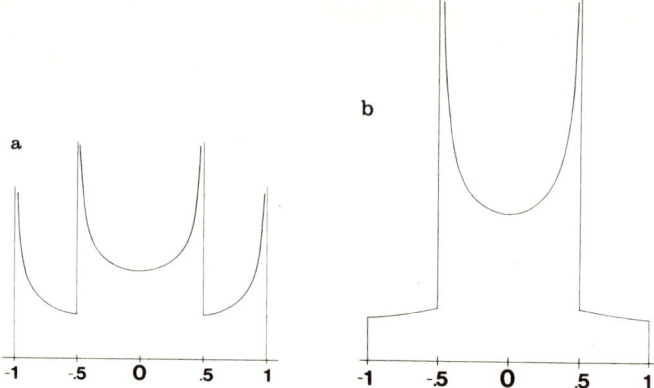

Figure 5. Calculated NMR absorption spectra for a liquid crystal where a) the director is randomly oriented in a plane containing the magnetic field direction and b) the director is completely randomly oriented. The frequency scale is in units of $\nu/(\nu_Q S)$.

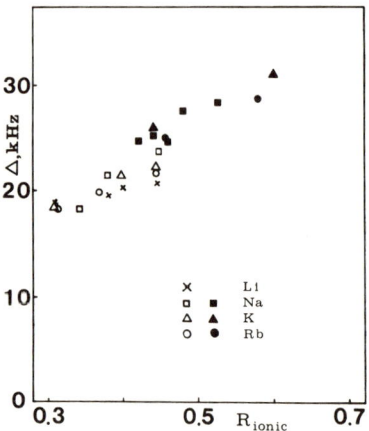

Figure 6. Observed deuteron quadrupole splittings at 27°C of decanol–OD deuterons for lamellar mesophase samples composed of D_2O, alkali octanoate, and decanol. The splittings are given as a function of R_{ionic}, the molar ratio of octanoate to octanoate plus decanol, at two mole fractions of D_2O. In the case of lithium $X_{D_2O} = 0.83$ and for the other alkali ions $X_{D_2O} = 0.77$ (filled symbols) or $X_{D_2O} = 0.83$ (open symbols).

that the polar head of decanol has a greater freedom of motion than the part of the hydrocarbon chain that is closest to the surface. One simple model to rationalize this finding is to assume that the $C_\alpha - C_\beta$ bond (C_α is the carbon adjacent to the hydroxyl group) is fixed in parallel to the director. With a free isotropic rotation about the $C_\beta - C_\alpha$ and $C_\alpha - O$ bonds one obtains S values in accordance with the experimental results for the decanol deuterons.

The increase in S with increasing R_{ionic} can be interpreted as resulting from a stronger interaction between $-COO^-$ and $-OH$ than between two $-OH$ groups, possibly through hydrogen bonding which perhaps leads to complex formation in the interface as has been proposed by Fontell et al. (16). Another alternative is that a higher concentration of charged species at the hydrophilic-hydrophobic interface tends to stabilize the interface.

Our observation that the degree of amphiphile orientation is, within our experimental error, independent of the alkali ion presents support for previous proposals (see e.g. Ref. 17) that the counterions are hydrated over the whole range of existence of the lamellar phase. As will be shown in the next section the degree of water orientation is, on the other hand, quite different for different counterions.

Water Orientation

For samples containing no exchangeable deuterons on the amphiphile or when the exchange is slow it is possible to directly obtain the splitting characteristic of the water deuterons from the NMR spectrum. Also for the case of intermediately fast deuteron exchange it is possible from the change in the spectrum on accelerating the exchange by e.g. a pH alteration to separate out the amphiphile and water quadrupole splittings.

The water deuteron quadrupole splittings are usually small. In the lamellar mesophases of the alkali n-octanoate-decanol-D_2O systems they are of the order of 1-2 kHz over a wide composition range. This corresponds to an order parameter of the order of 10^{-2}.

One way of rationalizing the water deuteron splittings is to assume a two-site model for the water, the two sites corresponding to "free" water (denoted f) and water bound to the amphiphilic surfaces (b). At least in the region of a lamellar phase where one has an one-dimensional swelling (18) on water addition one would expect that it is only the amount of free water that changes when the total water content is changed. One has then for the water deuteron splitting

$$\Delta = | P_f(\nu_Q S)_f + P_b(\nu_Q S)_b | = \left| \frac{X_{D_2O} - X_{D_2O,b}}{X_{D_2O}} (\nu_Q S)_f + \frac{X_{D_2O,b}}{X_{D_2O}} (\nu_Q S)_b \right| = \left| (\nu_Q S)_f + \frac{nX_{am}}{X_{D_2O}} ((\nu_Q S)_b - (\nu_Q S)_f) \right| \quad (10)$$

Here X_{D_2O} and X_{am} are the mole fractions of D_2O and total amphiphile, respectively. $X_{D_2O,b}$ is the mole fraction of bound water and n the average hydration number of the amphiphile.

If the model considered is valid we expect according to Equation (10) to obtain straight lines if Δ is plotted versus X_{am}/X_{D_2O}. It can be seen from Figure 7 that this occurs for several systems. In these cases the intercept is zero within the experimental error indicating that the deuteron splitting for the free water molecules is zero. Accordingly, no long-range ordering effect in the water layers with respect to the amphiphilic layers is indicated in our data. The non-linear plot obtained for the sodium octyl sulphate-decanol-D_2O system indicates that there is a change in the polar group hydration over the concentration interval investigated.

Table I. Values of n · Δ_b (cf. text) Obtained from Plots as in Figure 7 for Different Lamellar Mesophases.

System	n · Δ_b, kHz
Aerosol OT - D_2O	26.6
Lithium octanoate - decanol - D_2O	9.8
Sodium octanoate - decanol - D_2O	8.2
Potassium octanoate - decanol - D_2O	7.4
Rubidium octanoate - decanol - D_2O	7.9
Sodium octanoate - octanoic acid - D_2O	2.5

In Table I we give the slopes of plots according to Equation (10) for those cases where straight lines are obtained. These slopes constitute the products of n and Δ_b which can not be separated unless additional information on either quantity is available. Even if we assume n = 1 we obtain an order parameter of D_2O which is smaller than that obtained for the amphiphilic deuterons. Therefore, it can be concluded that the degree of ordering of the bound water molecules is considerably smaller than that for the amphiphilic polar head.

In the systems alkali octanoate-decanol-D_2O, spectra were obtained at the same temperature in both the slow and fast exchange limits. It was found that on fast exchange, which was achieved by addition of a small amount of alkali hydroxide, the splitting is given by

$$\Delta = |P_{D_2O} (\nu_Q S)_{D_2O} + P_{C_{10}OH} (\nu_Q S)_{C_{10}OH}|$$

where P_{D_2O} and $P_{C_{10}OH}$ are the fractions of water and decanol deuterons, respectively, and the $(\nu_Q S)$'s the splittings observed in the slow exchange limit. On the strength of Equation (7) we may conclude that the order parameter has the same sign for the water and the decanol deuterons. An analogous conclusion may be drawn from the data published for the octylammonium chloride-D_2O system (19).

It can be seen from Table I that the degree of water orientation may be quite different for different counterions. This is more clearly shown in Figure 8 where the observed deuteron quadrupole splitting is given as a function of R_{ionic} at a constant mole fraction of D_2O for lamellar mesophase samples composed of alkali octanoate, decanol and D_2O. Since decanol only gives a small contribution to the observed splitting and since this contribution is independent of counterion (see above) the differences between counterions in Figure 8 can be attributed to differences in water orientation. It can be seen that the degree of water orientation is in the sequence $Li^+ > Rb^+ \simeq K^+ > Na^+$. At present we are unable to provide a definite explanation for the irregular variation of the degree of water orientation with the ionic radius.

From studies on various types of soap systems (20-24) it can be concluded that a considerable fraction of the counterions are more or less firmly attached to the amphiphilic lamellae and that the bound counterions are hydrated (see also above). If the water of hydration of the bound counterions is extensively oriented with respect to the lamellae we expect the degree of water orientation to follow the hydration numbers, i.e. $Li^+ > Na^+ > K^+ > Rb^+$. Were, on the other hand, either the whole hydration complex or the water molecules in the hydration sphere rotating isotropically we would expect the reverse sequence for the degree of water orientation. Possibly a combination of these two effects may be the explanation of our observations. Thus only the strongly hydrated lithium ion may be oriented with its water of hydration.

It has recently been proposed that hydrogen bonding between the water of counterion hydration and the surfactant end-group may be an important factor for the counterion binding in surfactant systems (25). This is supported by NMR and diffusion studies (21, 26, 27) which show that the mode of counterion binding may be quite different for different soap end-groups. Of the alkali ions the hydrogen bonding should be strongest between the water of counterion hydration and the COO^--groups in the case of lithium and this may explain the high degree of water orientation observed in the lithium octanoate-decanol-D_2O system. However, this can not be the only effect since we have observed also in

266

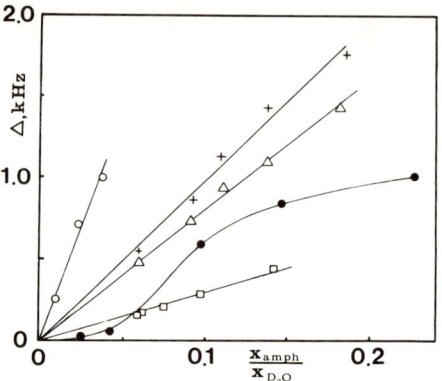

Figure 7. Observed water deuteron quadrupole splittings at 20°C for different lamellar amphiphile–D_2O mesophases. The splittings are given as a function of the molar ratio of total amphiphile to D_2O. The systems studied are Aerosol OT–D_2O (○), lithium octanoate–decanol–D_2O (+), sodium octanoate–decanol–D_2O (△), sodium octyl sulphate–decanol–D_2O (●), and sodium octanoate–octanoic acid–D_2O (□).

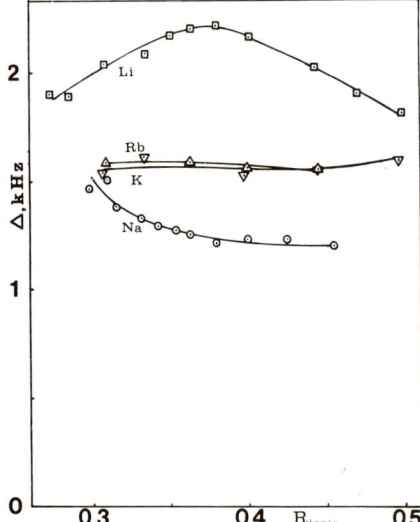

Figure 8. Observed deuteron quadrupole splittings at 20°C for lamellar mesophase samples composed of alkali octanoate, decanol, and D_2O. The splittings are given as a function of R_{ionic}, the molar ratio of ionic amphiphile to total amphiphile, at a constant mole fraction of water being 0.83.

the systems alkali octyl sulphate-decanol-D_2O that the water deuteron splittings are considerably greater with lithium than with sodium as counterion. Further studies on different systems are in progress with the hope of elucidating the mode of water orientation in amphiphilic liquid crystals.

A further possibility to investigate the orientation of water in a mesophase is to compare the deuteron (Δ_D) and ^{17}O (Δ_O) quadrupole splittings of the same sample. The electrical field gradients at the deuterons and at the oxygen have different relative orientations and the splittings can be used to determine the two independent orientation parameters for the water molecule. One has

$$\Delta_O = |\nu_Q^O \{ \tfrac{1}{2}(3\cos^2\theta_{DM} - 1) + \eta_O \sin^2\theta_{DM} \cos 2\phi_{DM} \}|$$

$$\Delta_D = |\nu_Q^D \{ \tfrac{1}{2}(3\cos^2\theta_{DM} - 1)(3\cos^2\theta - 1) + \sin^2\theta_{DM} \cos 2\phi_{DM} \sin^2\theta \}| \qquad (11)$$

Here it has been assumed that the asymmetry for the deuteron quadrupole coupling is negligible. The angle θ is the angle between the principal axes of the ^{17}O and ^2H electrical field gradient tensors and the angles θ_{DM} and ϕ_{DM} specify the transformation from the coordinate system defined by the molecular symmetry and the director coordinate system.

Experimental

Heavy water was purchased from Norsk Hydro, Norway or CIBA-Geigy, Switzerland, and had an isotopic enrichment of at least 99.7%. Sodium octanoate, decanol and octanoic acid were obtained from the British Drug Houses (BDH), England, sodium octyl sulphate from Schuchardt, Germany and Aerosol OT (sodium di-(2-ethylhexyl)-sulfosuccinate) from FLUKA, Switzerland. All these chemicals were of at least 97% purity. The other alkali octanoates were prepared as described in Ref. 5. Lithium octyl sulphate was prepared by adding chlorosulphonic acid to octanol in ether solution followed by neutralization with LiOH in an ice-ethanol mixture. The product was dried in vacuo. Samples were prepared by weighing the appropriate chemicals to total weight of ca 1 g into 10 mm tubes which were sealed off immediately. Homogenisation was achieved by heating the samples above the melting point of the liquid crystal. The alkali octyl sulphates hydrolyze at elevated temperatures (80° - 90°C) so these samples were shaken vigorously at lower temperatures with a 5 mm glass bead in the tube to facilitate mixing.

NMR-spectra were obtained by means of a Varian V-4200 wide-line spectrometer equipped with a V-3603 12 inch magnet. The magnetic field and the radio frequency were 1.403 T and 9.1786 MHz, respectively and were controlled by a Varian Mark II Fieldial unit and a crystal oscillator circuit. A Varian V-4540 Temperature Controller kept the sample temperature at a constant value which was 20°C where not otherwise stated. The amplitudes of the RF-field and field modulation were set sufficiently low not to cause any gross distortion of the obtained derivative spectra.

Because of signal intensity the decanolic deuteron splittings had to be measured for 4 g samples, which did not fit into the temperature controller. The sample temperature then was 27 ± 2°C. In this case the amplitudes of the RF and modulation fields had to be rather high. This did not seem to alter the values of the splittings significantly.

Summary

The general theory for the description of static quadrupolar effects in deuteron magnetic resonance of lyotropic liquid crystalline systems is presented. Especially the influence on the NMR spectrum of orientation effects on both the molecular and the macroscopic levels is considered. Furthermore, effects on spectral shape of chemical exchange phenomena and double quantum transitions are briefly described. Experimentally determined amphiphile and water deuteron quadrupole splittings for soap-water mesophases are discussed in terms of the theoretical results. For lamellar mesophase samples composed of alkali soap, decanol and heavy water it is possible to obtain both the decanol -OD and the water deuteron quadrupole splittings. The degree of decanol orientation is considerably greater than that of water and, furthermore, the decanol -OD deuteron splitting is only moderately affected by composition changes. The water deuteron splitting on the other hand changes considerably with counterion and with sample composition. The water deuteron splittings are compared with a simple model which assumes two types of water namely free and bound water molecules. It is shown that if the splittings of each site are independent of water content the degree of orientation of free and bound water molecules can be separately obtained. It is found that no long-range ordering of the water molecules with respect to the amphiphilic lamellae seems to exist.

Literature Cited

1. Ekwall, P., Advan.Liquid Cryst., in press.
2. Johansson, Å. and Lindman, B., in "Liquid Crystals and Plastic Crystals" G.W. Gray and P.A. Winsor, Ed., Vol. 2, p. 192, Ellis Horwood Publishers, Chichester, U.K., 1974.

3. Wennerström, H., Lindblom, G. and Lindman, B., Chem.Scr. (1974) 6, 97.
4. Lindblom, G., Acta Chem.Scand. (1972) 26, 1745.
5. Persson, N.-O., Wennerström, H. and Lindman, B., Acta Chem. Scand. (1973) 27, 1667.
6. Persson, N.-O. and Johansson, Å., Acta Chem.Scand. (1971) 25, 2118.
7. Lawson, K.D. and Flautt, T.J., J.Phys.Chem. (1968) 72, 2066.
8. Charvolin, J. and Rigny, P., J.Phys. (Paris) Colloq. (1969) 30, C4-76.
9. Wennerström, H., Persson, N.-O. and Lindman, B., J.Magn. Resonance (1974) 13, 348.
10. de Vries, J.J. and Berendsen, H.J.C., Nature (London) (1969) 221, 1139.
11. Black, P.J., Lawson, K.D. and Flautt, T.J., Mol.Cryst. Liquid Cryst. (1969) 7, 201.
12. Long, Jr., R.C. and Goldstein, J.H., Mol.Cryst. Liquid Cryst. (1973) 23, 137.
13. Cohen, M.H. and Reif, F., Solid State Phys. (1957) 5, 321.
14. Casleton, K.H. and Kukolich, S.G., Chem.Phys.Lett. (1973) 22, 331.
15. Charvolin, J., Manneville, P. and Deloche, B., Chem.Phys. Lett. (1973) 23, 345.
16. Fontell, K., Mandell, L., Lehtinen, H. and Ekwall, P., Acta Polytech.Scand., Chem.Incl.Met.Ser. (1968) 74 (III) 1.
17. Lindman, B. and Ekwall, P., Mol.Cryst. (1968) 5, 79.
18. Fontell, K. in "Liquid Crystals and Plastic Crystals" G.W. Gray and P.A. Winsor, Ed., Vol. 2, p. 80, Ellis Horwood Publishers, Chichester, U.K., 1974.
19. Johansson, Å. and Drakenberg, T., Mol.Cryst. Liquid Cryst. (1971) 14, 23.
20. Lindman, B. and Brun, B., J. Colloid Interface Sci. (1973) 42, 388.
21. Gustavsson, H., Lindblom, G., Lindman, B., Persson, N.-O. and Wennerström, H. in "Liquid Crystals and Ordered Fluids" J.F. Johnson and R.S. Porter, Ed., Vol. II, p. 161, Plenum Press, New York, 1974.
22. Gustavsson, H. and Lindman, B., J.Chem.Soc., Chem. Commun. 1973, 93.
23. Lindblom, G. and Lindman, B., Mol.Cryst. Liquid Cryst. (1973) 22, 45.
24. Lindman, B. and Danielsson, I., J. Colloid Interface Sci. (1972) 39, 349.
25. Mukerjee, P., (1972), personal communication.
26. Lindman, B., Kamenka, N. and Brun, B., C.R.Acad.Sci., Ser. C (1974) 278, 393.
27. Gustavsson, H., and Lindman, B., to be published.

19

Influence of Electrolyte on Phase Equilibria and Phase Structure in the Binary System of Di-2-Ethylhexyl Sulphosuccinate and Water

KRISTER FONTELL

Fysikalisk kemi 2, Lunds Tekniska Högskola, Lund, Sweden

Introduction

    Addition of electrolytes is known to influence the phase equilibria and phase structures of systems containing amphiphiles and water. From the 1920's until the 1940's McBain and associates, including the Volds, studied extensively the total composition range of systems of common soaps, electrolytes and water at various temperatures (1-11). They showed that there are large similarities between different systems in that the observed differences were not in kind but in degree. The phases observed in ternary systems derive from phases existing in the binary soap - water systems. One additional phase, kettle soap, was believed to occur in the ternary systems but its existence has later been questioned by Vincent and Skoulios (12). However, except for the time of McBain there have been remarkable few publications describing phase diagrams covering the complete concentration range. Most studies have been directed towards understanding changes in the properties caused by the electrolyte within the composition region of a single phase, especially premicellar or micellar solution phases at low amphiphile contents.
    Systems containing the amphiphile sodium di-2-ethylhexyl sulphosuccinate (Aerosol OT [x]) have been previously studied both from technical and theoretical aspects (13-19). The molecule has a forklike conformation with two branched hydrocarbon chains held together by a rather large polar sulphosuccinate group. The molecular structure is somewhat similar to that of a short chain lecithin.
    The addition of sodium chloride to aqueous systems of Aerosol OT has recently been reported to cause "coacervate formation" (20). The term "coacervation" has been used to describe the process of separation into two liquid layers caused by addition of halide ions to an aqueous lipide system. The term was ori-

---

[x]The isomer sodium di-octylsulphosuccinate also goes occasionally under the same trade name.

ginally introduced by Bungenberg de Jong for cases where it was not certain that a true phase separation occurred (21).

In the above mentioned investigation the sodium chloride was added to the binary Aerosol OT - water mixtures without distinguishing the phase behaviour of the parent system and it was ignored in the discussion. In addition, some of the findings were difficult to reconcile with experiences gained from our previous studies of the binary system of Aerosol OT and water (16, 22). It was therefore deemed of interest to study the phase diagram of the ternary system Aerosol OT - sodium chloride - water over its whole composition range.

Experimental

The study was performed at $20^{\circ}C$. The experimental methodology was the same as in our previous studies of phase equilibria and phase structures in amphiphile systems (23) and was essentially the same as used by McBain and associates (6-8). Aerosol OT was obtained from Fluka AG, Switzerland, and used as such after drying in vacuo at $78^{\circ}C$ over phosphorous pentoxide ; sodium chloride was of Baker "Analar Grade" and the water was twice-distilled "conductivity water". The components of the individual specimens were weighed into ampoules which were subsequently sealed. The samples were shaken at room temperature and after a homogeneous system been obtained were stored at $20^{\circ}C$ before inspection for any phase separation. Separation into two or three phases was sometimes instantaneous but in other instances was observed after prolonged storage. The separation process could in principle be speeded up by using centrifuging (23) but previous studies of the phase equilibria in binary Aerosol OT - water systems have shown that the separation of the phases in water-rich specimens may be influenced by the gravitational forces of the centrifugal field (22). The separated phases were analysed for the sodium chloride content by argentometric titration; for the Aerosol OT content by the titrimetric method of Epton (24); and for the water content by the Karl Fisher method.

Results and Discussion

Several studies of the phase equilibria of the binary system Aerosol OT - water have been published (16, 25). The Aerosol OT dissolves at $20^{\circ}C$ to a very slight extent in water, ca. 1.3 %. The anhydrous compound is liquid crystalline. It is birefringent and exhibits in the polarizing microscope the texture of a middle phase. The two-dimensional hexagonal structure of a middle phase has been confirmed by X-ray diffraction (16, 17). The building units of the structure are thus rod-like units of the reversed nature, that is the polar groups and the sodium ions form rods which are hexagonally arranged in a non-polar hydrocarbon matrix.

This hexagonal structure may take up about 18 % of water, the water entering into the polar cores of the rods (16). On further addition the structure breaks down and after passing a two-phase zone the region for a viscous isotropic phase is entered. The X-ray diffraction findings for this phase are reminiscent of those Luzzati and coworkers have obtained for a viscous isotropic phase exhibited by anhydrous strontium myristate at 232°C and there is thus a strong probability that the internal structure is the same (26-28). Luzzati has been able to determine for his phase its space group, Ia3d, and has put forward a proposal for its structure. The structure units would be short rodaggregates containing the polar groups of the amphiphile, the sodium ions and the water molecules of the specimen; the rods are joined in threes at each end and form two independent mutually interpenetrating networks while the hydrocarbon parts of the amphiphile form a surrounding non-polar matrix. It is the interplay of these two net-works which gives the system its body-centered cubic structure.

The region of existence of the viscous isotropic phase encompasses only about 3 % of water. On increasing the water content one enters, after passing a two-phase zone, the region for the lamellar liquid crystalline phase. The water content of this phase varies between wide limits, from about 30 % up to about 90 % of water. Throughout the phase there is an ideal swelling, that is the amphiphilic lamellae are unaltered when the water content is increased while the thickness of the intercalated water layers increases from 2 nm to above 10 nm (20-100 Å) (22).

The overall phase diagram of the ternary system Aerosol OT - sodium chloride - water at 20°C is sketched in Figure 1. The part of the diagram with high contents of amphiphile and/or halide was not studied in detail. As it is difficult to visualize the phase equilibria involved in the Aerosol OT corner Figure 2 attempts to give a clearer impression but this is obtained on the expense of the relation between the units on the three concentration axis.

The reversed middle phase, F, has a very low capacity for taking up sodium chloride. The region of existence of the viscous isotropic phase, I, extends with sodium chloride contents of 1-2 % as a narrow wedge towards higher contents of water, almost 40 %. The lamellar phase, D, can take up about 1 % of sodium chloride. On exceeding that amount one enters after passing a two-phase zone the region for an isotropic liquid phase, L_2. The region of existence of this phase is a narrow band running almost parallel with the base of the triangular phase diagram. The width of this region is only about 0.1 % of sodium chloride, while the contents of the two other components may vary widely, 40-96 % of water, and 3-35 % of Aerosol OT. On further increase of sodium chloride, there occurs separation between the liquid phase L_2 and an aqueous sodium chloride solution L_1' containing minimal amounts of Aerosol OT. On still further increasing the sodium chloride

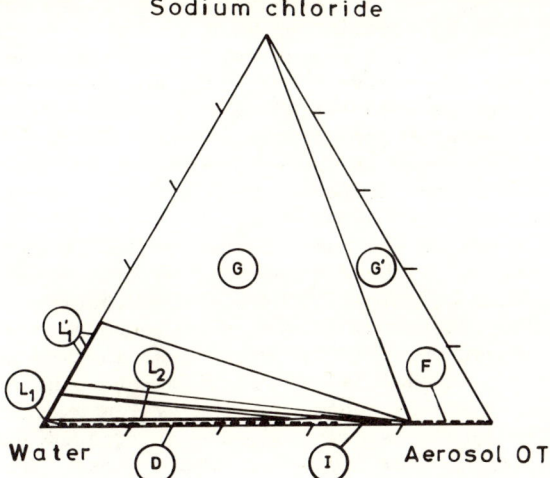

Figure 1. Tentative overall phase diagram of the ternary system di-2-ethylhexylsulphosuccinate–sodium chloride–water at 20°C. The concentrations are in weight %. Some two- and three-phase zones are not marked for readability. Key: Aerosol OT = sodium di-2-ethylhexylsulphosuccinate; L_1, L_1', L_2 = isotropic liquid solution phases; D = lamellar liquid crystalline phase; I = viscous isotropic cubic phase; F = reversed hexagonal liquid crystalline phase; G = three-phase zone $L_1 + F + NaCl(s)$; G' = two-phase zone $F + NaCl(s)$.

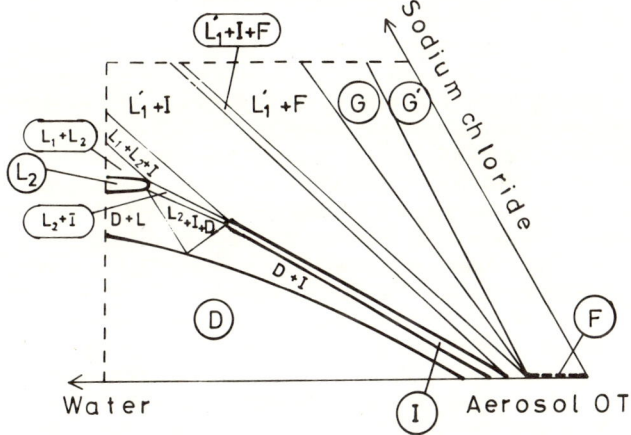

Figure 2. An attempt to visualize the principles for the phase equilibria in the Aerosol OT corner of the phase diagram of Figure 1. Units on the axis are distorted for readability.

concentration other two- and three-phase zones are entered. The saturated brine is in equilibrium with sodium chloride crystals and the reversed hexagonal phase. In the intermediate region the viscous isotropic and reversed hexagonal phases do not separate in pure state, but as dispersions of rather solid consistency. The gross appearance of the samples is very similar but the phases may be distinguished in the polarizing microscope, by X-ray diffraction, and by the different densities of the phases with respect to the separated brine.

The conditions in the water corner of the ternary diagram are shown in Figure 3. Above an Aerosol OT content of about 0.02 % the aqueous solutions L_1 have a low capacity for dissolving sodium chloride, it amounts at most to about 0.02 %. On exceeding that amount one obtains turbid systems that are extremely difficult, if not impossible, to separate into single phases. But when the sodium chloride content of the system exceeds about 1 % there is an instantaneous division into an aqueous sodium chloride solution and another isotropic liquid phase, the previously mentioned L_2 phase.

The exact location of the apexes of the intermediate three-phase triangle $L_1 - D - L_2$ (Figure 3) are difficult to determine because the spread of the analytical results for the sodium chloride content is larger than the width of the zone and furthermore there is an additional error in the determination of the Aerosol OT content of the D phase apex.

The viscosity of the lamellar liquid crystalline phase decreases on addition of the halide. The region of existence of the phase extends to higher contents of water on addition of sodium chloride, up to 95 % of water. The basic X-ray repeat distance remains unaltered for a sample containing the maximum amount of sodium chloride compared to one with the same Aerosol OT content and no salt. (However, specimens from the region containing high amounts of water have dimensions 15 nm (150 Å), which were outside the capability of the present X-ray diffraction set-up). These findings may be compared with those of Balmbra and coworkers (29) for the ternary system of decyltrimethylammonium decylsulphate, sodium bromide and water. They have similarly observed that addition of the halide extends the region of existence of the lamellar liquid crystalline phase towards large amounts of water. However, in their system the addition of the halide caused a decrease in the basic repeat distance.

The molecular composition at the water-rich end of the isotropic liquid phase L_2 is 1 mole of Aerosol OT: 2.8 moles of sodium chloride: 600 moles of water while the composition at the water-poor end is about 1:0.275:24. The phase is in equilibrium with aqueous sodium chloride solutions of concentrations between 0.17 and 1.4 molar (1-8 %) and with the lamellar liquid crystalline phase D, whose Aerosol OT content ranges from about 4 to 70 % while the sodium chloride content is at most 1 %. At the water-poor end the L_2 phase is in equilibrium with the optical

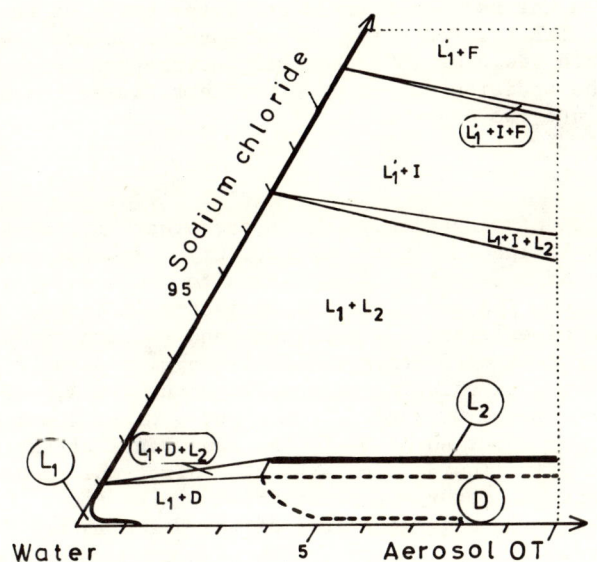

Figure 3. The water corner of the phase diagram in Figure 1

isotropic viscous phase I. The macroscopic appearance of the phase L_2 is a somewhat oily liquid with viscosity ranging from about 4 to 100 cP. The phase also exhibits thixotropic behaviour. At high contents of water the phase has a bluish tint, which suggests that the internal building units in this part of its region of existence are rather large.

Due to the similarity in the molecular structure of Aerosol OT with that of a short-chain lecithin, one possible phase structure would be of vesicular type. That is, the building units would be "microdroplets" consisting of a bilayer of amphiphile surrounding a polar core of aqueous sodium chloride solution and these microdroplets in turn being dispersed in an aqueous medium. If the suggestion is correct it should be possible to obtain information about differences in the packing of the hydrocarbon chains in the inner and outer parts of the vesicular bilayer by NMR studies using shift reagents as been demonstrated for lecithin vesicles (30). However, attempts have so far failed because the addition of the shift reagent caused the sample to separate into two phases.

Summary

Addition of sodium chloride to the binary system sodium di-2-ethylhexylsulphosuccinate (Aerosol OT) and water causes the occurrence of a new phase in addiiton to those already existing in the system. The new phase has an isotropic liquid structure. Its region of existence is a narrow band running parallel with the water - Aerosol OT axis of the triangular phase diagram. Its content of sodium chloride amounts to about 1.6 %, while the content of Aerosol OT ranges from about 3 to 58 % and the content of water from about 40 to 95 %. The appearance of the phase is a somewhat oily liquid which exhibits thixotropic behaviour. At high contents of water the phase has a bluish tint. The structure of the colloidal aggregates of the phase is still not established.

Literature Cited

1. McBain, J.W. and Langdon, G.M., J.Chem.Soc. (1925) 127 852
2. McBain, J.W. and Pitter, A.V., J.Chem.Soc. (1926) 129 893
3. McBain, J.W. and Field, M.C., J.Phys.Chem. (1926) 30 1543
4. McBain, J.W., International Critical Tables (1929) 5 446
5. McBain, J.W., Lazarus, L.M. and Pitter, A.V., Z.Phys.Chem. A (1930) 147 87
6. McBain, J.W., Brock, G.C., Vold, R.D. and Vold, M.J., J.Amer.Chem.Soc. (1938) 60 1870

7. McBain, J.W., Vold, R.D. and Vold, M.J., J.Amer.Chem.Soc. (1938) __60__ 1866
8. McBain, J.W., Gardner, K. and Vold, R.D., Ind.Eng.Chem. (1936) __36__ 808
9. Vold, R.D. and Ferguson, R.H., J.Amer.Chem.Soc. (1938) __60__ 2066
10. McBain, J.W. and Lee, W.W., Ind.Eng.Chem. (1943) __35__ 917
11. McBain, J.W., Thorburn, R.C. and McGee, C.G., Oil and Soap (1943) __21__ 227
12. Vincent, J.M. and Skoulios, A., J.Amer. Oil Chem.Soc. (1936) __40__ 20
13. Peri, J.B., J. Colloid Interface Sci., (1969) __29__ 6
14. Kitahara, A., Watanabe, K., Ko-No, K. and Ishikawa, T., J. Colloid Interface Sci., (1969) __29__ 48
15. Frank, S.G. and Zografi, G., J. Colloid Interface Sci., (1969) __29__ 27
16. Ekwall, P., Mandell, L. and Fontell, K., J. Colloid Interface Sci., (1970) __33__ 215
17. Balmbra, R.R., Clunie, J.S. and Goodman, J.F., Proc.Roy.Soc. (London) Ser. A. (1965) __285__ 534
18. Gilchrist, C.A., Rogers, J., Steel, G., Vaal, E.G. and Winsor, P.A., J. Colloid Interface Sci., (1967) __32__ 415
19. Park, D., Rogers, J., Toft, R.W. and Winsor, P.A., J. Colloid Interface Sci., (1970) __32__ 81
20. Acharya, R., Ecanow, A. and Balagot, R., J. Colloid Interface Sci., (1972) __40__ 125
21. Bungenberg de Jong, H.C., "La Coarcervation, les Coarcervats et leur Importance en Biologie", Actualities Scientitiques et Industrielles Vol. 397, Herman et Cie, Paris, 1936
22. Fontell, K., J.Colloid Interface Sci. (1973) __44__ 318
23. Mandell, L., Fontell, K., Lehtinen, H. and Ekwall, P., Acta Polytechn.Scand.Chem.Incl.Met. Series (1968) __74__ I - III
24. Epton, S.R., Trans.Faraday Soc. (1948) __44__ 226
25. Rogers, J. and Winsor, P.A., Nature (London) (1968) __216__ 477
26. Luzzati, V. and Spegt, P.A., Nature (London) (1967) __215__ 701
27. Luzzati, V., Tardieu, A., Gulik-Krzywicki, T., Rivas, E. and Reiss-Husson, F., Nature (London) (1968) __220__ 485
28. Fontell, K., J. Colloid Interface Sci., (1973) __43__ 156
29. Balmbra, R.R., Clunie, J.S., Goodman, J.F. and Ingram, B.T., J. Colloid Interface Sci. (1973) __42__ 226
30. Bystrov, V.F., Dubrovina, N.I., Barsukov, L.I. and Bergelson, L.D., Chem.Phys. Lipids (1971) __6__ 343

20

Dissolution Mechanism of Water-Soluble Polymers (Partially Saponified Polyvinyl Acetates)

HIRONOBU KUNIEDA and KŌZO SHINŌDA

Department of Chemistry, Yokohama National University,
Ooka-2, Minamiku, Yokohama, Japan

Introduction

Regular solutions (including ideal solutions and athermal solutions), electrolyte solutions and regular polymer solutions were already fairly well explored (1, 2, 3). The types of solvents and intermolecular forces involved in these solutions covers wide varieties, namely, from water to hydrocarbon and from London dispersion forces to ion-ion interactions. Nevertheless, solvent and solute mix randomly in all of these solutions and the theories also assume random mixing of components. Such an idealization seems adequate in these solutions, but it is the most simple and naive idealization.

These solutions, however, are rather exceptional solutions among so many solutions that we encounter in nature, in biological systems, in the process of manufacturing or in enviromental problems. Hence, in our opinion, the study on solutions of more sophisticated, delicate and realistic systems is craved for. Most striking feature of these solutions may be the dissolution due to the orientation, arrangement and structure formation of solute molecules, which are otherwise practically insoluble by random mixing. One typical example is a micellar solution of surfactant, in which surfactant molecules orient, arrange and form micelles. The solution which is explored in recent two decades applying solution theory (4). The saturation concentration of molecular (ionic) dispersion is very small, yet the solubility is unexpectedly large. Because, it dissolves forming micelles. Polymer which consists of hydrophilic groups and lypophilic groups may dissolve by orientation, rearrangement and structure formation of molecules. This type of dissolution seems very important, yet not well explored.

It is with this image that we began this study.

Experimental

Materials. Partially saponified polyvinyl acetate (PVA-Ac) were obtained from Nihon Gosei Kagaku Kogyo Co.. Dilute aqueous solutions of PVA-Ac (2-5 wt%) were heated and separated into two phases above the cloud point. Impurities which is more water-soluble, such as salts, were eliminated by repeated decantation of water phase. Then the samples were dried by freeze-drying, so that they will more easily mix with water. PVA-Ac fractionated by adding water to acetone solution did not show appreciable change in the phase diagram.

Extra pure grade KCl, NaCl, $CaCl_2$, Na_2SO_4, KSCN, n-butanol, ethyl acetate, D-sorbitol of Wako Pure Chemicals Co. were used. KCl was further purified by recrystallization three times from water, since it was used as standard solution for isopiestic vapor pressure measurements.

Procedures.

Phase diagram. Various amounts of PVA-Ac, water, (salt or organic compound) were sealed in ampoules. Cloud point were determined by raising and lowering temperature at a rate of 0.5°C per minute. In the case of very dilute solutions, say less than 1 wt%, the ampoules were left at constant temperature from several hours to 3 days till the precipitation occurred, because it was difficult to perceive that the solution became cloudy. Otherwise the apparent cloud point would have decreased with concentration initially.

Vapor-pressure measurements. The vapor pressure of water in aqueous solutions of PVA-Ac was determined by the isopiestic method ([5]). A cylindrical glass dish containing PVA-Ac solution was put on a glass-made supporter placed in a cylindrical, separable glass vessel and reference solution was added at the bottom of glass vessel. A series of such samples was kept in an air thermostat in reduced pressure for two to six months and the equilibrium concentration was determined by weight. KCl solutions were used as reference solutions

The water activities were interpolated from the activity-concentration relation given by Robinson and Stokes applying Debye-Hückel's formula ([2]).

Results and Discussion

Phase Diagram of PVA-Ac–H_2O System. PVA-Ac was prepared by alkali saponification of polyvinyl acetate in aqueous acetone solution. It is considered that PVA-Ac resembles the block copolymers of vinyl acetate vinyl alcohol (6, 7). In order to be soluble in water, saponification degree of PVA-Ac has to be more than 72 % or so. If the saponification degree is optimum, PVA-Ac dissolves in water at lower temperature over all composition but splits into two phases at higher temperature. The phase diagram of PVA-Ac–H_2O as a function of temperature is shown in Figure 1. The curve ABCD represents the mutual solubility of water and PVA-Ac. The shape of the curve as well as the lower critical solution temperature implies that the solution is quite nonregular. It is clear from Figure 1 that the liquid-liquid solubility curve is affected sensitively to the saponification degree, but not to the polymerization degree. Although the solution is so viscous and difficult to dissolve above 20-25 wt% (without the aid of added alcohol), it is soluble in water over all composition below the cloud point (region BC). The solution becomes cloudy and splits into two phases at temperature above the cloud point curve. The one is a polymer phase containing a large amount of water, and the other is the water phase containing practically no polymer. The water activity of this phase is close to unity as shown later. The steep slope of BC curve may be resulted from the distribution of the saponification degree of polymers.

The solubility of polymer in water phase is conventionally shown by

$$\ln a_2 = \ln \Phi_2 + \Phi_1(1 - V_2/V_1) + V_2\Phi_1^2 B'/RT \qquad (1)$$

On the other hand, the solubility of water in polymer phase is conventionally shown by

$$\ln a_1 = \ln \Phi_1 + \Phi_2(1 - V_1/V_2) + V_1\Phi_2^2 B'/RT \qquad (2)$$

where a is the relative activity, V_i the molecular volume of respective component in solution, Φ_i the volume fraction and B' the enthalpy of solution per unit volume (3). It is expected from Equation (1) that the solubility of polymer may be very small, because the molecular volume V_2 and therefore the heat of solution of polymer is very large. Inversely, the molar volume of solvent V_1 is so small that water dissolve infinitely in polymer, provided the heat of solution per mole is smaller than $\frac{1}{2}RT$.

The cloud point curve BC intersects to the solubility curve AB. Similar discontinuity in solubility curve is observed in nonionic surfactant-water system (8, 11). In the case of polymer, curve AB approaches to the water axis much closer than nonionic surfactant solution, because polymer does not dissociate into small molecules. Whereas the saturation concentration of surfactant phase is equal to the saturation concentration of single molecules. The solubility of water in polymer phase is large (99.99 mole% or 76 wt%) already at point C. The dissolution of water into polymer proceeds with temperature depression due to the increase of ice-berg formation and hydration of water surrounding solute molecule, and finally water and polymer mix each other completely below the temperature at B. The schematic diagram of the dissolution process of polymer which consists of hydrophilic and lyophilic groups, in water is shown in Figure 2. It is considered that water dissolves infinitely into polymer phase and pseudophase inversion occurs in polymer phase. If it happens, water becomes continuous phase. Polymer molecules will orient, rearrange so as to decrease the free energy of mixing. This phenomena is akin to the micellar dissolution of surfactant due to the increase of water solubility into nonionic surfactant phase at low temperature (11). If the saponification degree of PVA-Ac is smaller, i.e., less hydrophilic, the curve BC will shift to lower temperature and CD curve will shift to higher concentration, because the hydration of water per unit weight of polymer will decrease. Only swelling of water in polymer may occur below a certain saponification degree (∼70 %), and insoluble in water. Although molecules are aggregated to micelles by physical bond in the case of surfactant, and monomers are connected by covalent bond in polymer, the mechanism of dissolution in water is similar. Important conditions for complete dissolution of these substances are 1) the finite aggregation of these molecules in water and 2) infinite mixing with water.

Relative Activity of Water in PVA-Ac—H_2O System. The relative activity of water in aqueous solution of PVA-Ac (saponification degree 75.6 %, polymerization degree 2200) at 25 °C is plotted in Figure 3. Because of the large molecular weight, polymer solution whose weight per cent ranges from 0 to 50 wt% corresponds to 0 − 0.00015 in mole fraction unit. The dotted line expresses the relative activity of water in ideal dilute solution. The curve represents the relative activity of water determined by isopiestic method. The relative activity of water is practically equal to 1 up to 40

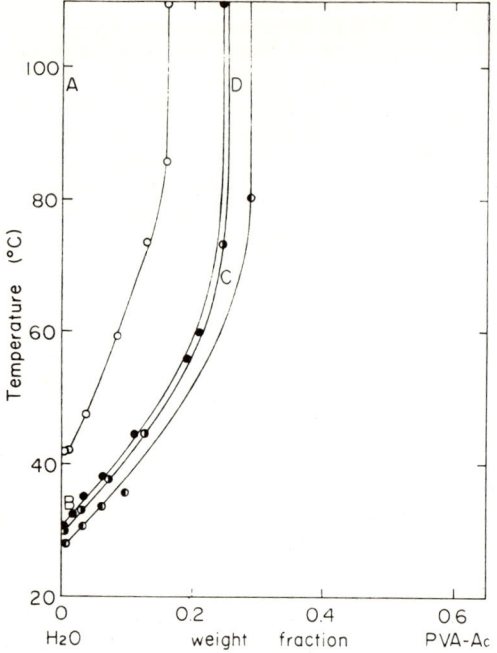

Figure 1. Phase diagram of PVA-Ac–H_2O as functions of temperature, saponification degree (D.S.), and polymerization degree (D.P.). ○: *D.P. 2200, 80% saponified.* ●: *D.P. 2200, 75.6% saponified.* ◐: *D.P. 1200, 75.1% saponified.* ◑: *D.P. 550, 74.1% saponified.*

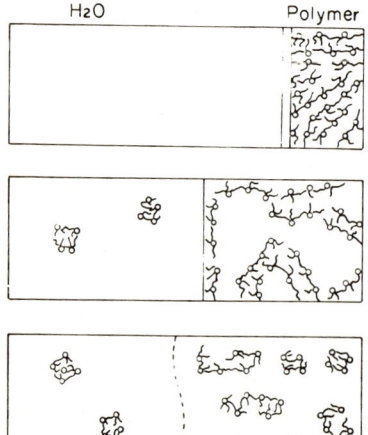

Figure 2. Schematic of dissolution process of polymer in water consisting of hydrophilic and lyophilic groups

wt% solution, i.e., deviate to positive side from that of ideal solution. Applying Gibbs-Duhem equation,

$$X_1\left(\frac{\partial \ln a_1}{\partial X_1}\right) + X_2\left(\frac{\partial \ln a_2}{\partial X_1}\right) = 0 \qquad (3)$$

we know that the activity of polymer also does not change with concentration. The relative activity of water decreases rapidly above 40 wt% solution. It can be concluded from this result that the solution is akin to two phase mixture of pure water and water swollen polymer phase. Hence, we know one phase solution below the cloud point (25 °C) and two phase solution above the cloud point (60 °C) in Figure 1 resemble each other thermodynamically, but not optically. The difference between two states is that the hydrated polymer aggregates infinitely above the cloud point, whereas the aggregation is finite below the cloud point. The mutual solubility of liquids in ordinary solution changes gradually with temperature. On the contrary, polymer (PVA-Ac) is practically insoluble in water above the cloud point, but it mixes with water over all composition slightly below the cloud point. It is concluded that the change from the complete solubility to insolubility with the small change of temperature is characteristic to polymers. Similar phenomenon can be observed also to the minute change in hydrophilic-lypophilic balance of polymer molecules, because the cloud point of PVA-Ac is moved to higher or lower temperature depending on the saponification degree. It is also depressed or raised in the presence of third substances. The effect of added inorganic salts and organic additives on the cloud point of 2 wt% aqueous solution of PVA-Ac (saponification degree 75.6 % and polymerization degree 2200) is shown in Figures 4 and 5. The trends of the change of cloud point in PVA-Ac solution is similar to those in aqueous solution of nonionic surfactant ([12]). It is then concluded that the change from complete solubility to insolubility of polymer with water may occur by the addition of third substances at a constant temperature. Complete insolubility and solubility are important, because biological tissue has to be fairly hydrophilic and yet completely insoluble.

Conditions of the Complete Dissolution of Water-Soluble Polymers. In the process of the dissolution of a water-soluble polymer, many factors come into play.
1. Liquid state of polymer in the presence of solvent.
   Dissolution of polymer with solvent is, in reali-

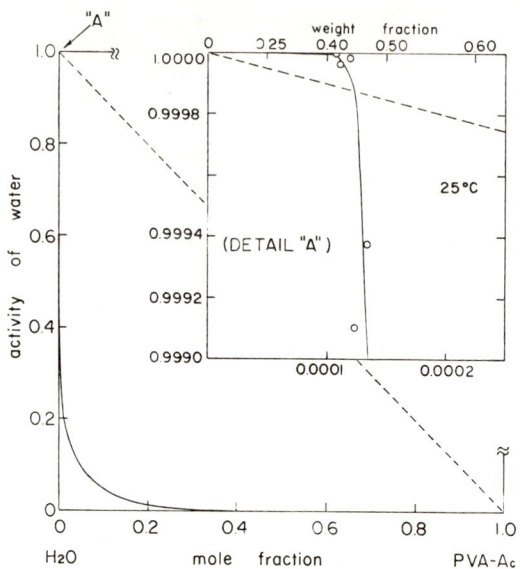

Figure 3. Relative activity of water in aqueous solution of PVA-Ac (saponification degree, 75.6%; polymerization degree, 2200) at 25°C

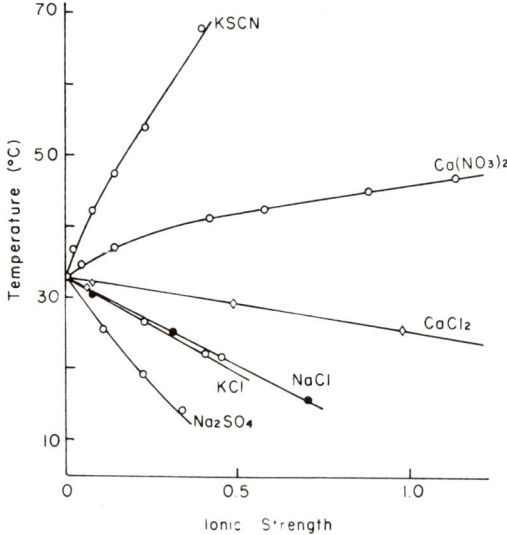

Figure 4. Effect of added inorganic salts on the cloud point of 2 wt % aqueous solution of PVA-Ac (saponification degree, 75.6%; polymerization degree, 2200)

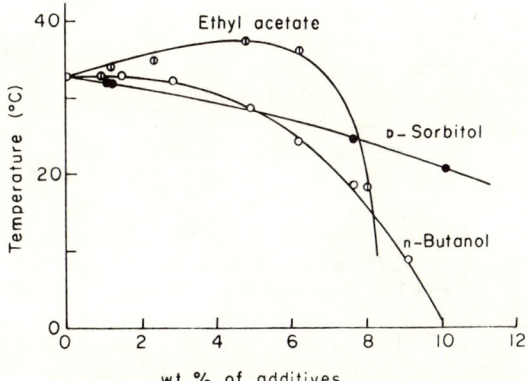

Figure 5. Effect of organic additives on the cloud point of 2 wt % aqueous solution of PVA-Ac (saponification degree, 75.6%; polymerization degree, 2200)

ty the dissolution of solvent into polymer, because the solubility of polymer in solvent is negligibly small. It is necessary that a polymer is in a liquid state in the presence of solvent, so that solvent can mix with polymer.

2. Linear molecule.

If a polymer is three dimensional it will swell solvent, but the swelling does not proceeds infinitely.

3. Flexible molecular structure.

Polymers which possess both hydrophilic and lypophilic groups have to orient and rearrange so as to decrease the energy of mixing with water. In this context block polymer may be preferable. Enthalpy of mixing, V_1B', has to be smaller than $\frac{1}{2}RT$.

4. Hydrophile-Lipophile Balance (HLB) and Pseudo-Phase Inversion.

The most important factor in the process of dissolution of water-soluble polymer seems to be 1) the swelling of solvent in polymer and 2) inversion of the continuous medium from polymer to water. This means that in the first instance the polymer is the continuous medium, but in the next step the water-soluble polymer molecules orient themselves so as to decrease the energy of mixing, hydrophilic groups being oriented outside and hydrophobic groups inside and finally water becomes the continuous medium due to the increase of dissolved water. Once this pseudo-phase inversion occurs in the polymer phase, both phase may easily mix.

Literature Cited

1. Hildebrand, J. H., et al., "Regular Solutions" Prentice-Hall, Englewood Cliffs. N.J. 1962, "Regular and Related Solutions" Van Nostrand Reinhold Co., New York, N.Y. 1970.
2. Robinson, R. A., and Stokes, R. H., "Electrolyte Solutions" Butterworths, London, 1959.
3. Flory, P. J., "Principles of Polymer Chemistry" Cornell Univ. Press, Ithaca, N.Y. 1953.
4. Shinoda, K., et al., "Colloidal Surfactants" Academic Press, New York, 1963.
5. Robinson, R. A., and Sinclair, D. A., J. Am. Chem. Soc., (1934) 56 1830.
6. Nagai, E., Kobunshi Kagaku, (1955) 12 195.
7. Sakurada, I., Kobunshi, (1968) 17 21.
8. Balmbra, P. R., Clunie, J. S., Corkill, J. M., and Goodman, J. F., Trans. Faraday Soc., (1962) 58 1661.
9. Corkill, J. M., Goodman, J. F., and Harrold, S. P., ibid. (1964) 60 202.
10. Carless, J. E., Challis, R. A., and Mulley, B. A.,

J. Colloid Interface Sci., (1964) 19 201.
11. Shinoda, K., ibid. (1970) 34 278.
12. Maclay, W. N., ibid. (1956) 11 272.

21

Thermodynamic Aspects of the Mixing of Mono-Amides and Poly-Peptides with Water

P. ASSARSSON, N. Y. CHEN, and F. R. EIRICH
Department of Chemistry, Polytechnic Institute of New York, Brooklyn, N.Y. 11201

Introduction, Review of Earlier Work, and Experimental

The solvation of amides is a subject of considerable interest partly because of the wide use of the low molecular weight members as solvents for organic and inorganic compounds, and also because the higher members can be considered as models for the physical properties and solution functions of the peptide bond in polypeptides and proteins. We studied the interaction of water with p-(vinyl pyrrolidone) PVP and of alkyl-substituted simple amides several years ago and some of the main results were subsequently published (1). In the following we wish to report on a number of yet unpublished data and, further, on new work on the heats of mixing of three amino acids, glycine (gly), proline (pro), and hydroxyproline (hypro) and some of their polymers and copolymers.

The compounds in question are shown on Figure 1. It is interesting to note that N-acetamide, with an endothermic heat of solution, is only partially soluble in water. All other amides studied are miscible with water in all proportions, three to four carbons in the alkyl groups being the upper limit for complete solubility (2,3). One may conclude that the breaking of inter amide polar bonds in the bulk phase and the forming of hydrogen bonds between amide dipoles and water are the predominant features governing this solubility pattern. The preference of the amide group in N-methyl acetamide to associate preferentially with water rather than with itself has been proven by spectroscopic investigation in the near infrared by following the N-H absorption band in the 1.5 micron region (4). Also likely to be pertinent for the solvation pattern is the fact that the atomic coordinates for the amide bond as investigated by Pauling (5) and confirmed by x-rays, shows the atoms to lie practically planar, and the N-C bond to have a partial double bond character.

Our substituted, and in particular the di-substituted, amides were studied over the whole range of concentration in water by measuring the absolute viscosities of the solutions, their density and volume changes, the heats of mixing and the heat capacities by micro-calorimetry, and the phase diagrams by further following the freezing point depressions. Supplementary infrared and NMR spectroscopic data were also obtained, but will not be discussed in this context.

Figures 2, 3, and 4 show some representative data obtained for the substituted amides. In general, di-substitution leads to higher exothermic heats of mixing than mono-substitution, while increasing length of the carbon alkyl chains causes the heat of mixing to become less negative. We concluded from these data that a definite molecular interaction exists between the peptide dipole and water.

The common feature among the amides and pyrrolidones investigated, see Figures 1-4, is the evidence, from thermodynamic and viscosity data, that two molecules of water are being bound per NCO-group. The logical sites for this association through hydrogen bonding are the unshared electron pairs of the carbonyl oxygen, a postulate which is in line with all spectroscopic studies and NMR data on the reactivity of the amide group (6). Even the fully substituted amides appear to be able, to some degree, to hold a third water molecule in the immediate solvation sphere by less energetic interaction with the unshared electrons of the nitrogen. The fact that one does not find evidence of a third water molecule associated with the N-substituted pyrrolidones is probably caused by the special geometry of the ring which leads also to the greater densities of these compounds. It is not possible, on the basis of known bond angles and bond distances of the water molecule, to say whether the two water molecules which are postulated to be hydrogen-bonded to the carbonyl might join into a ring structure.

Heat Capacities and Thermodynamic Functions

a <u>Mono-Amides</u>. We would like to report now briefly on the heat capacities and partial excess thermodynamic functions of water in the same substituted amides obtained from freezing point depressions ($\overline{\Delta F}_{excess}$) and calorimetry ($\overline{\Delta H}_{excess}$). Two typical diagrams obtained are shown in Figures 5 and 6. It is seen that the free energies change very little up to about 85 mole % of water (15 mole % of the amide), whereas the heats of mixing become increasingly negative over the same range. We have not followed the heats of mixing over the whole range, but qualitative evidence indicates that they remain negative. It is interesting that the entropy changes are also negative, so that some amount of the binding which was evidenced by the phase diagrams, or at least ordering, of water, is likely to extend into the higher dilutions of the amides in water.

Of particular interest are the data for the changes in heat capacities which we obtained over the whole range of compositions,

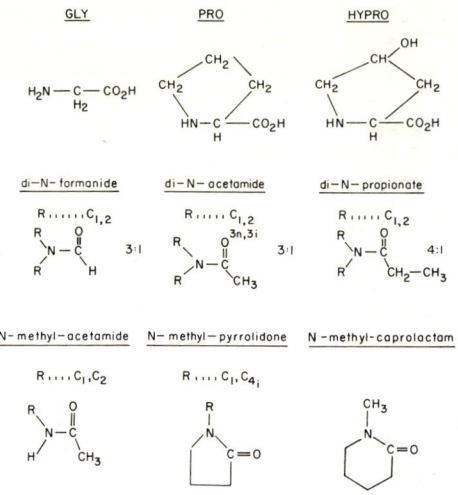

Figure 1.

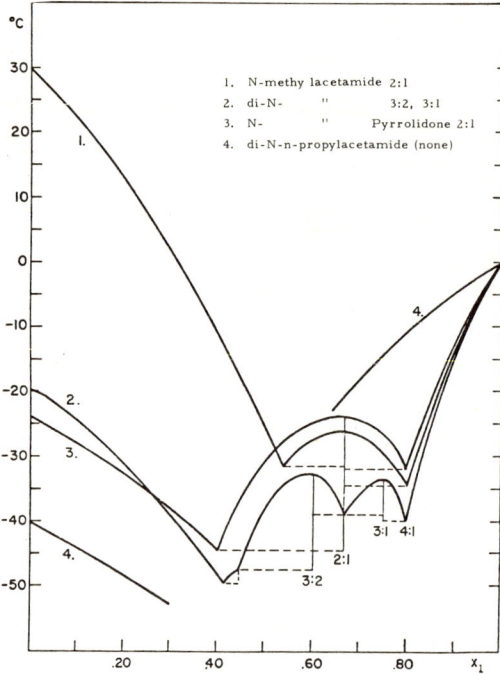

Figure 2. Phase diagrams of amide–water systems

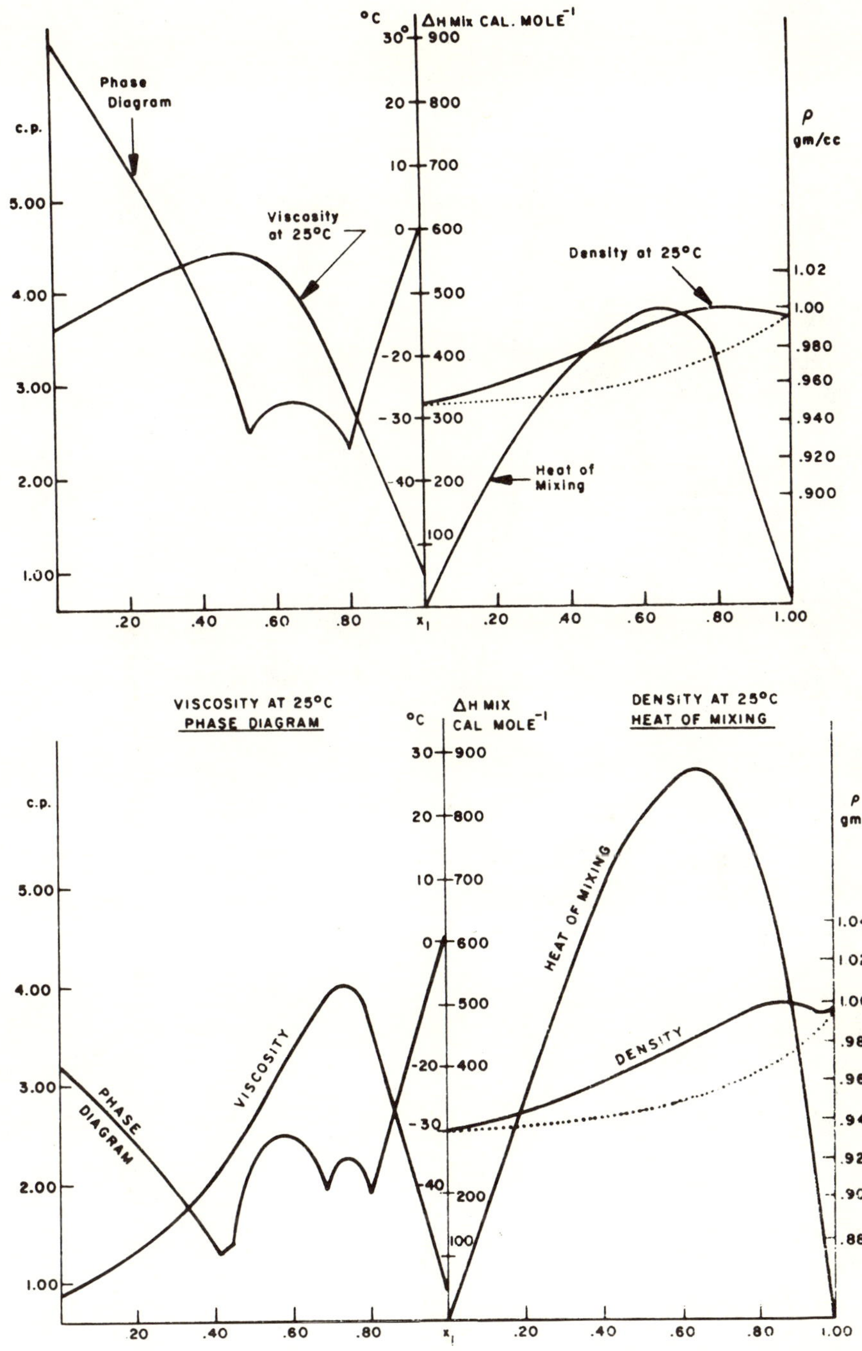

Figure 3. The a) (top) N-methylacetamide and the b) (bottom) N,N-dimethylacetamide–water systems

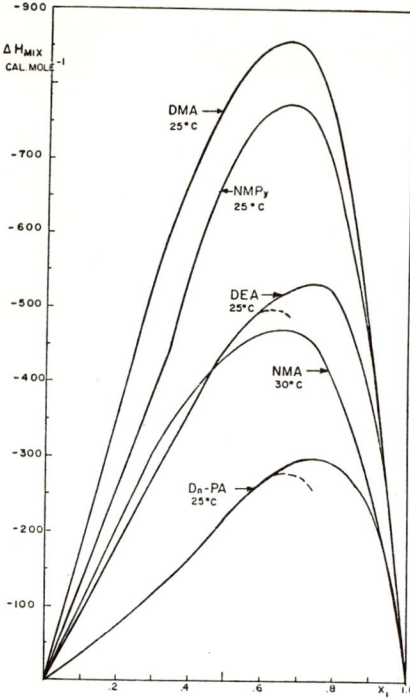

Figure 4. Heats of mixing of amides in water

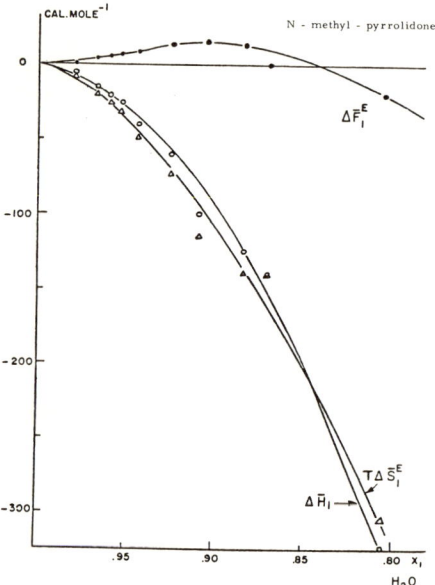

Figure 5. Partial excess functions of water in NMPy at 25°C

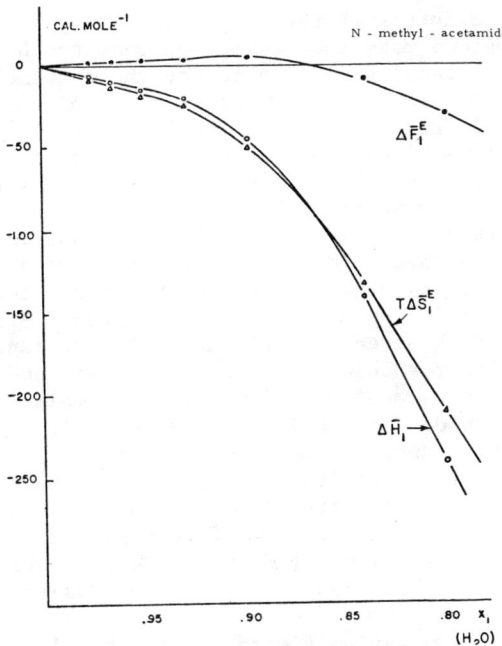

Figure 6. Partial excess functions of water in NMA at 30°C

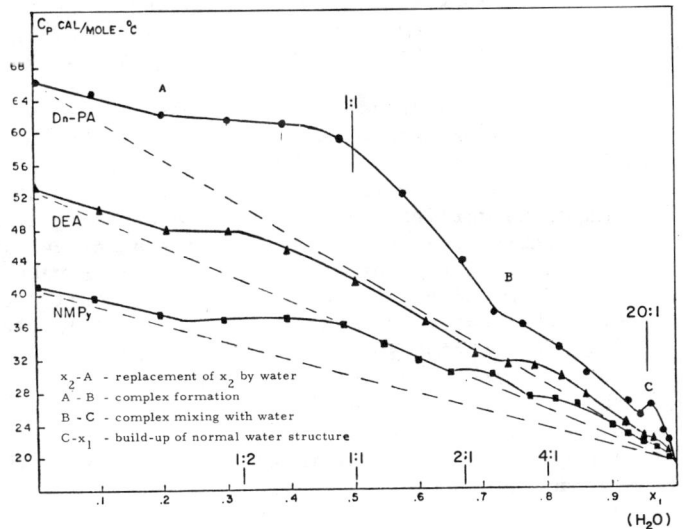

Figure 7. Heat capacities of amide–water systems at 25°C

see Figure 7. We interpret the many interesting features of the plots as follows: we make the assumption that the heat capacities of a mixture of molecules (partly due to the intramolecular degrees of freedom and partly to the vibrational and translational degrees of freedom of the molecules in the mixtures as a whole) are additive unless the external degrees of freedom are changed by non-ideal grouping or pairing of the mixing molecules. The straight lines connecting the heat capacity values of the pure amides, on the left side, with the heat capacity of pure water, on the right hand side, are drawn to represent such an ideality. In fact, N-Methyl acetamide, not shown here, exhibits such a linear relation up to 67 mole % of water. Strong deviations and changes in directions of the real heat capacity curves from this "ideality" should, therefore, indicate non-average changes in the arrangement and packing of the molecules in the mixture. Positive deviations in the heat capacity should stem from a rise in the number of molecules which are capable of increasing the numbers of their intermolecular degrees of freedom. In the mixtures under investigation we deduce that primarily the inter-molecular degrees of freedom of vibration and rotation, but not of translation, are increased, since we find that the densities of the mixtures rise less, while the viscosities rise more, than ideally.

Specifically, we see on Figure 7 that the first introduction of water up to a molar ratio of 1:4 into the partly polar, partly hydrocarbon, moiety of the pure amides changes the liequid structure only to a minor degree. From this point, the number of the molecules per volume acquiring previously inactive degrees of freedom, i.e. amides as well as the introduced water, rises substantially up to a mole ratio of about 1:1. From there, the heat capacity still stays in excess of additivity, though slightly declining, until a mole ratio of 3 molecules of water to 2 amides is reached.

We interpret the deviations up to this 3:2 ratio as being the result first of a breakup of amide association by the water, followed by hydrogen bonding up to 1 molecule of water per amide. At still higher amounts of water and consequent H-bonding, the excess heat capacity declines further, with a striking jump around the 20:1 ratio of water to amide. This signifies most likely the beginning of the formation of the regular water structure or, coming from the side of pure water, the completion of the breakdown of the water structure by the amide (7). In most studies of amide or peptide solutions in water in the literature it is this range of concentrations which has been covered. Mixtures near the 1:1 ratio have not been studied. Yet, in some biological systems in which we know that bound water is an essential part of the native conformations, this may be the very range of importance. Another aspect which we would like to emphasize is that the coincidence of breaks, or maxima, in the mixing functions for a variety of substituted amides at 67% of water, indicates a returning form of interaction between the amide group and water, somewhat modified by the substiuents on the nitrogen or on the alpha carbon, consisting of coupling 2 water molecules per substituted peptide bond.

<u>b Poly-Amides</u>. In unsubstituted amino acids, and in particular in polypeptides, the hydrogen of the amide group will, of course, participate in bonding to other peptide groups and to water. We have extended our studies, at first, to the three mentioned amino acids gly, pro, and hypro, and to some of their homo-oligomers and polymer and copolymers. N-substituted glycines were also studied by means of viscosity, density, and freezing point measurements. While this work is still in progress, we want to report here on some heat of mixing data obtained by a calorimeter with a precision of better than 1/100 of a degree. In conjunction with the heat capacity of the instrument, this amounts to a precision of about 0.2 cal. Because of the rather low solubility of some of the oligomers and polymers, low number average molecular weights from about 4,000 to 9,000 were used so that the polymers would dissolve in a sufficiently short period of time commensurate with the heat insulation of the instrument, and in sufficient amounts to yield data within the limits of our precision. The monomer concentrations up to about 4%, and of the polymers up to approximately 1% in water were studied. Based on an average amino acid molecular weight, the solutions were therefore approximately 1/1000 molar in amino acid residues. The polymer molecular weights were determined by osmometry and the composition of the copolymers by optical rotation analysis.

Among other basic difficulties encountered were the unknown degrees of crystallinity of the monomeric samples undergoing solution, (the polymers were found to be non-crystalline by x-ray and birefringence analysis), which caused endothermic solution enthalpies which varied with the state of crystallinity. We instituted therefore dilution studies in which we extrapolate to zero water, a method which yields also the heats of fusion for which we can not find data in the literature. At this moment we want to report qualitatively that the heats of mixing of the simple amino acids are less negative than those of the disubstituted amides, and that heat of fusion for the three amino acids gly, pro, hypro are of the order of 3,000-4,000 cal.

Figures 8 and 9 show the heats of mixing of two of the homopolymers, pro and hypro. The polymer of gly is too insoluble to obtain data. The Figures show also curves of copolymers of proline and glycine, of hydroxyproline and glycine, of proline and hydroxyproline, and a terpolymer of gly-pro-hypro at the approximate mole ratio of 3:4:5. The most important results are that all the heats of mixing are negative, substantially more so than those expected for the monomers once they will be fully evaluated. From data not here presented there is little effect

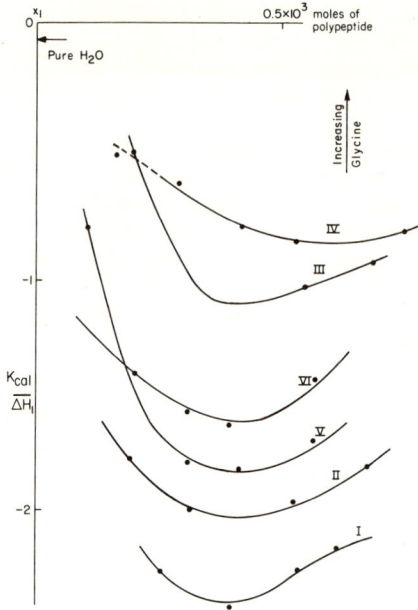

Figure 8. Enthalpies of mixing. I: polyproline. II: polyglycine–proline 1:20. III: polyglycine–proline 1:5.4. IV: polyglycine–proline 1:0.7. V: polyglycine–hydroxyproline–proline 3:4:5. VI: polyhydroxyproline.

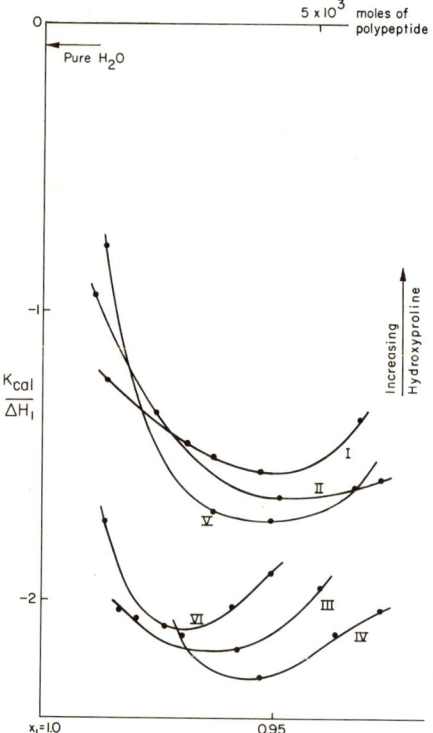

Figure 9. Enthalpies of mixing. I: polyhydroxyproline. II: polyglycine–hydroxyproline–proline 3:4:5. III: polyhydroxyproline–proline 3:1. IV: polyproline. V: polyhydroxyproline–proline 6:1. VI: polyhydroxyproline–glycine 4:3.

of molecular weights in our molecular weight range. The heats of mixing for these unsubstituted polypeptide groups are also substantially more negative than those for the N-substituted peptides.

Discussion

By and large, during the introduction of dry, unsubstituted, polypeptides into water three factors will affect the heat of mixing: the disruption of the inter-peptide forces, the disruption of the water structure, and the rebuilding of a different water structure around the solute. In the light of our earlier solvation studies on peptides the exchange of bulk hydrogen bonding to heterogeneous H-bonding, i.e., from the A-A to the A-B type, should not make a major difference. The increased negativity of the enthalpies should thus be attributed to the restructuring of water in the sense of iceberg formation which is well known to be the cause of the negative heats of mixing of the short chain alcohols (8).

The finding that the heat of mixing of polypro is substantially more negative than that of polyhypro is surprising in view of the hydrogen bonding to be expected between water and the hydroxy groups of the hypro. Again, one might explain this in terms of the rough equivalence of exchanging A-A by A-E hydrogen bond, and in terms of the more extensive hydrophobic bonding and structuring of water around polyproline. Another possibility would be differences in the polypeptide conformations (9,10). A more trivial factor may be that not all of the protective acetyl groups on hypro were removed.

Looking at the copolymers between gly and pro, Figure 8, we see that the heats of mixing are all less negative than for polypro. The close likeness of the proportions of the monomer feed in the reaction mixture to the monomer residue ratio in the polymer indicates an approximately random distribution, or at least the formation of blocks or of periodicities of similar length. Thus, the decrease in the negative heats of mixing for increasing gly proportions in the glypro copolymer indicates less hydrophobic bonding to this polypeptide. In polyglycine, as stated, the absence of nonpolar groups together with high chain regularity leads to a high heat of fusion which prevents polymer solution. For the prohypro copolymers, Figure 9, the heats of mixing lie approximately between the curves for the homopolymers, so that no new factors seem to enter in the solvation behavior of these copolymers. Unexpectedly, though, the copolymer of glyhypro shows a more negative enthalpy than polyhypro, in contrast to the effect of the gly-component in the gly-pro copolymer.

Summary

Our data on the thermodynamics of mixing of N-substituted amides with water confirm our earlier conclusion (1) of a strong binding of two water molecules to di-N-substituted amides, with a third more lightly bonded. This indicates that the carbonyl group is capable of holding two water molecules by H-bonding, the third presumably being attached to the nitrogen. Applying this conclusion to N-substituted amides or peptides, these should also be solvated by an inner shell of three water molecules. In dilute solution, according to our heat capacity data, these simple amides affect approximately 15 water molecules in an outer zone. Since, in the competition between inter-peptide and peptide-water forces, the latter appear to be substantially favored, the water molecules remain bonded up to the 1:1 molar ratio. For unsubstituted polymeric peptides, the regular array of hydrogen bonds along the polymeric molecules favors the inter-peptide H-bonding, as indicated by the low solubilities of many unsubstituted polypeptides. For the polypeptides of the pro-, hypro-, and gly- series here studied, we find great exothermicity and therefore more extensive hydrogen bonding and formation of water structures than those accompanying the solution of simpler peptides and amides.

Literature Cited

1 Assarsson, P. and Eirich, F. R., in "Molecular Association in Biological and Related Systems", Adv. Chem. Ser. No.84, American Chemical Society, Washington, D. C., 1968.
2 Wood, F., Ramsden, D. K., and King, G., Nature, (1966) 212, (5062), 606
3 Needham, T. E., Jr., Diss. Abst. Inter. B., (1970) 31 (6) 3303
4 Klotz, J. and Franzen, J., J. Am. Chem. Soc. (1960), 82, 5241
5 Pauling, L., Proc. Nat. Acad. Sci. U.S., (1932), 18, 293
6 Costain, C. C. and Dowling, J. M., J. Chem. Phys. (1960), 32, 158
7 Klotz, I. M. and Farnham, S. B., Biochem., (1968) 7 (11) 3879
8 Franks, F., Ann. N. Y. Acad. Sci., (1965), 125, 277
9 Forsythe, K. H. and Hopfinger, A. J., Macromolecule, (1973), 6(3) 423
10 Segal, D. M., J. Mol. Biol., (1969), 43(3) 497

22

Equilibrium Dialysis and Viscometric Studies on the Interaction of Surfactants with Proteins

W. U. MALIK and V. P. SAXENA

Department of Chemistry, University of Roorkee, Roorkee, India

Introduction

    Ever since the discovery of Kuhn and coworkers (1) in 1940 that certain cationic soaps exert powerful disinfecting action on bacteria, presumably owing to their combining capacity with the cell protein and that their action is proportional to their surface activity, chemists have evinced keen interest in studying the role of synthetic detergents in biological systems and processes. Recent studies have shown that the nonionic polysorbate 80 causes a disruption of the membrane structure (2) and the nonionic Triton X-100 exerts a similar effect on lysosomes, mitochondria, and erythrocytes (3,4). The solubilizing action of detergents on cholesterol has been reported to check the development of atheroma (5). The formation of bile salt-monoglyceride mixed micelles has been shown to be a significant factor in lipid absorption (6). In view of the important role played by the synthetic detergents in various metabolic processes and their evergrowing use in the food and pharmaceutical industries, the need to develop a knowledge of their biological effects acquires great significance providing enough justification for carrying out systematic studies on the chemical interaction of detergents with biological effects acquires great significance providing enough justification for carrying out systematic studies on the chemical interaction of detergents with biological substances. In the present communication we have discussed the results of our investigations on some protein-detergent systems employing equilibrium dialysis and viscometric techniques. Substituted proteins, viz., p-toluene sulphonylated gelatin (hereafter called TSG) and iodinated casein (hereafter called IC) were used for the equilibrium dialysis studies which provided evidence for their statistical binding with the anionic detergents: sodium dodecyl sulphate (hereafter called SDS) and sodium octyl sulphate (hereafter called (SOS). The viscometric studies were made with ovalbumin (hereafter called OV), a globular protein, and transfusion gelatin (hereafter called TG), a fibrillar protein with the anionic detergent sodium lauryl sulphate (hereafter called (SLS) and the cationic detergent cetylpyridinium bromide (hereafter called CPB) with a view to gain an insight into the nature of particle-particle interaction.

Experimental

   Materials (a) Proteins. p-toluene sulphonylated gelatin was prepared by adding p-toluene sulphonyl chloride (10g) in small lots to a solution of gelatin (30g) made alkaline (pH 11) with potassium hydroxide and maintaining the pH throughout at this value by the occasional addition of the alkali. The reaction mixture was stirred for 3 hours and then centrifuged. The centrifugate was acidified with acetic acid until the pH was 3.5 when a white coagulum formed. This was separated and cut into small pieces which were subjected to prolonged washing with cold methyl alchol (1 litre) in a continous extractor for a week when a chloride free product was obtained. It was dried in vacuo and ground to a powder. Its sulphur content was found to be 1.985%. Iodinated casein was prepared by adding chloramine-T (6g) in small lots to a solution of cow's whole milk casein (20g) in water (800ml) containing dissolved sodium bicarbonate (6g) and potassium iodide (2.5g). The reaction mixture was stirred for an hour and the protein precipitated at pH 3.8 by the addition of hydrochloric acid. The curd was subjected several times to dissolution in dilute sodium hydroxide and precipitated with hydrochloric acid until the inorganic iodide was completely removed. The product was dried at 80°C. It was found to contain 7.73% of iodine. Ovalbumin was prepared from egg white according to the procedure of Sørensen and Høyrup (7) by extraction with a saturated ammonium sulphate solution and crystallizing at a pH of about 4.6 brought down by adding 0.2N sulphuric acid. Transfusion gelatin was obtained through the courtesy of the Director, National Chemical Laboratory, Poona, India in the form of 6% solution.
   (b) Detergents. All the four detergents (SDS, SOS, SLS, and CPB) were British Drug House products. These were further purified by recrystallization from acetone.

   Preparation of solutions. Solutions of TSG and IC were prepared by soaking the protein in doubly distilled water for several hours and then stirring the mixture mechanically after adding a known amount of 0.1 M KOH to bring the pH to 6.5. Concentrations of these solutions were determined by drying a known aliquot to constant weight in an air oven at 105°C and applying the correction for potassium hydroxide to get the absolute weight of the protein. Ovalbumin solution was prepared by shaking it with water at room temperature. The solutions of ovalbumin and transfusion gelatin were purified by dialysis and the concentrations of the dialysed solutions determined by drying a known volume at 105-110°C to constant weight in an air oven.

Procedure for equilibrium dialysis. 20 ml aliquots of the protein-detergent mixture in which the concentrations of SDS and SOS varied from 0.2×10^{-3}M to 20.0×10^{-3}M at a fixed conc. (6×10^{-5}g) of the protein were made by mixing requisite volumes of 6% protein, 0.05M detergent, and distilled water, and brought up to the desired pH. These were stored for two days at 25°C. 5 ml portion from each aliquot was then pipetted into dialysis bags made from 15cm long strips of Visking Nojax sausage casing (1.8cm in diameter), freed from sulphur by repeated boiling with distilled water, by tying a square knot on one end. Most of the air inside the bags was forced out by pressing which were then closed with two overhead knots, one above another. Each bag was carefully suspended in a test tube containing 5 ml of double distilled water brought up to the same pH, which was then stoppered. The tubes were placed in a stand supported inside a water thermostat and gently shaken for 48 hours by means of an electrical shaker. The dialysis bags were then withdrawn from the test tubes and the amount of the detergent present in the dialysate was estimated spectrophotometrically using the cationic dye pararosaniline hydrochloride.

To 1 ml solution of pararosaniline hydrochloride (4×10^{-4}M) was added an appropriate volume of the test solution (not exceeding 4 ml). The total volume was made up to 5 ml with distilled water. 5 ml of a mixed solvent (equal volumes of chloroform and ethyl acetate) were added and the solution shaken manually about 50 times. It was then centrifuged for about a minute at 5000 r.p.m., for complete separation of the organic and aqueous phases, the former containing the coloured complex at the bottom. Its spectral absorption was measured on Klett Summersion Photoelectric Colorimeter using green filter against a reference tube filled with the solvent.

Procedure for viscometric studies. Three sets of protein-detergent aliquots of 20 ml each were prepared for viscosity measurements as given below: (i) In the first set, the protein concentration was fixed (0.5%) while the concentration of the detergent varied from 2-50 millimoles per litre. The viscosities of the aliquots of this set were measured at pH 2.0, 3.0, 6.0 and 8.0 in the case of anionic detergent-protein system and at pH 2.0, 3.0, 8.0 and 9.0 in the case of cationic detergent-protein system. (ii) In the second set, the concentration of the protein was varied, from 0.8% to 1.4% in the case of OV and from 0.6% to 1.2% in the case of TG using the same concentrations of detergent solutions. Viscosity measurements were made at pH 8.0 for the anionic detergent-protein system and at pH 3.0 for the cationic detergent-protein system. (iii) The third set consisted of solutions having different protein-detergent ratios. Viscosity measurements were made at pH 8.0. The viscosities were measured by Scarpa's method as modified by Prasad et al.(8). The viscometric constant, k, was calculated from the experimentally determined values of t_1 and t_2, the time of rise and the time of fall respectively, for a given volume of solution under constant pressure (for double distilled water). The viscosity, η, of the solution is related to t_1 and t_2 under constant pressure as

$$\eta = k \frac{t_1 t_2}{t_1 + t_2}$$

The viscosity of the test solution was determined by finding t_1 and t_2 for a known volume of solution (20ml). All measurements were carried out at 30 ± 0.1°C in a thermostatic water bath.

Calculations

(i) <u>Equilibrium Dialysis</u>. The extent of binding was calculated by Klotz's equation (9):

$$\alpha = \frac{\epsilon_{app} - \epsilon_\beta}{\epsilon_F - \epsilon_\beta}$$

where ϵ_{app} is the apparent molar extinction coefficient, ϵ_β the molar extinction coefficient of the bound detergent, ϵ_F the molar extinction coefficient of free detergent and α is the fraction of the free detergent. The value of Vm, the number of moles of the detergent bound per mole of protein, is given by the expression:

$$V_m = A_p / P_t$$

where Ap is the number of moles of protein bound detergent and P_t is the total number of protein molecules.

(ii) <u>Viscosity</u>. The intrinsic viscosity, $[\eta]$, at different detergent concentration was obtained by plotting η_{sp}/c vs c. (η_{sp} is the specific viscosity and c the conc. of the protein in g/ml) and extrapolatting the curves to zero concentration. The interaction index k, was calculated using Kramer's equation which has been found to be valid for such systems over a wide range of concentrations. The equation is

$$\ln \eta_{rel}/c = [\eta] - k[\eta]^2 c$$

where' η_{rel}' is the relative viscosity.

Results and Discussion

__Equilibrium dialysis studies.__ Typical data for the binding of SDS with TSG and of SDS with IC are given in Table I and II. Typical curves for the plots of V_m vs log C_F and plots of $1/V_m$ vs $1/C^F$ are given in Figures 1, 2 and 3. Values of V^m obtained from extrapolation are given in Table III. The course of binding of SDS and SOS with p-toluene sulphonylated gelatin and iodinated casein as the detergent is added in progressively increasing concentration to a fixed amount of the protein at definite values of pH (7.7 and 9.5) has been followed by plotting the average number of moles of the detergent bound per 10^5g protein, V_m, against the log of concentration of the unbound detergent, log C_F (Figure 1). The general shape of the curves obtained with different detergent protein combinations at different pH values is the same. An examination of the plots of V_m vs log C_F reveals that in each case the mode of binding of the surface active molecules to the protein varies in three distinct patterns as the concentration of the former is gradually increased. These patterns of binding have been roughly demarcated by dividing the plots into three regions A, B, and C, with a view to facilitating a better appreciation of the mechanism of detergent-protein interaction at various stages. In the region A, which is the region of relatively high proportion of protein to detergent, the plot of V_m vs log C_F follows a linear course. This is indicative of a more or less statistical distribution of the detergent over the entire available protein molecules in this region. Similar observations have been reported by Klotz et al. (9) in dye-protein interaction. The applicability of the simple statistical theory in the present studies may be easily tested by plotting $1/V_m$ against $1/C_F$ (Figure 2, 3) when a plot is obtained which is linear up to a certain limit. The values of the reciprocal of the intercepts on the ordinate obtained after extrapolations of the straight lines with various detergent-protein combinations are given in TableIII. These values represent the maximum number of binding sites available, i.e., the max.values of V_m, in this region. Beyond the region A, the course of the plots of V_m vs log C_F sharply deviates from the initial linear course in the direction of higher values of V_m, reflecting a distinct change in the mode of binding which no longer appears to be wholly statistical. It is reasonable to assume that after combination with the number of moles of detergent as a maximum for statistical binding (vide Table III), a large number of ions get bound essentially as a unit. Possibly after occupying the readily accessible sites (basic groups) on the surface of the protein molecule, the extra detergent molecules penetrate the original tightly folded protein molecule causing it to undergo a change in physical organization in which form it offers hithertofore inaccessible sites for combination and, therefore, exhibits greater binding capacity for the detergent. With the structural disordering of the protein molecule, the potential

Table I. Binding of Sodium Dodecyl Sulphate (SDS) with p-Toluene Sulphonylated gelatin (TSG).

pH = 7.7; Conc. of TSG = 6×10^{-5}g; Temperature = $25 \pm 0.1°C$; Period of equil. = 48 hours

Conc. of SDS (C) $\times 10^{-5}$M	Conc. of free SDS(C_F) $\times 10^{-5}$M	Conc. of bound SDS(C_B) $\times 10^{-5}$M	Moles SDS bound per 10^5g TSG, V_m	$\log C_F$	$1/V_m$	$\frac{1}{C_F} \times 10^5$
20	12	8	1.33	$\bar{1}.0792$	0.75	0.0833
30	18	12	2.0	$\bar{1}.2553$	0.50	0.0555
40	24	16	2.66	$\bar{1}.3802$	0.375	0.0417
60	36	24	4.0	$\bar{1}.5563$	0.25	0.0273
80	51.5	28.5	4.75	$\bar{1}.7118$	0.21	0.0194
100	64	36	6.0	$\bar{1}.8062$	0.166	0.0156
120	75	45	7.5	$\bar{1}.8751$	0.133	0.0133
200	116	84	14.0	3.0645	0.07	0.0090
300	150	150	25.0	3.1761	0.04	0.0066
400	175	225	37.5	3.2455	0.026	0.0056
600	264	336	56.0	3.4216		
800	342	458	78.0	3.5340		
1000	400	600	100.0	3.6021		
1200	450	750	125.0	3.6532		
1600	520	1080	180.0	3.7160		
2000	590	1410	235.0	3.7709		

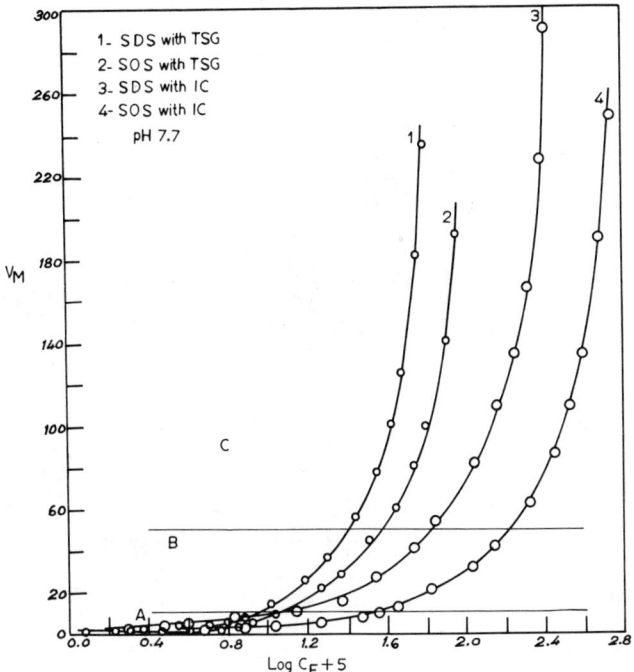

Figure 1. Plot of V_m vs. log C_F

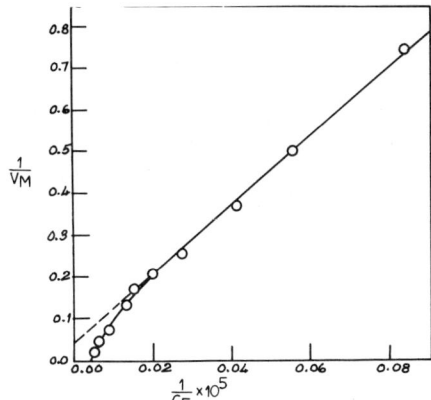

Figure 2. Plot of $1/V_m$ vs. $1/C_F$ (SDS with TSG) at pH 7.7

Table II. Binding of Sodium Dodecyl Sulphate (SDS) with Iodinated Casein (IC).
pH = 7.7; Conc. of IC=6 x 10^{-5}g; Temperature = 25 ± 0.1°C; Period of equil.=48 hours

Conc. of SDS (C) x10^5M	Conc. of free SDS,(C_F) x10^{-5}M	Conc. of bound SDS, (CB) x10^{-5}M	Moles SDS bound per 10^5g IC, V_m	log C_F	$1/V_m$	$\frac{1}{C_F}$ x10^5
20	2	18	3	$\bar{5}$.3010	0.333	0.500
30	3	27	4.5	$\bar{5}$.4771	0.222	0.333
40	4	36	6	$\bar{5}$.6021	0.167	0.250
60	7	53	8.8	$\bar{5}$.8451	0.114	0.143
80	14	66	11	$\bar{4}$.1461	0.090	0.071
100	22	78	13	$\bar{4}$.3423	0.077	0.0454
120	24	96	16	$\bar{4}$.3802	0.062	0.0416
200	35	165	27.5	$\bar{4}$.5441	0.037	0.0286
300	54	246	41	$\bar{4}$.7324	0.0244	0.0185
400	70	330	55	$\bar{4}$.8451	0.0182	0.0143
600	114	486	81	3.0569	0.0135	0.0064
800	140	660	110	3.1461		
1000	178	822	137	3.2504		
1200	210	990	165	3.3222		
1600	238	1362	227	3.3766		
2000	260	1740	290	3.4150		

barrier to the entry of more detergent is progressively reduced resulting in the breaking up of internal linkages. However, it is evident that the protein is not fully unfolded in this region.

In the region C, V_m increases apparently without limit and its values far exceed the number of total positively charged groups on the protein molecule. Obviously, unlike the region B, the binding here is no longer wholly electrostatic in nature and involves nonspecific forces also. Such extra bound detergent ions have been reported by Yang and Foster ([10]) from their studies on the interaction of sodium dodecylbenzene sulphonate, by Lundgren et al. ([11]), from electrophoretic investigations with ovalbumin, and by Strauss et al. ([12]) from electrical transference studies with serum albumin. The excessive binding of the detergent with the protein may be attributed to the non-polar attraction of additional detergent to that already statistically bound ([11]). Presumably, the binding of one detergent ion at a site on the protein molecule favours the binding of additional detergent ions in its immediate vicinity through Van der Waals interaction of the neighbouring hydrophobic detergent groups. This would eventually result in the formation of mixed micelles. The extra bound detergent causes further unfolding of the protein mechanically due to the large detergent ions penetrating into the folds of the once highly condensed polypeptide chain and electrostatically due to the repulsion of the similarly charged atoms on the adjacent folds of the chain. This permits the bound aggregates to have a close approach to the hydrophobic residues of amino acids along the polypeptide chain with the result that the denaturation of protein takes place. It may be noted that over the entire range of the detergent concentration employed, the extent of binding to the protein is more with sodium dodecyl sulphate than with sodium octyl sulphate (Table I, II). A general increase in the extent of binding with the increasing size of the detergent anion has been observed by many workers. Boyer et al. ([13]) have reported increased binding in interactions with albumin in the order: butyrate, caproate, caprylate, and caprate, while similar results were obtained by Karush and Sonenberg ([14]) in studies with octyl, decyl, and dodecyl sulphates. The increased binding ability of longer chain detergent has been attributed to stronger Van der Waal's forces arising from increased molecular size. The values of V_m, i.e., the number of surfactant (SDS or SOS) moles bound with a definite weight of the protein at pH 9.5 are significantly lower in the region A of the curve than those at pH 7.7. However, the values at both the pH come closer in the region B and still closer in the region C. These variations are in accordance with the fact that the total number of cationic groups decrease with increase in pH. Since the magnitude of binding of the detergent in the region A depends on the total number of the cationic groups available on the protein, it shall be less at higher pH. However, in the regions B and C, the binding involves hithertofore inaccessible sites and

Table III. Values of V_m, the Average Number of Moles of SDS or SOS Bound per 10^5 g of Protein Up to Which the Statistical Binding Occurs. Values Obtained From Extrapolation.

Detergent	p-toluene sulphonylated gelatin.		Iodinated casein	
	pH 7.7	pH 9.5	pH 7.7	pH 9.5
SDS	25.0	20.0	18.0	15.4
SOS	30.0	25.0	20.0	16.7

Table IV. Ionizable groups in p-Toluene Sulphonylated Gelatin and Native Gelatin.

Groups	Reasonable analytical values in native gelatin per 10^5g.	Observed in case of p-toluene sulphonylated gelatin per 10^5g.
α -Carboxyl β -Carboxyl	120	110
Imidazole	5	4
α-Amino -Amino	32	0
Guanidinium	49	42
Total cationic	86	46

Table V. Ionizable Groups in Native Casein and IC.

Groups	Reasonable analytical values in native casein per 10^5g.	Observed in case of IC per 10^5g.
α-carboxyl β-carboxyl	96	90
Imidazole	20	18
Amino	56	52
Guanidinium	24	22
Tyrosine/Tryptophan	5.83	4.67
Total cationic	100	92

tends to become increasingly independent of the cationic groups of the protein molecules. A comparison of the binding pattern of p-toluene sulphonylated gelatin and iodinated casein at pH 7.7 with that of the corresponding native proteins, viz., gelatin and casein (15) at the same pH provides strong evidence for statistical binding of the detergent to the protein. The values of the number of moles of the detergents bound with the substituted proteins are lower than those with the corresponding native proteins in the region A by a factor of approximately 0.6 in the case of p-toluene sulphonylated gelatin and approximately 0.9 in the case of iodinated casein. These factors are nearly the same by which the total number of the available cationic groups on these proteins are less than on the native proteins (16) (Table IV,V). In the regions B and C, where the binding of the detergent becomes increasingly independent of the surface cationic groups, the binding values with the substituted proteins tend to approach closer and closer to those obtained with the native proteins.

Viscometric studies. Typical data for SLS-OV system and SLS-TG system are given in Tables VI and VII and depicted in Figures 4-7. Viscosity data of SLS-TGS and SLS-OV systems of varying detergent-protein ratio at pH 8.0 are depicted in Figure 8. Investigations were carried out on both the acidic and basic sides of the isoelectric point. The results show an initial decrease in viscosity with both anionic detergent-cationic protein (below the isoelectric point) and cationic detergent-anionic protein (above the isoelectric point) systems (Figure 4, 5) until precipitation takes place. The precipitation zone is shown by a blank space in the curves. On adding a large amount of the detergent, the precipitate formed redissolves and an increase in viscosity is observed. In the case of anionic detergent-anionic protein neither initial decrease is observed nor the zone of precipitation is realised. On the other hand the viscosity continuously increases (Figure 4, 5). The proteing molecule which initially existed in the expanded state on either side of the isoelectric point gets tightened up with the progressive combination with detergent ions and consequently the viscosity decreases. Boyer et al. (13) have reported similar observations in their studies with serum albumin in the presence of sodium caprylate and explained them on the basis of the tightening up of protein molecule. With further addition of the detergent, the stage is finally reached when precipitation occurs owing to the orientation of the hydrophobic part of the detergent ions in solution. With the addition of further quantities of the detergent, a second adsorption layer is formed by Van der Waal's attraction forces between carbon chains (17-20) which makes the molecule hydrophilic, as such the precipitate again disperses causing the viscosity of the solution to rise.

Table-VI. Sodium Lauryl Sulphate-Ovalbumin System.

Concentration of detergent in millimoles/litre	Intrinsic viscosity	Interaction Index
0.0	40.5	6.30
10.0	52.2	5.20
15.0	60.0	4.10
20.0	72.0	3.50
25.0	86.5	2.70
30.0	110.0	2.20
35.0	124.0	1.90
40.0	136.0	1.60

Table-VII. Sodium Lauryl Sulphate-Transfusion Gelatin System.

Concentration of detergent in millimoles/litre	Intrinsic viscosity	Interaction Index
0.0	35.55	6.55
10.0	42.10	5.58
15.0	54.15	4.42
20.0	60.52	3.75
25.0	69.41	3.40
30.0	78.00	2.74
35.0	99.76	2.44
40.0	120.00	1.87

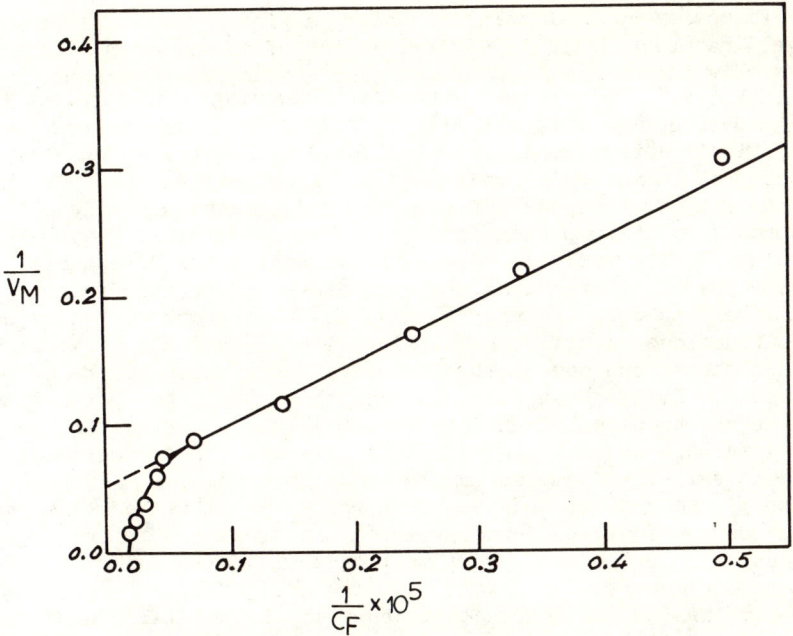

Figure 3. Plot of $1/V_m$ vs. $1/C_F$ (SDS with IC) at pH 7.7

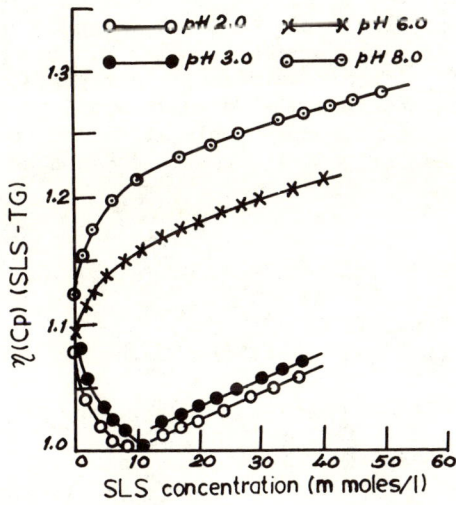

Figure 4. Viscosity variation of TG in the presence of SLS at different pH values

Viscosity data of soln. in which varying amount of the detergent has been added to a fixed amount of the protein (Figure 6, 7) show that the viscosity, as also the intrinsic viscosity (Table VI, VII), increases with the increasing concentration of the detergent. The overall effect of the anionic or the cationic detergent is to bring about a progressive unfolding of the protein molecule resulting in its solubilization. In the initial stages, when smaller amounts of detergent are added, the electrostatic force predominates and causes its binding with the protein. In the presence of larger amounts of the detergents, a second type of interaction involving non-electrostatic forces is more likely to occur. Lundgren (11) has also proposed two types of combinations identifying the first one as stoichiometric binding and the second one as the secondary association of the extra detergent. The size and structure of the hydrocarbon portion as well as the presence of an ionic group in the detergent are two likely factors controlling such interactions. The former contributes to salt-like combination between the oppositely charged groups of the protein and detergent while the latter confers stability on the ionic bond and promotes the binding of extra detergent. Support to the above mentioned view also comes from the data on the interaction index, 'K' (Table VI, VII). A rapid fall in the values of the interaction index in the initial stages indicates salt like combinations while a smaller decrease in its values at higher detergent concentrations exhibits a lesser amount of particle-particle interaction forces. It may be, therefore, concluded that the secondary association of extra detergent is in the form of loose combination due to non polar attraction with the already electrostatically bound detergent. The viscosity variations of the protein detergent system of set (iii) have been plotted in Figure 8. These plots exhibit maxima at the detergent-protein ratio of 1:3 corresponding to a complex of constant composition. Similar results have been reported by Lundgren et al. (11) from electrophoretic studies. However, such plots could not be obtained with dilute solutions. The formation of the protein detergent complex only at higher concentrations suggest the presence of rod-like particles in such systems (22).

Literature Cited.

1. Kuhn, R., Bielig, H. and Dann, O., J. Ber. Deut Chem. Ges., (1940) 738, 1080.
2. Kay, Cancer Res., (1965) 25, 764.
3. Du Duve, Wattiaux and Wibo, Biochem.Pharmacol., (1962) 9, 97.
4. Weismann, Biochem.Pharmacol., (1965) 14, 525.
5. Keeser, Arch. Int. Pharmacodyn., (1951) 87, 371.
6. Dawson and Rhodes, Ed., Metabolism and Physiological Significance of Lipids, John Wiley, New York, 1964.

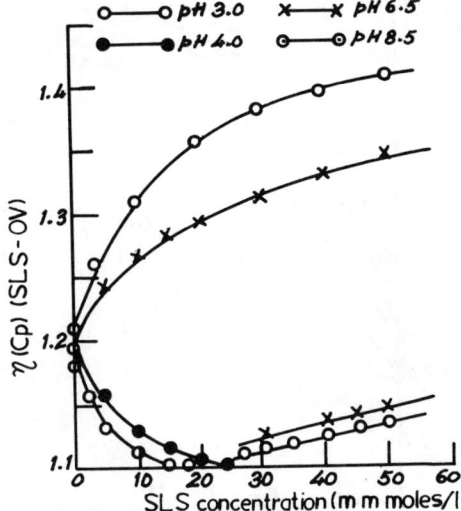

Figure 5. Viscosity variation of OV in the presence of SLS at different pH values

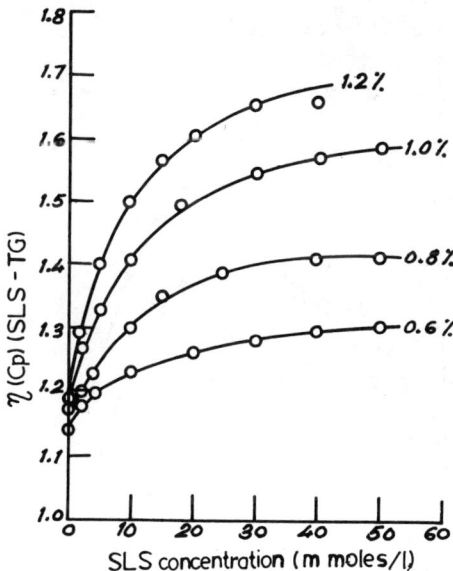

Figure 6. Viscosity changes in the TG–SLS system with varying concentrations of protein and detergent at a fixed pH of 8.0

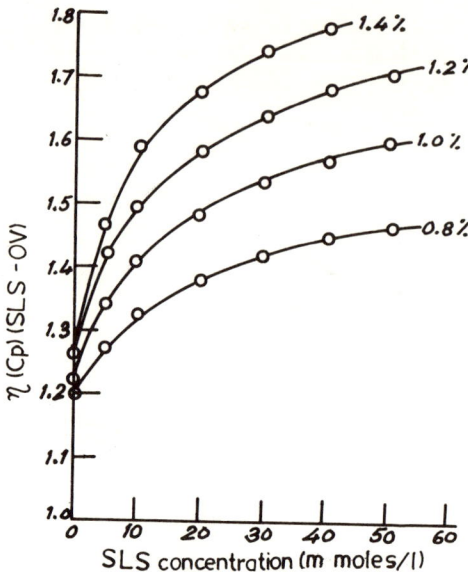

Figure 7. Viscosity changes in the OS–SLS system with varying concentrations of protein and detergent at a fixed pH of 8.0

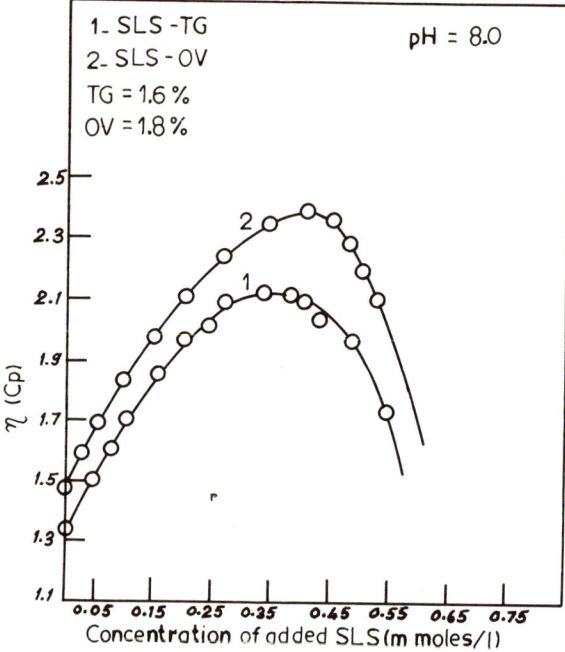

Figure 8. Viscosity of TG–SLS and OV–SLS system of varying protein–detergent ratio

7. Sørenson, S.P.L. and Høyrup, H., Compt. Rend Trav. Lab., Carlsberg, (1917) 12, 164.
8. Prasad, M., Mehta, S.M. and Desai, J.B., J.Phys.Chem., (1932) 36, 1384.
9. Klotz,I.M., Walker, F.M. and Pivan, R.B., J.Amer.Chem. Soc., (1946) 68, 1486.
10. Yang,J.T. and Foster, J.F.,J.Amer.Chem.Soc., (1953) 75,5560.
11. Lundren,H.P., Elam,D.W. and Connel, R.A.O., J.Biol.Chem., (1943) 143, 183.
12. Strauss,G. and Strauss, U.P., J.Phys.Chem., (1958) 62,1321.
13. Boyer,P.D., Ballou,G.A. and Luck,J.M., J.Biol.Chem., (1946) 162, 199.
14. Karush,F. and Sonenberg, M.,J.Amer.Chem.Soc., (1949) 71, 1369.
15. Ashraf,S.M., Ph.D.thesis, University of Roorkee, India, 1969
16. Randhawa,H.S.,Ph.D.Thesis,University of Roorkee, India, 1969
17. Pankhurst,K.G.A. and Smith,R.C.M., Trans.Faraday Soc., (1944) 40, 565.
18. Pankhurst,K.G.A. and Smith,R.C.M., Trans.Faraday Soc., (1945) 41 630.
19. Pankhurst,K.G.A. and Smith,R.C.M., Trans.Faraday Soc., (1947) 43, 506.
20. Pankhurst,K.G.A. and Smith,R.C.M., Trans.Faraday Soc., (1947) 43, 511.
21. Wurzschmitt,B., Z.Anal.Chem., (1950) 130, 105.
22. Putnam,F.W. and Neurath,H.N., J.Biol.Chem., (1945) 160,397.

23

Derivation of Equations to Account for the Variation of pH and Specific Conductivity of Gum Arabic Acid Sol with Dilution

B. N. GHOSH

University College of Science & Technology, Calcutta—700009, India

Introduction

The electrochemical properties of gum arabic acid, purified by electrodialysis, have been investigated by a number of authors. By potentiometric titration with an alkali Thomas and Murray (1) have found it to behave as a fairly strong monobasic acid with an equivalent weight of 1175. They have also noticed that neutralisation is complete at pH 7 and equilibrium is attained in a short time indicating that all the acidic groups are exposed to attack by alkali. It is quite possible that disaggregation of the particles of the acid starts at a pH much lower than 7 thereby exposing all the acidic groups by the time the final stage of titration is reached. Briggs (2) and also Pauli and Ripper (3) have investigated the effect of dilution on the pH and specific conductivity of sols of the acid. The osmotic pressure of arabic acid sols, under conditions of Donnan equilibrium, has been investigated by Briggs (4) and Klaarenbeck (5). The dissociation constant of the acid has been found by Briggs (2) to vary with its concentration and degree of neutrlisation. Overbeek (6) reports that for a given degree of neutralisation, the pH and the dissociation constant of the acid depend much on the concentration of an added neutral salt.

In this paper it will be shown that equations can be derived to account for the variation of pH and specific conductivity of the acid with dilution on the basis of a theory proposed by the author (7) to explain the 'Pallmann effect'. The main assumptions on which the theory is based are stated below:

I. Each micelle or colloidal particle is surrounded by an electrical double layer the solution side of which is made up of two parts, (i) the fixed Stern (8)-Mukherjee (9) layer and the diffuse (i.e., the Gouy-Chapman) layer.

II. When a colloidal particle moves sufficently close to a reversible electrode placed in the sol so that the diffuse double layer is in contact with it then only the counter ions present in that part of the diffuse double layer which is in contact with the electrode can record their activity on it (i.e. the electrode).

III. Starting from the outermost boundary of the diffuse double layer the number of counter ions per unit volume increases according to the Boltzman distribution law as the surface of the particle is approached.

Derivation of Equation for H^+ Ion Activity

For the sake of simplicity let us assume that all the particles or micelles of gum arabic are of the same size and x their number per unit volume at a concentration C of the gum, the concentration being expressed in grams of gum per kg of water. Briggs has suggested that gum arabic acid may have its c.m.c. below 2.5g of gum per kg of water. Gum arabic acid is, however, known as a polyelectrolyte and so it is not likely to have any c.m.c. Let θ be the fraction, per unit area, of the surface of the reversible electrode in contact with the diffuse double layers of the particles whose number is x per unit volume at a concentration C of the gum. Then the rate at which the bare portion $(1-\theta)$ per unit area of the electrode surface comes in contact with the double layers of the particles is $k_1(1-\theta)x$ and the rate at which they are removed from unit area of the surface is $k_2\theta$, where k_1 and k_2 are rate constants and depend on viscosity and temperature.
At equilibrium;

$$k_1 (1 - \theta) x = k_2 \theta \tag{1}$$

Now assuming $x = k_o C$ where k_o is a proportionality constant and putting $K = k_2/k_1 k_o$ we get from Equation (1) the following expression

$$\theta = \frac{x}{k_2/k_1 + x} = \frac{C}{K + C} \tag{2}$$

It may be noted that θ is the ratio of two areas and is independent of the unit in which C is expressed. Let H_c be the H^+ ion activity of the sol indicated by the reversible electrode at the concentration C of the gum and let H_i be the H^+ ion activity of the intermicellar solution. As H_c is made up of two parts, one part contributed by the H^+ ions present in the diffuse double layers of the particles and the other part is contributed by the H^+ ions present in the intermicellar solution so it follows that

$$H_c = \theta H_s + (1 - \theta) H_i \tag{3}$$

where H_s is a proportionality constant. It may be pointed out that according to assumption III the concentration of the H^+

ions in the diffuse double layer close to the plane of contact with the electrode is much greater than that in the intermicellar solution. The activity H_i of the H^+ ions in the intermicellar solution is very small and is probably maintained by the release of some acid of low molecular weight from its combination with some of the constituents of the gum. Equation (3) can also be written as follows,

$$H_c - H_i = \theta (H_s - H_i) \tag{3a}$$

Since H_i is very small compared to H_s and does not change appreciably with the dilution of the sol so $(H_s - H_i)$ may be treated as a constant and put equal to H_m. Substituting the value of θ in Equation (3a) we may write it as follows,

$$1/(H_c - H_i) = K/CH_m + 1/H_m \tag{3b}$$

Derivation of Equation for Specific Conductivity

Imagine a plane between two parallel electrodes of a conductivity cell containing a sol of gum arabic acid of concentration C and let the plane be parallel to the electrodes. By adopting a similar line of argument as in the case of H^+ ion activity it can be shown that a fraction θ per unit area of the plane is in contact with the diffuse double layers of the particules and the remaining fraction $(1 - \theta)$ is in contact with the intermicellar solution. Let X_c represent the specific conductivity of the sol at a concentration C of the gum and X_i that of the intermicellar solution. As X_c is made up of two parts, one contributed by the ions present in the diffuse double layers of the particles and the other contributed by the ions in the intermicellar solution so it follows that

$$X_c = \theta X_s + (1 - \theta) X_i \tag{4}$$

where X_s is a proportionality constant. Putting $(X_s - X_i) = X_m$ and substituting the value of θ in Equation (4) we can write it as follows: -

$$(X_c - X_i) = X_m C/(K + C) \tag{4a}$$

or $\quad 1/(X_c - X_i) = K/CX_m + 1/X_m$

As X_i does not change appreciably with dilution so X_m may be treated as a constant.

Verification of the Equations

Briggs has determined at 25°C the H^+ ion activity of sols of arabic acid over a wide range of concentration of the acid. The sample of acid was purified by electrodialysis and its ash content was 0.05% (dry weight basis). His data are recorded in Table I in which H_c and hence pH have been calculated using Equation (3) taking $\theta = C/(K + C)$, where C denotes grams of arabic acid per kg of water.

The variation of specific conductivity with the concentration of arabic acid in the sol has been investigated at 25°C by Briggs and Pauli and Ripper. Their data are recorded in Table II. X_c has been calculated by using Equation (4).

The constants in the Equations (3) and (4) have been found in the following way. It may be pointed out that H_i and X_i are very small compared to H_c and X_c respectively when the concentration C of the colloid in the sol is high. According to Equation (3b) a plot of $1/(H_c - H_i)$ against $1/C$ should be linear. Therefore, using different probable values of H_i, $1/(H_c - H_i)$ has been plotted against $1/C$ and that value of H_i is selected which makes the plot linear. The value of H_i thus found is 8.0×10^{-5}. From the slope and intercept of the straight line, H_m and K are found.

Since $H_m - H_s - H_i$, therefore, from the known values of H_m and H_i, the value of H_s can be easily found. In this way the values of H_s, H_i and K have been found to be 1.55×10^{-2}, 8.0×10^{-5} and 86.5 respectively.

In the same way as discussed above, the values of X_s, X_i and K in Equation (4) have been found to be 4.8×10^{-3}, 2.8×10^{-5} and 86.5 respectively for the sample of arabic acid used by Briggs. In the sample of arabic acid used by Pauli and Ripper only X_i has been found to be different, the other two constants remain unchanged. It may be mentioned that the ash content of Pauli and Ripper's sample of arabic acid is 0.02% which is less than that of the sample used by Briggs. X_i for Pauli and Ripper's sample is 1.6×10^{-5}.

It will be noticed from the data recorded in Table 1 that the observed values of pH agree satisfactorily with those calculated by using Equation (3) over the range of concentration 100 to 2.5 grams of gum acid per kg of water. It may also be noted that in Equation (3) only H_c and θ vary with dilution and the other terms H_s, H_i and K remain constant.

From the data recorded in Table II it will appear that the agreement between the observed and calculated values of specific conductivity is fairly satisfactory. The constants X_s and K in the samples used by Briggs and Pauli and Ripper remain unchanged only X_i, which is very small in magnitude, changes. It is probable that X_i depends to some extent on the ash content of the sample. Pauli and Ripper's sample of arabic acid has lower ash content and lower X_i than those of Briggs.

Table I. Briggs Data on pH

C		Values of pH	
		Observed	Calculated
100	0.536	2.08	2.08
80	0.481	2.13	2.12
60	0.410	2.20	2.19
50	0.366	2.25	2.24
40	0.316	2.30	2.30
25	0.224	2.44	2.45
20	0.188	2.52	2.53
10	0.104	2.75	2.78
5	0.055	3.00	3.03
2.5	0.028	3.26	3.29

Table II. Briggs Data on Specific Conductivity

C	Values of (X_c) 10^5 reciprocal ohms	
	Observed	Calculated
100	260.4	258.7
80	231.6	231.9
60	196.2	198.2
50	175.1	177.6
40	151.0	153.7
25	108.2	109.8
20	89.8	92.4
10	52.5	52.3
5	29.3	28.9
2.5	16.1	16.2

Pauli and Ripper's Data on Specific Conductivity

20.0	93.0	91.5
18.2	83.6	84.5
14.6	69.9	70.5
9.9	51.4	50.4
8.3	44.0	43.4
4.2	24.0	23.8

Summary

On the basis of certain assumptions the author has deduced equations which can account for the effect of dilution on pH and specific conductivity of gum arabic acid sol in a fairly satisfactory manner.

Literature Cited

1. Thomas, A.W. and Murry, H.A., Phys. Chem., (1928), 32, 676.
2. Briggs, D.R., J. Phys. Chem., (1934), 38, 867.
3. Pauli, W. and Ripper, E., Kolloid-Z., (1933), 62, 164.
4. Briggs, D.R., J. Phys. Chem., (1934), 38, 1145.
5. Klaarenbeek, F.W., "Colloid Science", Vol. I, p. 191, Elsevier Publishing Company, London, 1963.
6. Overbeek, J. Th. G., Bull. Soc. Chim. Belg., (1948), 57, 252.
7. Ghosh, B.N., J. Indian Chem. Soc., (1971), 48, 185.
8. Stern, O., Z. Electrochem., (1924), 30, 508.
9. Mukherjee, J.N., Trans. Faraday Soc., (1920-21), 16, 103.

24

Electrokinetic Approach to the Selection of Blood Compatible Materials and Anticoagulant Drugs

SUPRAMANIAM SRINIVASAN*

Department of Applied Science, Brookhaven National Laboratory, Upton, N.Y. 11973

BARRY R. WEISS

Electrochemical and Biophysical Laboratories, Department of Surgery and Surgical Research, State University of New York, Downstate Medical Center, Brooklyn, N.Y. 11203

Introduction

The clotting of blood in vivo or in vitro is an interfacial phenomenon initiated at the blood vessel wall or prosthetic material-blood interface. The total molar concentration of the ionic constituents in blood is about 0.16M. From a colloidal chemical standpoint, one can expect the dispersion of blood cells in this environment to depend markedly on their surface charge characteristics. Furthermore, the adsorption of blood proteins on the blood vessel wall or on prosthetic materials, their reactions at the interface with blood, the adhesion of blood cells on the wall, the aggregation of blood cells directly or via bridges (e.g., fibrin strands) are intermediate steps in the overall thrombosis or clotting reaction.

A knowledge of the mechanism of the overall thrombosis reaction is essential in the selection and evaluation of (i) materials for vascular and heart valve implants and (ii) drugs which prevent or inhibit the disease. Since the intermediate enzymatic reactions which lead to thrombosis are interfacial in nature and occur in electrolytic environments, one may expect many of them, to be electrochemical in origin. There is considerable evidence for this view presented in a number of original and review articles (1-5).

In this chapter are presented (i) a brief description of the electrokinetic methods used to determine the surface charge characteristics of (a) materials potentially useful as vascular prostheses (b) the vascular constituents (blood vessel wall and blood cells) and alterations produced by chemicals; (ii) correlations between the surface charge characteristics of materials and their pro or antithrombogenic characteristics; and (iii) correlations between the effects of chemical compounds on the surface

---

* At the time of presentation of this paper, Supramaniam Srinivasan was on a sabbatical leave program with the General Products Division, International Business Machines Corporation, Boulder, Colorado 80302.

charge characteristics of the vascular constituents and their pro or anti-coagulant properties.

Electrokinetic Methods Used in Cardiovascular Research

The surface charge characteristics of materials, potentially useful for vascular prostheses, were determined by streaming potential or electroosmosis measurements _in vitro_ in physiologic fluids (0.154M sodium chloride or Krebs solution) (1). Streaming potential measurements were also carried out _in vivo_ on prostheses fabricated from some insulator materials, implanted in canine carotid and femoral arteries (1). For this measurement, two silver/silver chloride reference electrodes were inserted through side branches closest to the two ends of the prosthesis and the potential measurements were made between the two electrodes with and without obstruction of blood flow. The difference between the two readings is the streaming potential. A schematic representation of the experimental set-up for such measurements is shown in Figure 1.

The surface charge characteristics of the blood vessel wall in physiological solutions were also determined _in vitro_ by streaming potential and electro-osmosis measurements. For the streaming potential measurements, canine carotid or femoral arteries were excised from the animal and connected between two reservoirs. The streaming potentials were measured as a function of pressure drop across the tube during the flow of solution. For the electro-osmosis measurements, sections of canine aorta or vena cava were used and the rate of flow of fluid transport across the membrane, in the presence of an electric current across it, measured.

Electrophoretic techniques are considerably more convenient than sedimentation potential measurements for obtaining information concerning the surface charge characteristics of blood cells (red cells, white cells and platelets). Electrophoretic mobilities of cells have been determined in 5% dextrose, 0.154N NaCl or Krebs solution.

In order to determine the effects of chemical compounds on the surface charge characteristics of the vascular components, electrokinetic measurements have been carried out in solutions which contained one of these chemicals in the desired concentration (6). Control experiments were also performed without addition of the chemical to the solution.

Correlations Between the Surface Charge and Pro or Antithrombogenic Characteristics of Insulator Materials

Studies on the effects of electrochemical factors on blood coagulation date back to 1824 when it was reported that blood _in vitro_ is deposited on the positive electrode but not on the negative electrode (7). The application of electric currents in the production of thrombosis, aneurysms and hemostasis was also carried out in the 19th century (8). In 1928, it was proposed that materials with substantially negative zeta potentials favor adsorption of positively charged blood elements, thereby concentrating the necessary elements for initiation of blood coagulation (9). Plasma fractions and coagulation affecting chemicals create uniquely different surface charge effects on materials (10). No direct correlations between zeta potentials and thrombengic propensities of materials were found in some early work (11). However, in a later work by this same group, it was concluded that zeta potential measurements offer a convenient way of studying adsorption of blood components on plastics and that this study may lead to an understanding of the interaction of blood with plastics (12). Another group has suggested that positively charged surfaces are prone to thrombosis (13). Recently this group concluded that streaming potential techniques may serve as a quality control in the preparation of nonthrombogenic materials (14). Graphite benzalkonium heparin (GBH) surfaces have a negative zeta potential in Ringer's solution and retain their negative potential even after exposure to plasma proteins (15). These surfaces are moderately wettable, highly conductive, antithrombogenic and have good lubricant properties.

The surface energy of a material has been suggested as an important criterion in the selection of blood compatible materials (16, 17). The related parameters, surface tension and contact angle, have also been determined for some materials and correlated with their antithrombogenic properties. Platelet adhesion decreases with decreasing critical surface tension of materials. Lower energy surfaces are less thrombogenic. The effects of surface active groups (hydroxyl, sulfate, sulfhydryl, carboxyl, amino, sulfamido and quarternary ammonium salt) on clotting activity have been determined (18). Electrets with a negative surface charge tend to be anti-thrombogenic and the extent of platelet adhesion on them is less than on non-polarized polymers (19).

The zeta potentials from streaming potential measurements in 10^{-4}N NaCl solution were in the range -32 to -42 mV for non-electrified estane tubes and in the range of -50 to -71 mV for electrified estane tubes. The electrified estane tubes were considerably more thrombo-resistant than the non-electrified ones, as ascertained from in vivo implantation studies (20). Polymer electrets formed by high voltage corona discharge contain a surface layer of charged ions and are referred to as homocharged electrets. These ions slowly drain off the polymer reducing the external electric field of the electret. Further, homocharged electrets do not possess a sufficient volume polarization or heterocharge which would lead to more stable electret potentials (21). Heterocharged electrets are prepared by placing the material in a non-ionizing electric field at an elevated temperature (22). The volume polarization induced by the electric field is frozen in the material as it is cooled down to room temperature. It has been shown (22) that volume polarization in cellulose acetate is considerably increased by addition of 20 wt % polystyrene sulfonic acid (Figure 2). Such electrets are quite stable in aqueous solution. Introducing volume polarization into a material should enhance its antithrombogenic nature.

Polyurethane samples with varying concentrations of acetylene black carbon were prepared and zeta potentials of these samples were obtained as a function of carbon concentrations (23, Figure 3). There appears to be a correlation between the zeta potentials and X-ray diffraction studies in that all carbon impregnated samples which exhibited a negative zeta potential also had a strong crystalline pattern (Figure 4). Above 21.13% carbon, there is a reversal in the sign of the zeta potential and a loss of crystallinity (cf. Figures 3 & 4). Of the samples tested for antithrombogenicity from implantation studies, only the material containing 17.24% carbon was significantly antithrombogenic (23).

Heparin is one of the most potent anti-coagulants. Several attempts have been made to bond heparin to plastic surfaces. In one type of study, heparin is bonded ionically to a plastic via a bridge e.g. a polymerizable quarternary ammonium salt such as methyl iodide salt of dimethyl aminoethyl methacrylate. In Table I are shown the zeta potentials of the untreated, quarternized and heparinized surfaces as well as the blood clotting times in tubes of the base polymers (24). Though many of these surfaces, after heparinization were anti-thrombogenic they were not stable because of heparin elution from the surface. This was also found to be the case with ionically bonded heparin surfaces prepared with hydroxy-3-methacryloyl-oxy-propyl-trimethyl ammonium chloride (GMAC) or with tri dodecyl methyl ammonium chloride (TDMAC) as the bridge compounds between the heparin and the base polymer. Attempts were made to bond heparin covalently to surfaces (25). Though these surfaces are reasonably stable in electrolytic environments, there is a loss in heparin activity (in respect to anti-thrombogenic character) when bonded covalently to surfaces. Gluteraldehyde cross-linked heparin surfaces seem to be anti-thrombogenic but there is no long term implantation data available (26).

In a recent article, the dependence of blood compatibility on surface charge, surface free energy and hydrophilicity of materials was examined (27). It was concluded that though each of these parameters is involved, they are not by themselves adequate to explain biocompatibility. A wide variety of insulator materials have been investigated with respect to surface charge and surface free energy characteristics and blood compatibility. In Table II, these results are summarized.

The extent of cell (red cells, white cells and platelets) adhesion to materials is an indication of the bio-compatibility of materials. This parameter is found along with results of implantation studies in Table II.

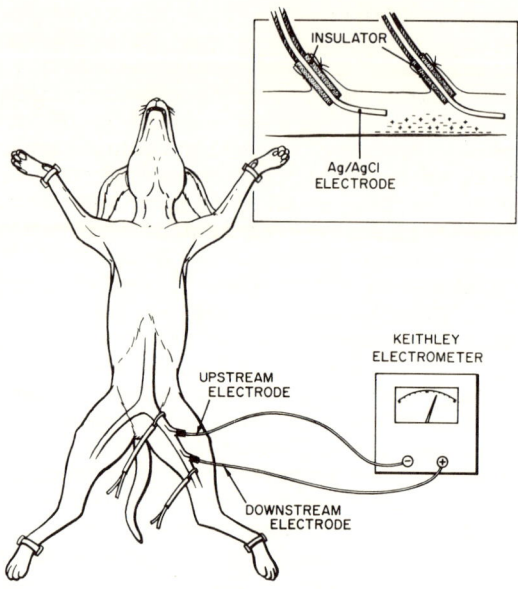

Electrochemical Engineering and Bioscience

Figure 1. Experimental arrangement for measurement of streaming potentials in canine blood vessels, in vivo (6)

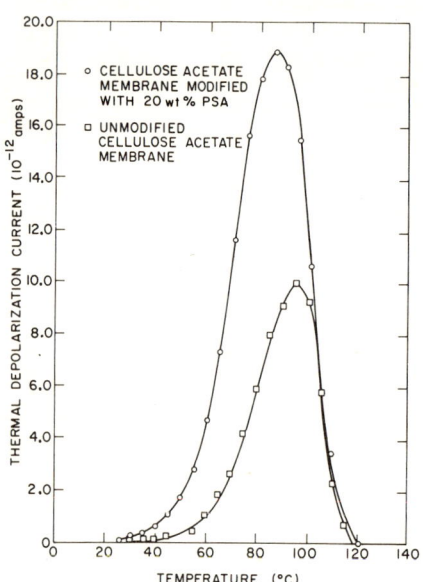

Journal of Biomedical Materials Research

Figure 2. Thermal depolarization of heterocharged cellulose acetate and of cellulose acetate containing 20 weight % polystyrene sulfonic acid (22)

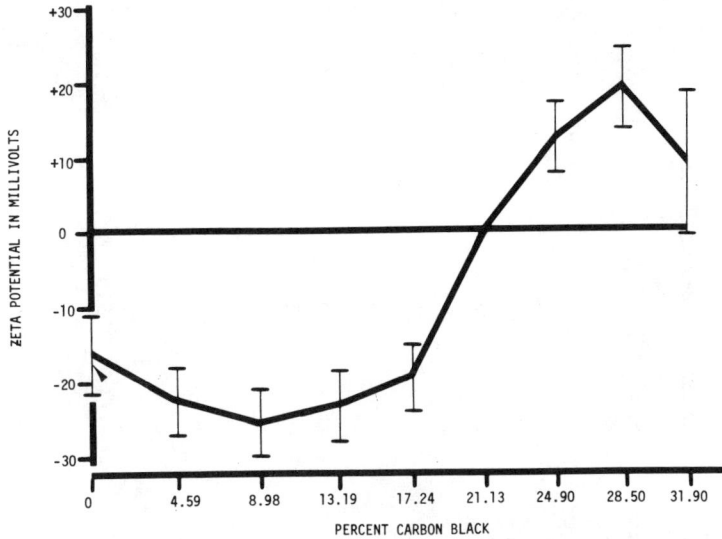

Figure 3. A plot of zeta potentials of the polyurethane samples as a function of carbon content. The zeta potential of pure carbon is noted by the arrow along the vertical axis. Vertical bars represent 95% confidence limits (23).

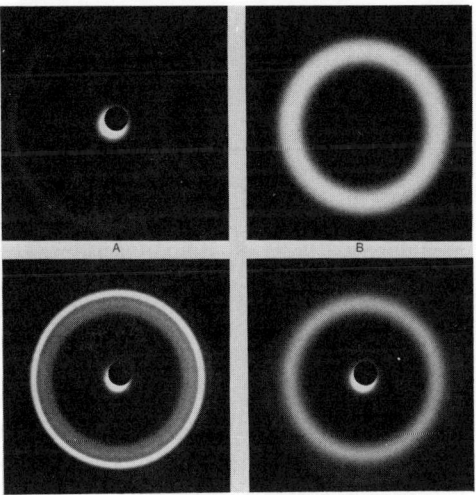

Figure 4. X-ray diffractogram produced by (A) pure acetylene black carbon, (B) pure NDV polyurethane, (C) crystalline pattern produced by 4.59 to 17.24% carbon, and (D) amorphous pattern caused by increasing the carbon content to 21.13% or greater (23)

Table I. Zeta Potentials of Untreated, Quaternized and Heparinized Polymer Surfaces. Blood Clotting Times on Base Polymers are also Presented.

Polymer	Zeta Potentials (mV)			Clotting Times (min)
	Base Polymer	After Quaternization	After Heparinization	Base Polymer
Polystyrene	-13	+11	-19	12
Polyethylene	-16	+9	-14	11
Silicone rubber	-11	+3	-6	20
Prolypropylene	-16	+9	-14	12
Cellophane	--	--	--	6
Vinyl pyridine-butadiene rubber	-5	+9	-7	12
Teflon	-10	+63	-9	10
Natural rubber	-11	+12	-9	8
Epoxy	-4	+15	-9	13
Polyvinyl fluoride	-11	+50	-6	10
Polyvinylidene fluoride	-10	+35	-8	12
Hydrin rubber	0	--	-5	9
Ethylene/propylene rubber	-6	+2	-7	13
Styrene butadiene rubber	-13	-2	-10	12
Fluorinated silicone rubber	-8	+21	-12	8
Polyethylene terephthalate film	-8	+42	-7	10
Glass	-25	--	--	3.5

From "Annals of New York Academy of Sciences," Volume 146 (24).

Table II. Correlations Between Surface Characteristics (Zeta Potentials in Krebs Solution, Surface Tension) and Antithrombogenic Characteristics of Some Materials

	Material	Zeta Potential in Krebs Solution, mV	Critical Surface Tension, dyn/cm	Antithrombogenic Behavior from Implantation in Canine Vena Cava		Adherence no. of cells/cm^2 of		
				2 Hours	2 Weeks	RBC	WBC	Platelets
1.	Copolymer - Urethane Stanford Research Institute	-6.7	29	Excellent	Moderate	20,000	60,000	2,320,000
2.	Estane B. F. Goodrich, Inc.		29	Good	Poor	30,000	60,00	2,080,000
3.	Fluorosilicone Elastomer 100% FS	-5.8	19	Moderate	Moderate	320,000	250,000	3,680,000
4.	Fluorosilicone Elastomer 65/35 Mole % FS/DMS	-8.5	26	Good	Good	100,000	200,000	2,560,000
5.	Fluorosilicone Elastomer 38/62 Mole % FS/DMS (3) to (5) from Dow Corning	-8.5	24	Good-Moderate	Good-Moderate	150,000	80,000	2,240,000
6.	TDMAC Heparinized Silicone Rubber Battelle Memorial Institute	-15.9	--	Excellent	Excellent	10,000	0	150,000
7.	APTES Heparinized Silicone Rubber Carnegie Mellon Institute	--	--	Very Good	Very Good	180,000	0	90,000
8.	Polyacrylamide Hydrogel(neutral) Polysciences, Inc.	--	--	Excellent	Very Good	70,000	0	256,000
9.	Ioplex 101 (0.5 meq. anionic) Amicon, Inc.	-7.8	--	Excellent	Very Good	510,000	0	320,000
10.	LTI Pyrolytic Carbon Gulf Atomic	--	50	Excellent	Excellent	Light	None	Moderate
11.	Polycarbonate/Silicone Copolymer (Mem 213 - General Electric)	--	--	Moderate-Poor	--	30,000	60,000	1,550,000

Correlation Between the Effects of Drugs on the Surface Charge Characteristics of the Vascular System and Their Anti or Pro-thrombogenic Properties

It has been shown that antithrombogenic agents increase the negative charge densities of the blood vessel wall and blood cells while thrombogenic agents have the opposite influence (1,28). Investigation into the relationship between the effect of a drug on surface charge and its antithrombogenicity is of value in: (i) understanding the electrochemical mechanisms involved in intravascular thrombosis; (ii) elucidating the nature of drug action in the prevention of thrombosis; and (iii) the development of in vitro screening techniques for the selection of antithrombogenic drugs.

The knowledge that a greater negative surface charge density on blood vessel walls and blood cells is conducive to antithrombogenicity lends credence to the concept of thrombosis as an electrochemical, potential-dependent phenomena (1-5). Thus, from this viewpoint, the antithrombogenic action of a particular drug can be thoughtof as being due to its effect on the surface charge of the vascular system. This provides a useful criterion for the selection of antithrombogenic drugs.

Electrostatic Repulsive Force in Blood Cell Interactions. Rouleaux formation has been shown to result from red blood cell (RBC) aggregation (RCA) in the presence of macromolecules (29). It has also been demonstrated that the mechanism of macromolecule-induced RCA is related to the action of the macromolecule as a bridge between cell surfaces (30). The normal red blood cell is negatively charged due to the predominant ionized carboxyl groups of N-acetyl-neuraminic acid (which make up over 60% of the RBC charged groups) (31, 32). The relative contribution of charge of the RBC (erythrocyte) compared to lymphocyte and thrombocyte (platelet) is clearly seen from the histogram plot in Figure 5. It is the like negative charges of adjacent cells which result in the electrostatic repulsion that tends to cause disaggregation. Jan and Chien (29) studied the effect of enzymatic removal of N-acetyl neuraminic acid from RBC on the presence and extent of aggregation in the presence of dextran, a neutral polysaccharide macromolecule. Dextran is used widely as a plasma expander (i.e. to increase blood volume) and thus it is of interest to study its effect on the blood elements. Neuraminidase treated RBC's show greater aggregation in dextrans of molecular weights 20,000 (D x 20), 40,000 (D x 40) and 80,000 (D x 80) when compared to normal RBC's. Also, with neuraminidase treated RBC, there is none of the disaggregation characteristic of the normal RBC at high dextran concentration, when aggregation occurs (Figure 6). This suggests that the disaggregation of the normal RBC is related to increased electrostatic repulsion resulting from the dextran. The repulsive force would be minimal in the case of the neuraminidase-treated cells since the N-acetylneuraminic acid (which

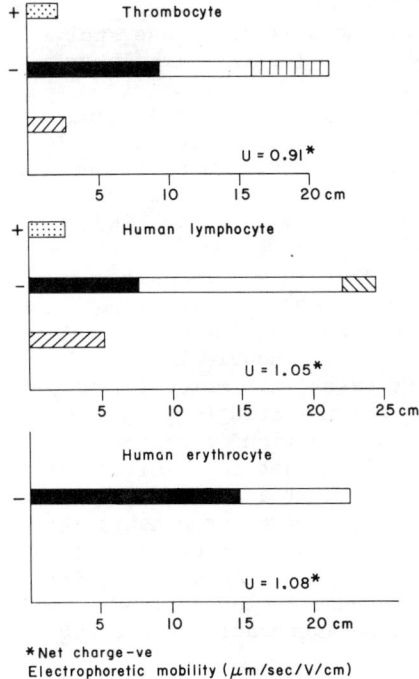

Figure 5. Histogram plots showing the relative distribution of the groups in arbitrary units. ▦—NH_2; ■—α carboxyl groups (NANA), RDE susceptible; □—unidentified; ⬚—phosphate, alkaline phosphatase susceptible; ◩—phosphate groups—RNase susceptible; and ▨—SH group. Net charge −ve. Electrophoretic mobility ($\mu m/sec/V/cm$) (32).

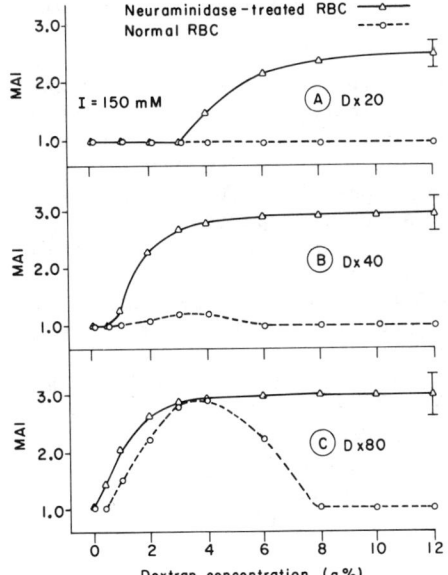

Figure 6. Effects of dextrans on aggregation of normal RBC's (dotted lines) and neuraminidase-treated RBC's (solid lines) in saline. A, B, and C shows results in dextrans 20, 40, and 80, respectively. The vertical bars represent SEM (29).

fosters the charged carboxyl groups) has been removed.

Danon (33) reported on the use of the polymeric base, polylysine (plys). A low molecular weight plys (degree of polymerization, 16) required a concentration of 6.5×10^{-8} µg/cell to achieve aggregation. The concentration needed for aggregation decreases with increasing molecular weight, approaching 1.3×10^{-8} µg/cell for plys, degree of polymerization, 200. Thus, it appears that plys segments provide macromolecular bridges connecting one erythrocyte to the next. The polybase acts by actively binding the RBC and not by simply altering the repulsive force.

Brooks and Seaman (34) have shown that in simple salt solutions dextran adsorption is related to polymer concentration. The adsorption results in an increase in RBC zeta potential, probably as a consequence of expansion of the electric double layer. Figure 7 shows that zeta potential increases with molecular weight as well as concentration of dextran. Polymer bridging, the result of dextran adsorption, can cause aggregation although the role of electrostatic repulsive forces increases in significance with adsorbed polymer concentration. Thus, at a critical concentration of dextran (which is a function of molecular weight) the repulsive forces will become stronger than the attractive forces. The result will be disaggregation. The observation of the critical aggregation concentration lends credence to the role of electrostatic attraction and repulsion in aggregation and disaggregation phenomena.

Drugs, Surface Charge and Anti or Procoagulant Properties.
The electrostatic forces that are crucial to aggregation phenomenon, and therefore thrombosis have been examined. The remainder of this section will be devoted to some of the drugs that produce an alteration in surface charge and therefore initiate or inhibit thrombosis.

Heparin and Protamine. The most widely used antithrombogenic drug is heparin. It is thought to originate in connective tissue mast cells. Heparin is an acid mucopolysaccharide of high molecular weight (Figure 8). It is unique in the sulfamic linkages binding the sulfate groups to amino groups (35). Heparin, made up of about 40% esterified sulfuric acid, is apparently one of the strongest mammalian acids. At physiologic pH, it is believed that heparin is hydrolyzed to an acidic product. Thus, acidic groups adsorbed onto blood cells and blood vessel wall intima make a significant contribution to (negative) charge as demonstrated by electrokinetic studies.

The effect of heparin on the surface charge of platelets has been demonstrated. It has been shown that incubation of platelets with heparin causes an increase in electrophoretic mobility (EM) and therefore zeta potential (36). In vivo and in vitro streaming potentials have also been seen to increase with heparin addition, which reflect an increase in zeta potential (and therefore negative

Bibliotheca Anatomica

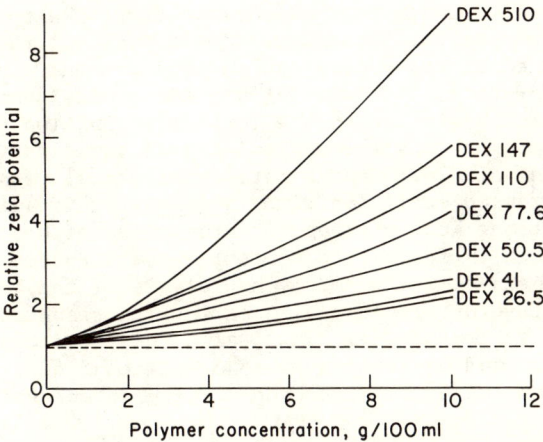

Figure 7. Relative zeta potential as a function of polymer concentration for normal human erythrocytes suspended in saline plus different molecular weights of dextran (DEX) (34)

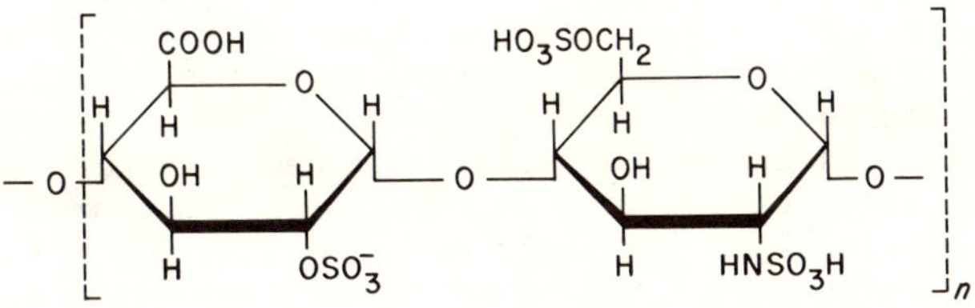

Figure 8. The structure of the tetrasaccharide unit of heparin

charge density) at the solid-solution interface (37). Correlation between this increased negative surface charge and antithrombogenicity has been provided by rat mesenteric vessel occulsion studies (38).

In this in vivo experiment, an extracted portion of the rat mesoappendix is placed over a transilluminated plastic stage. An electric current is applied, by means of fine platinum electrodes placed at opposite edges of the mesoappendix. The current (usually ranging from 10-50 μa) induces occlusion of the rat mesenteric vessels. The effect of a drug can be gauged either in terms of time needed for total occlusion or percent occlusion at a specific time. Thus, an in vivo technique for the evaluation of the relative antithrombogenicity/thrombogenicity of various drugs is now available. Infusion of heparin into experimental rats results in a greatly increased occlusion time (38, 39). This again demonstrates the strong antithrombogenic properties of heparin. The relation between sulfonic acid content and antithrombogenic activity has been demonstrated by hydrolysis of the ester linkage as a result of de-N-sulfation of heparin with resultant loss of anticoagulant activity (40). This, also, has been confirmed by rat mesenteric studies which show no significant difference in occlusion times between control and de-N-sulfated heparin infused rats (39).

As a final note on heparin activity, it has been shown that heparin selectively adsorbs over fibrinogen on cleaved mica surfaces (41). A surface covered with heparin will have a relatively high negative charge density. Work on adsorption of fibrinogen on mercury has shown heavy adsorption at positive and low negative potentials with no adsorption at potentials more negative than -1.6V (vs. S.C.E.) (42). Thus adsorption of fibrinogen onto vascular surfaces, the relation of which to thrombus formation has been documented elsewhere (43, 44), evidently a potential dependent phenomenon, may be prevented by the high negative charge imparted onto the surface by heparin adsorption. The mechanism remains unclear;though, in some recent work no adsorption or charge transfer processes were detected across the Pt-NaCl interface at the heparin level that brings about total inhibition of fibrinogen adsorption on mica (45).

Protamine, a common antagonist for heparin, has been shown to have the opposite effect of heparin, i.e. to decrease the zeta potential of platelets and blood vessel walls (36, 37).

Noradrenaline and ADP. Aggregating agents noradrenaline and adenosine diphosphate (ADP) exhibit a biphasic effect on platelet electrophoretic mobility, i.e., increase with low concentration and decrease with higher concentration (46-48). Figure 9 shows the dosage for maximum EM to be .05 mg/l. Figure 10 represents a theoretical schematic application to an in vivo system, with areas of optimal stability, reversible aggregation, and irreversible aggregation. Confirmation of this model by observation

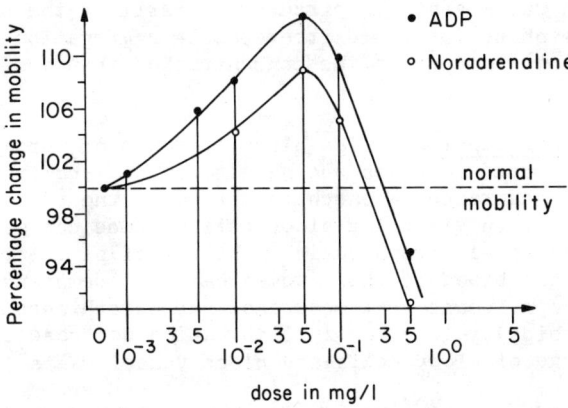

Figure 9. Percentage change in electrophoretic mobility of platelets from control valves as a function of concentration of (i) ADP, and (ii) noradrenaline (46)

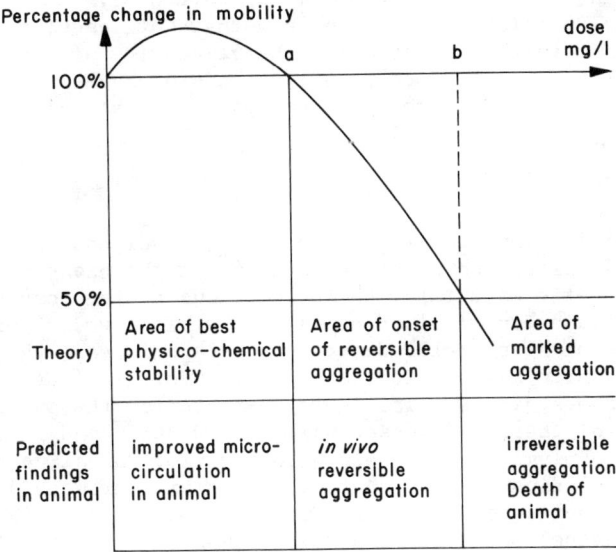

Figure 10. Qualitative representation of percentage change in electrophoretic mobility as a function of chemical concentration, showing regions of stability or aggregation of cells (46)

of effect on rat mesenteric circulation verifies the basic theory although the phase for rapid irreversible aggregation is encountered at a lower noradrenaline concentration than would be expected (46).

Oral Contraceptives. The high incidence of thromboembolic phenomena observed in women who have taken oral contraceptives (49, 50) has generated research to determine the role of these chemical agents in effecting blood cell surface charge. A significant change in electrophoretic mobility of platelets sensitized by ADP from the blood of these women has been demonstrated (51). There is also evidence that mestranol and norethindrone (contraceptive steroids), alone and in combination decrease the negative surface charge of blood cells and blood vessel walls (52).

Flavonoids. A mixture of some flavonoids, derivatives of Vitamin P, have been used for the treatment of venous disorders in Europe. The flavonoids increase the negative charge density of the blood vessel wall and blood cells significantly while also increasing the occlusion time of rat mesenteric vessels (53). Thus, the flavonoids are an atoxic group of drugs worthy of further investigation for their role in the prevention of thrombosis.

Table III presents typical changes in charge characteristics of the blood vessel wall and blood cells caused by heparin, protamine, venoruton (a flavonoid compound), aspirin, mestranol (an estrogen) and norethindrone (a progesterone). According to the results in this Table, it follows that anticoagulant drugs increase or maintain the negative charge density of the blood vessel wall and of blood cells while the procoagulants decrease this parameter, or in some cases causes a reversal of the sign of the surface charge.

Summary

Blood clotting, in vivo or in vitro, involves a series of reactions - adsorption of blood proteins on the blood vessel wall or on prosthetic materials, their reactions at the interface with blood, the adhesion of blood cells on the blood vessel wall, the aggregation of blood cells directly or via bridges (e.g., fibrin strands) - which may be electrochemical in origin. Using electrokinetic methods, it is possible to characterize the surface charge properties of the blood vessel wall, blood cells and prosthetic materials. Polymeric materials with a uniform negative charge density (including electrets) tend to inhibit thrombosis. The enzymatic blocking of N acetyl neuraminic acid (the main source of negative charge) on red cell surfaces causes their aggregation. With increasing molecular weight of related classes of compounds, lower concentrations of drugs are required to cause aggregation of blood cells. Anticoagulants increase or maintain the negative

Table III. Correlations Between the Surface Charge Effects and Pro-or Anticoagulant Effects of Some Drugs

Drug	Effects on surface charge* of		Pro-or anticoagulant (from rat mesentery exp.)
	Blood vessel wall	Blood cells	
Heparin	−	−	Highly anticoagulant
Protamine	+	+	Procoagulant
Venoruton (flavonoid)	−	−	Anticoagulant
Aspirin	−	−	Anticoagulant
Mestranol	+	+	Slightly pro-coagulant
Norethindrone	+	+	Slightly pro-coagulant

* − = increase of negative surface charge density, + decrease of negative surface charge density.

From "Electrochemical Engineering and Bioscience," (6).

charge density of the blood vessel wall and of blood cells while procoagulants have the opposite effect.

Acknowledgements

A considerable portion of the work on the selection and evaluation of potential antithrombogenic materials and anticoagulant drugs was carried out at the Electrochemical and Biophysical Laboratories, Department of Surgery and Surgical Research, State University of New York, Downstate Medical Center, Brooklyn, N.Y. 11203 in collaboration with Professor P. N. Sawyer and a number of other co-workers during the period from 1966 to 1973. The authors are also grateful to Dr. N. Ramasamy, of State University of New York, Downstate Medical Center, for helpful discussions and suggestions.

Literature Cited

1. Srinivasan, S., Sawyer, P.N., J. Ass. Adv. Med. Ins., (1969), 3, 116.

2. Srinivasan, S., Sawyer, P.N., J. Colloid Interface Sci., (1970), 32, 456.

3. Srinivasan S., Duic,L., Ramasamy, N., Sawyer, P.N., Stoner, G.E., Ber. Bunsen Ges., Phys. Chem., (1973), 77, 798.

4. Sawyer, P.N., Srinivasan, S., "Bibliotheca Anatomica," 12, p. 106, Karger, Basel, 1973.

5. Ramasamy, N., Keates, J.S., Srinivasan, S., Sawyer, P.N., Bioelectrochemistry and Bioenergetics, (1974), 1, 244.

6. Srinivasan, S., Sawyer, P.N., in "Electrochemical Engineering and Bioscience," pp. 17-36, H.T. Silverman, I.F. Miller and A.J. Salkind, Eds., The Electrochemical Society, Princeton, N.J., 1973.

7. Scudamore, C., "Essay on Blood," Longham, Hurst, Rees, Orme, Brown and Green, London, 1824.

8. Poore, G.V., "A Textbook of Electricity in Medicine and Surgery," D. Appleton and Co., N.Y., 1876.

9. Gortner, R.A., Briggs, D.R., Proc. Soc. Expl. Biol., (1928), 25, 820.

10. Wood, L.A., Horan, F.E., Sheppard, E., Wright, I.S.,in "Blood Clotting and Allied Problems," p. 89, J.E. Flynn, Ed., Josiah Macy Foundation, N.Y., 1950.

11. Leininger, R.I., Mirkovitch, V., Beck, R.E., Ardus, P.G., Kolff, W.J., Trans. Amer. Soc. Artif. Int. Organs, (1964), 10, 237.

12. Leininger, R.I., Epstein, M.M., Falb, R.D., Grode, G.A., ibid., (1966), 12, 151.

13. Milligan, H.L., Davis, J., Edmark, K.W., J. Biomed. Mat. Res., (1968), 2, 51.

14. Edmark, K.W., Davis, J., Milligan, H.L., Throm. Diathes. Haemorr., (1970), 24, 286.

15. Gott, V.L., Whiffen, J., Dutton, R.C., Leininger, R.I., Young, W.P., in "Biophysical Mechanism in Vascular Homeostasis and Intravascular Thrombosis," p. 297, P.N. Sawyer, Ed., Appleton Century Crofts, N.Y., 1965.

16. Baier, R.E., Dutton, R.C., J. Biomed. Med. Res., (1969), 3, 181.

17. Lyman, D.J., Brash, J.L., Klein, K.G., in "Artificial Heart Program Conference - Proceedings June 9-13, 1969, Washington, D.C.," Chapter 11, R.J. Hegyeli, Ed., U.S. Government Printing Office, Washington, D.C., 1969.

18. Grode, G.A., Falb, R.D., Anderson, S.J., ibid., Chapter 2.

19. Murphy, P., Lacroix, A., Merchant, S., ibid., Chapter 10.

20. Murphy, P.V., LaCroix, Merchant, S., Bernhard, W., J. Biomed. Mat. Res., Symposium, (1971), 1, 59.

21. Gross, B., J. Chem. Phys., (1949), 17, 866.

22. Gable, R.J., Wallace, R.A., J. Biomed. Mat. Res., (1974), 8, 91.

23. Taylor, B.C., Sharp, W.V., Wright, J.I., Ewing, K.L., Wilson, C.L., Trans. Amer. Soc. Artif. Int. Organs, (1971), 17, 22.

24. Leininger, R.I., Falb, R.D., Grode, G.A., in "Materials in Biomedical Engineering,", p. 11, S.N. Levine, Ed., Annals of the New York Academy of Sciences, Volume 146, 1968.

25. Grode, G.A., Merker, R.L., (1969), Personal communication.

26. Lagagren, H.R., Eriksson, J.C., Trans. Amer. Soc. Artif. Int. Organs, (1971), 17, 10.

27. Bruck, S.D., Trans. Amer. Soc. Artif. Int. Organs, (1972), 18, 1.

28. Sawyer, P.N., Srinivasan, S., Bull. N.Y. Acad. Med., (1972), 48, 235.

29. Jan, K.M., Chien, S., "Bibliotheca Anatomica," Karger, Basel, (1972), 11, 251.

30. Chien, S., Jan, K.M., Usami, S., Luce, S.A., Miller, L.H., Fremount, H., ibid., 29.

31. Seaman, G.V.F., Brooks, D.E., ibid., 251.

32. Mehrishi, J.N., ibid., 260.

33. Danon, D., ibid., 289.

34. Brooks, D.E., Seaman, G.V.F., ibid., 272.

35. Durant, G.V., Hendrickson, H.R., Montgomery, R., Arch. Biochem., (1962), 99, 418.

36. Stoltz, J.F., Stoltz, M., Alexander, P., Struff, F., Larcan, A., "Bibliotheca Anatomica," Karger, Basel, (1969), 10, 343.

37. Srinivasan, S., Aaron, R., Chopra, P.S., Lucas, T., Sawyer, P.N., Surg., (1968), 64, 827.

38. Lustrin, I., Breen, H., Reardon, J., Weslowski, S.A., Sawyer, P.N., Surg., (1965), 58, 857.

39. Aaron, R.K., Srinivasan, S., Burrowes, C.B., Sawyer, P.N., Thrombos. Diathes. Haemorrh., (1970), 23, 621.

40. Eika, C., Thromb. Res., (1973), 2, 349.

41. Stoner, G.E., Srinivasan, S., Gileadi, E., J. Phys. Chem. (1971), 75, 2107.

42. Stoner, G.E., Walker, L., J. Biomed. Mat. Res., (1969), 3, 645.

43. Baier, R.E., Dutton, R.C., ibid., 191.

44. Vroman, L., Adams, A.L., Thrombos. Diathes. Haemorrh. (1967), 18, 510.

45. Ramasamy, N., Weiss, B.R., Srinivasan, S., Sawyer, P.N., Extended Abstracts of 145th Meeting of the Electrochemical Society, San Francisco, May 12-17, 1974, 74 (1), 680.

46. Stoltz, J.F., Stoltz, M., Larcan, A., "Bibliotheca Anatomica," Karger, Basel, (1969), 10, 474.

47. Hampton, J.R., Mitchell, R.A., Brit. Med. J., (1966), 1, 1074.

48. Hawkins, I., Nature (New Biology) (1971), 233, 92.

49. Inman, W.H., et.al., Brit. Med. J. (1970), 2, 203.

50. Tyler, E.V., JAMA, (1963), 185, 131.

51. Bolton, C.H., Hampton, J.R., Mitchell, J.R., Lancet, (1968), 1, 1336.

52. Srinivasan, S., Solash, J., et. al., Contraception (1970), 9, 291.

53. Srinivasan, S., Lucas, T., Burrowes, C.B., Wanderman, N.A., Redner, A., Bernstein, S., Sawyer, P.N., "Sixth European Conference on Microcirculation, Aalborg, 1970," p. 394, Karger, Basel, 1971.

INDEX

INDEX

A

Acrylate esters 135
Additions, PVA 182, 187
Adducts, ethyleneoxide 237
Adenosine diphosphate (ADP) 334
Adhesion, cell 325
Adhesion, platelet 324
APP (adenosine diphosphate) 334
Adsorbates, macromolecular 121
Adsorbed layer 165
Adsorption
   of CTA^+, bilayer 177
   emulsifier 136
   equation, Gibbs 6
   ion 212
   of the ionic emulsifier 136
   isotherms 175, 177
      Gibbs 3
      Langmuir 49
      of metal octoates 218
      of NaDBS and $CTAB^+$ 179
      of PVA 181
   layer 155
   of SHS 138, 141
Aerogel 9, 271
Aggregates 48, 57, 61, 253
Aggregation
   of cells 335
   degree of stabilized emulsions 118
   equilibrium 229, 231
   fast 146
   finite 281
   of normal red blood cells and neuraminidase 331
   number 219, 229, 236
   particle 53
   rate constants for 165
   reversible 146
   of two colloidal spheres 55
Air interface, liquid– 44
Alcohol, fatty 137
Alcohol–heavy water mesophase, soap– 257
Alignment, macroscopic 257
Alkali saponification 280
Amide–water system 290, 293
Amides 288, 292
Amphiphile 261, 270
Amphiphilic lamellae 254
Anchored polymer chains 165
Anionic surfactant 24, 178
Anisometric crystals 50
Anisotropic aggregates 61
Anisotropic systems 254
Anisotropy 57
Anticoagulant drugs 322
Antithrombogenic characteristics 329
   of insulatory materials 323
Anti-thrombogenic properties of drugs 330

Aqueous
   dispersions 173
   phase nucleation 141
   poly(ethylene oxide) solutions 170
Association polyelectrolyte 183
Asymmetry parameter 256
Attraction 22, 134, 145

B

Barrier, potential energy 115, 151
Barrier, steric 166
Bilayer adsorption 177
Biliquid foams 18, 27, 28, 36
Bimolecular collisions 151
Bimolecular surface reaction 49
Biomembranes 36
Binary Aerosol OT–water mixtures 271
Binary system of di-2-ethylhexyl sulpho-succinate and water 270
Binding 303, 304, 306, 307
Bingham yield value 198, 201
Birefringence, magnetic 60
Block copolymer 169
Block polymer 286
Blood
   cells 330, 331
   clotting times 328
   compatible materials 322
   interface, prosthetic material–proteins 322
   vessels, canine 326
Body-centered cubic structure 272
Bond, metal–carboxylate 219
Bond, polar 213
Bonding, hydrophobic 177
Bovine serum albumin (BSA) 113
Bridging 24, 332
Brownian motion 131, 194
BSA (bovine serum albumin) 113
Bubbles
   bridges between 24
   floating 22
   independent 23
   monolayers of 22
   on surface 21
Bubbling gas 43
Bulk phase 22

C

$C_{12}E_6$, (n-dodecylhexaoxyethylene monomer) 149, 156
Cancer cells 36
Canine blood vessels 326
Capacities, heat 289, 293
Capillary pressure, Laplace 20
Carbon black (Sterling R) 211, 213

345

Carbon number ... 231
Carboxylate bond, metal– ... 219
Cardiovascular research ... 323
Casein ... 308
Cationic soaps ... 299
Cationic surfactant ... 24
Cell(s)
  adhesion ... 325
  aggregation of ... 331, 335
  biliquid foams, oil ... 28
  cancer ... 36
  interactions, blood ... 330
Cellulose acetate ... 326
Cetylpyridinium bromide (CPB) ... 300
Cetylpyridinium chloride (CPC) ... 71, 97
Cetyltrimethylammonium bromide (CTABr) ... 173
Chain(s)
  anchored polymer ... 165
  chain interactions ... 206
  detergent, longer ... 307
  interaction, polymer– ... 193
  ions, long ... 47
  stabilizing ... 165
Charge neutralization ... 6
Charge, surface ... 323, 330, 332, 336
Charged toner particles ... 211
Chemical potential, excess ... 226
Clotting times, blood ... 328
Cloud point ... 284, 285
Cmc (critical micelle concentration) ... 219, 236, 242, 243, 246
Coagulation ... 211
  of cupric oxide sols ... 42
  effect of electrolyte on ... 56
  effect of surface active agent on ... 58
  of electrolyte ... 40
  of α-FeOOH sol ... 44, 51, 54, 58
  of iron oxide sol ... 45
  mechanical ... 40
  rate of ... 40, 43, 48
  surface ... 40, 48, 55
  value ... 46
Coagulum ... 53
Coalescence
  of clear oil phase ... 81
  of coated emulsion droplets ... 112
  on demulsification, effect of ... 71
  influence of nonionic detergents on ... 123
  rate ... 119, 121
Coacervation ... 270
Coated emulsion droplets ... 112
Collision number ... 40
Collisions, bimolecular ... 151
Colloid
  chemistry, history of ... 1
  chemistry, teaching of ... 2
  lyophobic ... 111
  polymer ... 137, 139
Colloidal
  particle ... 316
  silver sols ... 149
  solutions ... 55
  spheres ... 55
  stability ... 139
  suspension ... 14, 193

Comb polymers ... 166
Complexes, polymer–surfactant ... 173
Conductance ... 216, 217
Conductivity of soaps ... 219
Conductivity, specific (see Specific conductivity)
Contact angle ... 29, 324
Contact electrification ... 212
Continuous phases ... 128, 130, 166
Contraceptives, oral ... 336
Copolymer ... 169, 193
Copying process, electrostatic ... 211
Core particle radius ... 204
Counterion ... 261, 317
Coupling term, quadrupole ... 254
CPB (cetylpyridinium bromide) ... 300
CPC (cetylpyridinium chloride) ... 97
Creaming ... 64, 81, 90
Creep compliance–time data ... 127
Critical flocculation temperature ... 193
Critical micelle concentration (cmc) ... 219, 236, 242, 243, 246
Critical voltage of emulsification (E.(crit.)) ... 99
Crystals ... 50, 55, 262
Crystalline middle soap, liquid ... 3
Crystalline phase, lamellar liquid ... 258, 272
Crystallinity, degrees of ... 295
CTA^+, bilayer adsorption ... 177
CTABr (cetyltrimethylammonium bromide) ... 173, 179
Cubic packing geometry ... 132
Cupric oxide sols ... 42, 44, 45
Curd fiber phase ... 4
Cytological patterns ... 30
Cytoplasmic movement ... 32

D

DDAO (N,N-dimethyl-n-dodecylamine oxide) ... 57
Decanol orientation ... 261
Deflocculation ... 129
Degrees of freedom, inter-molecular ... 294
Demulsification
  effect of flocculation and coalescence on ... 71
  of emulsions ... 70, 72, 73
  ultracentrifugal ... 64, 74
Depolarization, thermal ... 326
Deryagin-Landau-Verwey-Overbeek (DLVO) theory ... 145
Detergent(s)
  action ... 7
  with biological substances, interaction of ... 299
  longer chain ... 307
  nonionic ... 123
  ratio, protein– ... 314
  synthetic ... 299
Deuteron
  exchange ... 256
  NMR spectra ... 258, 260
  NMR studies ... 253
  quadrupole splittings ... 260, 262, 266

INDEX 347

Dextrans ... 331
Dialysis, equilibrium ... 299, 301, 302
Dielectric constant ... 46
Di-2-ethylhexyl sulphosuccinate ... 270
Diffraction, x-ray ... 7
Diffractogram, x-ray ... 327
N,N-dimethylacetamide ... 291
N,N-dimethyl-n-dodecylamine oxide, (DDAO) ... 57
Dipole, peptide ... 289
Disaggregation ... 332
Discontinuous phase ... 28
Disjoining pressure ... 19
Disperse phase ... 27
Dispersions
  aqueous ... 173
  ethirimol ... 173
  latex ... 167
  molecular ... 278
  nonaqueous ... 211
  sterically stabilized ... 165
Dissolution of water-soluble polymers ... 278, 282, 283
Ditynadallism, magnetic ... 60
DLVO theory, (Deryagin-Landau-Verwey-Overbeek) ... 110, 145
n-Dodecylhexaoxyethylene monomer $(C_{12}E_6)$ ... 149
Double layer ... 11, 174, 215, 316
Double quantum transitions ... 257
Doublets, separation of ... 205
Droplets, emulsion ... 110, 112, 113
Dropets, styrene monomer ... 135
Drops, water ... 129, 134
Drugs, anticoagulant ... 322, 330
D.T.A. data ... 129
Dying, solvent ... 107

E

E(crit.) (critical voltage of emulsification) ... 99, 100, 102, 103
Electret ... 324
Electric field at the interface ... 106
Electrical
  double layer ... 11, 316
  emulsification ... 97, 98, 100
  field gradients ... 267
Electrification, contact ... 212
Electrocapillary spinning ... 107
Electrochemical properties ... 316
Electrokinetic methods in cardiovascular research ... 323
Electrokinetic selection ... 322
Electrolyte ... 40, 56, 270
Electrophoretic development of latent images ... 212
Electrophoretic mobility ... 176, 215, 335
Electrostatic copying process ... 211
Electrostatic forces ... 110, 190, 330
Electroviscous effect, secondary ... 200
Emulgents, macromolecular ... 110
Emulsification
  (E(crit.)), critical voltage of ... 99
  electrical ... 97, 98, 100
  of KCL ... 99
Emulsified water drops ... 134
Emulsifier
  adsorption ... 136

Emulsifier (continued)
  ionic ... 136
  mixed ... 135
  stabilizing action of the ... 14
Emulsion(s)
  cetyl pyridinium chloride in ... 71
  characteristics of ... 116
  coalescence rates for ... 119
  composition of ... 128
  demulsification of ... 70, 72, 73
  droplets ... 110, 112, 113
  gum arabic ... 117
  kinetic studies of ... 86
  microgas ... 25
  Nujol–water–SDS ... 66
  oil-in-water ... 28
  olive oil–water–SDS ... 69
  o/w ... 80
  PMMA latex from ... 142, 143
  polymerization, "true" ... 135
  preparation of ... 126
  rheological evaluation of ... 127
  separation of oil from ... 85
  shelf-life of ... 92
  sodium alignate ... 117
  for solvent dyeing ... 107
  stability ... 105, 110
  stabilized ... 118
  styrene monomer ... 140
  ultracentrifugal technique in the study of ... 76
  in the ultracentrifuge, behavior of ... 80
  water-in-liquid paraffin ... 126
  w/o ... 105, 126, 133
Encapsulating phase ... 28
Endocytosis ... 33, 35
Energy
  barrier, potential ... 115
  of bimolecular collisions ... 151
  change free ... 227
  dissipation, total ... 193
  interaction ... 193, 221
  of micellization, standard free ... 234
  minimum, potential ... 166
  profiles, interaction ... 112
  for separation of doublets ... 205
  surface ... 324
Enhanced viscosity ... 206
Enthalpic parameters ... 225
Enthalpies of mixing ... 296
Entropic parameters ... 225
Equilibrium
  aggregation ... 229, 231
  compositions of micelles and monomers ... 241, 243
  dialysis ... 299, 301, 302
  phase ... 270
Equivalent conductance ... 217
Erythrocytes, human ... 333
Ethirimol ... 173, 175, 179
Ethyleneoxide adducts ... 237
2-Ethylhexanoic acid ... 211
Excess
  chemical potential ... 226
  osmotic pressure ... 226
  thermodynamic functions, partial ... 292, 294
Experimental limiting stability ratio ... 155

F

Fast aggregation	146
Fast flocculation	147
Fatty alcohol	137
α-FeOOH sol	44, 51, 54, 58

Field
- gradients, electrical ... 267
- magnetic ... 259
- RF- ... 257
- shear ... 205

Fiber phase, curd ... 4
Flat plates ... 187
Flavonoids ... 336
Flexible molecular structure ... 286
Floating bubbles ... 22
Flocculation
- –deflocculation of the water drops ... 129
- on demulsification, effect of ... 71
- fast ... 147
- kinetics ... 167, 176
- poly(ethylene oxide) concentrations for ... 171
- rate constant of ... 123
- reversible ... 113
- at secondary minima ... 111, 145, 160
- temperature, critical ... 193

Flotation of oil particles ... 91
Flotation rates ... 91
Flow behavior ... 193, 194, 201
Fluorinated surfactants, partially ... 239, 242
Fluorine NMR results ... 245
Fluorocarbons ... 239
FOA (perfluoro-octanoic acid) ... 240
Foam(s)
- biliquid ... 18, 27
- fractionation ... 24
- gas ... 19
- lamella ... 25
- macrocluster gas–liquid ... 18
- macrocluster system ... 18, 23
- oil cell biliquid ... 28
- polystyrene ... 132
- representation of ... 20
- two-dimensional biliquid ... 36

Force, double layer repulsion ... 174
Fractionation, foam ... 24
Free energy ... 227, 234
Freedom, inter-molecular degrees of ... 294

G

Gas, bubbling ... 43
Gas foams ... 18, 19
GBH surfaces (graphite benzalkonium heparin) ... 324
Gel structure ... 10
Gelatin, native ... 308
Gelatin, transfusion (see also TG) ... 300, 310
Geometry, cubic packing ... 132
Gibbs adsorption equation ... 6
Gibbs adsorption isotherm ... 3
Glass spheres, wettable ... 24
Glycine ... 288
Graphite benzalkonium heparin (GBH) surfaces ... 324
Gravimetric volume fractions ... 202
Grease, physical properties of a ... 9
Gum arabic acid sol ... 316
Gum arabic emulsion ... 117

H

H^+ ion activity ... 317
Heat
- capacities ... 289, 293
- of miscellization ... 235
- of mixing ... 288, 292
- of transition ... 5

Heavy water mesophase, soap–alcohol– ... 257
Hemimicelle formation ... 178
Henry's equation ... 215
Heparin ... 332, 333
Heptane solution, carbon black in ... 211
Heterocharged cellulose acetate ... 326
Hexadecanol (HD) ... 136
Hexagonal mesophase ... 253
Hexagonal structure ... 271
Histogram plots ... 331
History of colloid chemistry ... 1
Homogeneous nucleation ... 137, 139
Homopolymers ... 166
Human erythrocytes ... 333
Hydration ... 264, 265
Hydrocarbon
- -like character, degree of ... 245
- paraffinic ... 115
- phobicity of ... 239
- solubility of water in ... 249
- surfactant ... 229

Hydrodynamic forces ... 208
Hydrodynamic volume fraction ... 197, 201
Hydrophile–lipophile balance ... 286
Hydrophilic groups ... 282
Hydrophilic polymer ... 193
Hydrophobic
- bonding ... 177
- characteristics ... 173
- moiety ... 239
- sites ... 220

Hydroxyproline ... 288

I

IC (iodinated casein) ... 306, 308
Iceberg formation ... 281
Images, latent ... 212
Insolubility of polymer in water ... 283
Insulator materials ... 323
Inorganic crystals ... 55
Inorganic salts ... 284
Interaction energy ... 112, 193, 221
Interaction parameter, polymer–solvent ... 185
Interface(s)
- coalescence rates for ... 119
- discontinuity of electric field at the ... 106
- liquid–air ... 44
- liquid–liquid ... 45
- prosthetic material–blood ... 322

Interfacial tension ... 104
Intermolecular degrees of freedom ... 294
Inter-peptide forces ... 297
Inversion, pseudo-phase ... 286
Iodinated casein (IC) ... 306, 308
Ion
- activity, H^+ ... 317

Ion (continued)
- adsorption ... 212
- counter ... 261, 317
- long chain ... 47

Ionic
- emulsifier ... 136
- strength ... 136
- surfactants ... 166, 226

Ionizable groups ... 308
Iron oxide sol ... 45
Isopiestic method ... 281
Isotherm, adsorption (see also Adsorption isotherm) ... 175, 177
Isotropic liquid phase ... 272
Isotropic solution ... 3

K

Kinetic studies of emulsions ... 86
Kinetic theory of slow flocculation at the secondary minimum ... 145
Kinetics
- of adsorption of SHS ... 138, 141
- flocculation ... 167, 176
- Smith–Ewart ... 135
- ultracentrifugal demulsification ... 64

L

Lamella, foam ... 25
Lamellae, amphiphilic ... 254
Lamellar liquid crystalline phase ... 258, 272
Lamellar mesophase ... 253
Langmuir adsorption isotherm ... 49
Laplace capillary pressure ... 20
Latent images, electrophoretic development of ... 212
Latex
- characterization of ... 167
- dispersions ... 149, 167
- from emulsion, PMMA ... 142, 143
- particles ... 193
- polystyrene (PS) ... 51, 142, 156
- stability curves for ... 171

Layer
- adsorbed ... 165
- adsorption ... 155
- double (see Double layer)
- Stern ... 185

Lens on water ... 33
Lens, sprending ... 35
Lenses, adjacent ... 31
Limiting stability ratio, experimental ... 155
Linear molecule ... 286
Lipophile balance, hydrophile– ... 286
Liquid
- –air interface ... 44
- crystal ... 262
- crystalline middle soap ... 3
- crystalline phase, lamellar ... 258, 272
- foams, macrocluster gas– ... 18
- –liquid interface ... 45
- paraffin ... 126, 127
- phase, isotropic ... 272
- state of polymer ... 283

London constant ... 175
London equation ... 113
Long chain ions ... 47

Long-range order ... 253
Loops, polymer ... 197
Lyophobic colloid ... 111
Lyophobic suspensions ... 145
Lyophilic groups ... 282

M

Macroinstability ... 76
Macrocluster
- gas–liquid and biliquid foams ... 18
- system foam ... 23
- systems ... 26, 29, 31

Macromolecular
- adsorbates ... 121
- emulgents ... 110
- substances ... 113

Macroscopic alignment ... 257
Magnetic
- birefringence ... 60
- ditynadallism ... 60
- field ... 259

Manganese salts of 2-ethylhexanoic acid ... 211
Material–blood interface, prosthetic ... 322
McBain's paradox ... 4
Mechanical stability ... 51
Mechanical and surface coagulation ... 40
Media, nonaqueous ... 225
Media, non-polar ... 225
Membrane osmometry ... 233
Membranes, monolayer ... 26
Meniscus ... 24
Mesoappendix, rat ... 334
Mesophase, lamellar ... 253
Mesophase, soap–alcohol–heavy water ... 253, 257

Metal
- –carboxylate bond ... 219
- octoates ... 218
- soaps in oils ... 7

N-Methylacetamide ... 291
Methyl methacrylate (MMA) ... 136
Methylisobutylketone (MIBK) ... 97–100
Micelle(s) ... 136
- equilibrium compositions of ... 241, 243
- formation ... 219, 225, 227, 233, 240
- mixed ... 188, 307
- partially fluorinated surfactant ... 239
- of sodium dodecyl sulfate ... 248
- surface of ... 247

Micellar solution of surfactant ... 278
Micellization ... 234, 235
Micro-calorimetry ... 289
Microcrystalline wax ... 126, 127, 133
Microdroplets ... 276
Microelectrophoresis ... 214
Microflotation ... 48
Microgas emulsion ... 25
Microinstability ... 76
Middle soap, liquid crystalline ... 3
Minimum
- potential energy ... 166
- primary ... 145
- secondary (see Secondary minimum)

Mixed emulsifiers ... 135
Mixed micelles ... 188, 240, 307
Mixing
- enthalpies of ... 296

Mixing *(continued)*
  heats of 288, 292
  infinite 281
  of mono-amides and poly-peptides
    with water 288
Mixtures, binary Aerosol OT–water 271
MMA (methyl methacrylate) 136
Mobility, electrophoretic 176, 215, 335
Moiety, hydrophobic 239
Molecular dispersion 278
Molecular structure, flexible 286
Mono-amides with water, mixing 288
Monolayer membranes 26
Monolayer of bubbles 22
Monomer(s) 118
  droplet, styrene 135
  emulsion, styrene 140
  equilibrium compositions of 241, 243
  ($C_{12}E_6$), n-dodecylhexaoxyethylene... 149
Morphology of the aggregates 57

N

NaDBS (sodium dodecylbenzene
  sulfonate) 173, 179
Negative surface charge 336
Nematic character 61
Neuraminidase 331
Neutralization, charge 6
Noradrenaline 334
Nonaqueous
  colloidal suspensions 14
  dispersion 211
  media 225
Nonionic
  detergents 123
  polysorbate 80 299
  surfactant 233
  Triton X-100 299
Nonpolar media 225
NMR results, fluorine 245
NMR spectra, deuteron 253, 258, 260
Nucleation 137, 139, 141
Nujol–water–Triton X-100 emulsions 69

O

Occlusion time 334
Oil(s)
  cell biliquid foams 28
  from emulsions 85
  metal soaps in 7
  particles, flotation of 91
  phase, clear 81
  polyhedra 74
  separation 65
  soluble surface active molecules
    (OILSSA) 27
  -soluble surfactants 225
  -in-water emulsions 28
OILSSA (oil soluble surface active
  molecules) 27
Olive oil–water–SDS emulsions 69
Oral contraceptives 336
Order, long-range 253
Order parameter 255
Organic additives 285
Orientation effects 254
Orientation, stress 61
Osmium-stained styrene monomer
  emulsion 140
Osmometry, membrane 233
Osmotic pressure 180, 226
OS–SLS system, viscosity changes in ... 314
Ostwald flow behavior 194, 201
Ovalbumin (OV) 300, 310, 313
O/w emulsions 80

P

Packing geometry, cubic 132
Pallman effect 316
Paraffin 126, 127
Paraffinic hydrocarbon 115
Partial excess thermodynamic
  functions 292, 294
Partially saponified polyvinyl acetate
  (PVA-Ac) 278
Particle(s)
  aggregation 53
  charged toner 211
  latex 193
  oil 105
  primary 139
  radius, core 204
  selfcrowding effect 200
  size distribution 101, 139
Peptide dipole 289
Peptide forces, inter- 297
Perfluoro-octanoic acid (FOA) 240
Perfluoro surfactants 242, 245
Perikinetic secondary minimum
  flocculation 160
pH, Briggs data on 320
pH of gum arabic acid sol 316
Phase(s)
  bulk 22
  clear oil 81
  continuous 128, 130, 166
  curd fiber 4
  diagrams 8, 273, 280, 290
  discontinuous 28
  disperse 27
  encapsulating 28
  equilibria 270
  inversion, pseudo- 286
  isotropic liquid 272
  lamellar liquid crystalline 258, 272
  nucleation, aqueous 141
  of sodium palmitate 4
Phobicity 239
Photoelectrophoresis 211
Plane, shear 183
Plasma expander 330
Platelets 324, 335
Plates, flat 187
PMMA latex from emulsion 142, 143
Polar bond 213
Polar groups 173, 264, 272
Polarization, explosive 34
Poly-amides 295
Polyederschaum 23
Polyelectrolyte 174, 183
Poly(ethylene oxide) 170, 171

INDEX

Poly(ethylene oxide-*b*-methyl
   methacrylate) 193
Polyhedra, oil 74
Polyisoprene block copolymer, styrene– 169
Polymer(s)
   block 286
   bridging 332
   chain interaction 193
   chains, anchored 165
   colloids 137, 139
   comb 166
   concentrations 165
   hydrophilic 193
   liquid state of 283
   loops 197
   –solvent interaction parameter 185
   surfaces 328
   –surfactant complexes 173
   –surfactant mixed micelles 188
   in water 278, 282, 283
Polymeric stabilizers 15
Polymerization degree 282
Polymerization of acrylate esters 135
Poly(methyl methacrylate) sphere 193
Polymorphic forms 5
Polypeptides 288, 297
Polysorbate 80, nonionic 299
Polystyrene 185
   foam 132
   latex 51, 142, 149
   sulfonic acid 326
Polyurethanes 327
Polyvinyl acetates (PVA-Ac) 278
Poly(vinyl alcohol) (PVA) 173
Pores, opening and closing of 37
Potassium chloride, emulsification of 99
Potential
   energy barrier 115, 151
   energy minimum 166
   excess chemical 226
Potentials, streaming 326
Potentials zeta (*see also* Zeta
   potentials) 180, 182
Powder pattern 257
Precipitation zone 309
Pressure
   disjoining 19
   Laplace capillary 20
   osmotic 180, 226
   vapor 3
Primary minimum 145
Primary particles 139
Procoagulant properties 332
Proline 288
Prosthetic material–blood interface 322
Prostheses, vascular 322
Protamine 332
Protein(s) 299, 307, 314, 322
Prothrombogenic characteristics of
   insulator materials 323
Prothrombogenic properties of drugs 330
Proton transfer 212
PS latex–$C_{12}E_6$–water system 156
Pseudo-phase inversion 286
PVA(poly(vinyl alcohol)) 173
   -Ac, (partially saponified polyvinyl
   acetates) 278, 280, 281

PVA *(continued)*
   additions 182, 187
   adsorption isotherm of 181
   concentration 186
   variation of zeta potential with 181

Q

Quadrupole coupling term 254
Quadrupole splitting, deuteron 260, 262, 266
Quantum transitions, double 257

R

Radius, core particle 204
Rat mesoappendix 334
Rate constants for aggregation 165
Rate constants, zero order 66
Rayleigh scatterers 167
Refractive indices 175
Repulsive force 174, 208, 330
Reversible aggregation 146
Reversible flocculation 113
RF-field 257
Rheological
   data 127, 130, 140
   evaluation of the emulsions 127
   parameter 121
   studies of polymer chain interaction 193

S

Salt effect 136
Salts of 2-ethylhexanoic acid, manganese
   and zinc 211
Saponification 278, 280, 282
Scatterers, Rayleigh 167
SDS (sodium dodecyl sulfate) 97, 240, 248
   concentration on E(crit.), effect of 100
   emulsions, Nujol–water– 66
   emulsions, olive oil–water– 69
   with iodinated casein, binding of 306
   with TSG, binding of 304
Secondary electro-viscous effect 200
Secondary minimum
   depth 187
   flocculation at 111, 145, 160
   position of the 122
   reversible aggregation at the 146
Sedimentation 90
Selfcrowding effect, particle 200
Separation
   of clear oil 81
   of doublets 205
   of oil 65
Shear
   dependent term 131
   field 205
   forces 129
   plane 183
   rates 141
   stress 193
Shelf-life emulsions 92
SHS (sodium hexadecyl
   sulfate) 136, 138, 141
Silver sols, colloidal 149
Sites, hydrophobic 220
Slater–Kirkwood equation 113

Slater–Kirkwood–Moelwyn–Hughes
  equation 113
SLS (sodium lauryl
  sulphate) 300, 310, 313, 314
Smectic character 61
Smith–Ewart kinetics 135
Smoluchowski theory 147
Soap(s)
  boiling 7
  cationic 299
  conductivity of 129
  liquid crystalline middle 3
  on mechanical stability of polystyrene
    latex, effect of 51
  metal 7
  sodium 5, 7
  –water mesophases 253, 257
Sodium
  alginate emulsion 117
  chloride 271
  dodecyl sulfate (SDS) (see also
    SDS) 240, 248, 304
  dodecylbenzene sulfonate (NaDBS) 173
  hexadecyl sulfate (SHS) 136, 138, 141
  laurate 4
  lauryl sulphate (SLS) 300, 310, 313, 314
  octyl sulphate (SOS) 299
  palmitate 4
  soaps 5, 7
Sol(s)
  colloidal silver 149
  cupric oxide 42, 44, 45
  α-FeOOH 44, 51, 54, 58
  iron oxide 45
Solubility 249, 281, 283
Solute, water structure around the 297
Solution(s)
  colloidal 55
  isotropic 3
  of manganese and zinc salts of
    2-ethylhexanoic acid 211
  micellar 278
  poly(ethylene oxide) 170
  properties 220
Solvation 288
Solvent dyeing 107
Solvent interaction parameter, polymer– 185
SOS (sodium octyl sulphate) 299
Specific conductivity
  Briggs data on 320
  on E(crit.), influence of 102, 103
  equation for 318
  of gum arabic acid sol 316
  Pauli and Ripper's data on 320
Spectra, deuteron NMR 258, 260
Spermatozoon 36
Sphere(s)
  colloidal 55
  poly(methyl methacrylate) 193
  wettable glass 24
Spinning, electrocapillary 107
Splitting, angular dependence of 257
Splittings, deuteron quadrupole .. 260, 262, 266
Spreading behavior 29
Sprending lens 35

Stability 101
  of aqueous dispersions 173
  colloidal 139
  curves for latex 171
  emulsion 110
  measurements 183
  mechanical 51
  of nonaqueous colloidal suspensions 14
  order of 89
  sterically stabilized dispersions 165
  ratio 15, 155, 156, 186
  tests of w/o emulsions 105
  ultracentrifugal 83, 92
Stabilization of lyophobic suspensions 145
Stabilized emulsions 110, 113, 118
Stabilizers, polymeric 15
Stabilizing action of emulsifier 14
Stabilizing chains 165
Standard free energy of micellization 234
Steric barrier 166
Steric force 190
Sterically stabilized dispersions 165
Sterling R, carbon black 213, 218
Stern layer 185
Streaming potentials 326
Stress orientation 61
Stress, shear 193
Structure, phase 270
Styrene monomer droplet sizes 135
Styrene monomer emulsion,
  Osmium-stained 140
Styrene–polyisoprene block copolymer 169
Surface(s)
  active agent 27, 58
  active materials 107
  active molecules (OILSSA), oil soluble 27
  benzalkonium heparin (GBH) 324
  bubble on 21
  characteristics 329
  charge 323, 330, 332, 336
  coagulation 40, 48, 55
  energy 324
  of a micelle 247
  polymer 328
  reaction, bimolecular 49
  tension 241, 324
Surfactant(s)
  anionic 24, 178
  cationic 24
  complexes, polymer– 173
  hydrocarbon 229
  ionic 166, 226
  micellar solution of 278
  micelles 188, 239
  nonionic 233
  oil-soluble 225
  perfluorinated 242, 245
  with proteins, interaction of 299
  systems 265
  ω-trifluoro 249
  (WATSSA), water soluble 27
Suspension 14, 145, 193
Synthetic detergents 299

T

Teaching of colloid chemistry 2
Temperature, critical flocculation 193

INDEX

353

Tension, interfacial 104
Tension, surface 241, 324
Tetrabutylammonium chloride (TBAC) 97
Tetrasaccharide unit of heparin 333
TG (transfusion gelatin) 300, 310
   system, sodium lauryl
      sulphate– 310, 313, 314
   viscosity variation of 311
Thermal depolarization 326
Thermodynamic functions, partial
   excess 292, 294
Thermodynamics of micelle formation 233
Thermodynamics of mixing of mono-
   amides and poly-peptides with water 288
Theta conditions 197
Thixotropic behavior 276
Thrombosis 322
Time data, creep compliance– 127
p-Toluene sulfonylated gelatin (TSG) 304
Toner particles, charged 211
Transfer, proton 212
Transfusion gelatin (see also (TG)) 300
Transition, heats of 5
Transitions, double quantum 257
Triblock copolymer 193
ω-Trifluoro surfactants 249
Triton X-100 .... 69, 299
TSG (p-toluene sulfonylated gelatin) ..304, 308
Turbidmetric technique 167
Turbulence, role of 52

U

Ultracentrifugal
   demulsification 64, 74
   stability 83, 92
   technique 76, 78, 93
Ultracentrifugation, time of 68, 70
Ultracentrifuge 80

V

van der Waal's constant 120, 199, 204
van der Waal's forces 11, 145
Vapor pressure 3, 279
Vascular
   constituents 322
   prostheses 322
   system 330
Vessels, canine blood 326
Viscoelastic properties of w/o emulsions 126
Viscoelasticity of the adsorbed film 110
Viscosity
   apparent 200
   of aqueous poly(ethylene oxide)
      solutions 170
   calculations 302
   changes in OS–SLS system 314
   changes in TG–SLS system 313, 314
   enhanced 206
   of OV 313
   of TG 311
Viscometric studies 299, 301

Void effect 157
Voltage of emulsification E(crit.),
   critical 99
Volume fractions, gravimetric 202
Volume fraction, hydrodynamic 197, 201

W

Water
   amides in 292
   binary mixture of Aerosol OT– 271
   binary system of di-2-ethylhexyl
      sulphosuccinate and 270
   clearing of 82
   drops emulsified 134
   drops, flocculation–deflocculation of 129
   electrical emulsification of 98, 100
   emulsions, oil-in– 28
   of hydration 265
   in hydrocarbons 249
   lens on 33
   -in-liquid paraffin emulsions 126
   mesophases, soap– 253, 257
   mixing of mono-amides and
      poly-peptides with 288
   molecular interaction between peptide
      dipole and 289
   orientation 254, 263
   polymer in 282, 283
   relative activity of 281
   sodium laurate in 4
   soluble surfactant (WATSSA) 27
   structure 297
   systems, amide– 290, 293
   system, PS–latex–$C_{12}E_6$ 156
   system, PVA-Ac 280, 281
   –SDS emulsions, Nujol– 66
   –SDS emulsions, olive oil– 69
   -soluble polymer 278, 283
   –Triton X-100 emulsions, Nujol– 69
WATSSA (water soluble surfactant) 27
Wax, microcrystalline 126, 127, 133
Wettable glass spheres 24
Wetting 247
W/o emulsions 105, 126, 133

X

X-ray diffraction 7
X-ray diffractogram 327

Y

Yield value, Bingham 198, 201

Z

Zero order rate constants 66
Zeta potentials 180, 182
   metal octoate concentration 217, 218
   of polymer surfaces 328
   of polyurethane samples 327
   variation of 181
Zinc salts of 2-ethylhexanoic acid 211

QD
549
C62

MAR 19 1976